Lecture Notes in Mathematics

Edited by A. Dold and B. Eckmann

1156

Global Differential Geometry and Global Analysis 1984

Proceedings of a Conference
held in Berlin, June 10–14, 1984

Edited by D. Ferus, R. B. Gardner, S. Helgason and U. Simon

Springer-Verlag
Berlin Heidelberg New York Tokyo

Editors

Dirk Ferus
Fachbereich 3 – Mathematik, Technische Universität Berlin
Straße des 17. Juni 135, 1000 Berlin 12, Federal Republic of Germany

Robert B. Gardner
Department of Mathematics, University of North Carolina
Chapel Hill, NC 27514, USA

Sigurdur Helgason
Department of Mathematics, Massachusetts Institute of Technology
Cambridge, MA 02139, USA

Udo Simon
Fachbereich 3 – Mathematik, Technische Universität Berlin
Straße des 17. Juni 135, 1000 Berlin 12, Federal Republik of Germany

Mathematics Subject Classification (1980): 53-06, 53Cxx, 58-06

ISBN 3-540-15994-0 Springer-Verlag Berlin Heidelberg New York Tokyo
ISBN 0-387-15994-0 Springer-Verlag New York Heidelberg Berlin Tokyo

Printed in Germany

Printing and binding: Beltz Offsetdruck, Hemsbach/Bergstr.
2146/3140-543210

Preface

The Colloquium "Global Differential Geometry and Global Analysis 1984" was held June 10 - 14, 1984 at the Fachbereich Mathematik der Technischen Universität Berlin. One hundred differential geometers and analysts from Europe and Overseas joined for a series of survey lectures on major recent developments and for workshops on a broad variety of topics. This volume reflects the structure of the conference and includes both, survey articles and research papers. Most of the contributions, however, contain more material, especially more details of proofs, than presented at the conference.

The colloquium was supported by: Deutsche Forschungsgemeinschaft, Deutscher Akademischer Austauschdienst, Berliner Mathematische Gesellschaft, Gesellschaft der Freunde der TU Berlin, Technische Universität Berlin. To all of them we express our sincere thanks. We also thank Bernd Wegner for his valuable help with the preparation of this volume, and the Springer-Verlag for the opportunity to publish it as a Lecture Note.

The editors.

C o n t e n t s

both future and past complete and for each $p \in M$, every null geodesic emanating from p contains a past and future null cut point to p. Then Galloway shows that (M,g) splits isometrically as a Lorentzian product $(\mathbb{R} \times H, -dt^2 \oplus h)$ where (H,h) is a compact Riemannian manifold.

In the present paper, we consider a different class of space-times than those studied in [10] and we use quite different techniques to obtain the following splitting theorem.

Theorem 4.2: Let (M,g) be a globally hyperbolic space-time of dimension ≥ 2 with everywhere nonpositive timelike sectional curvatures $K \leq 0$ which contains a complete timelike line $\gamma: (-\infty,\infty) \to (M,g)$. Then (M,g) is geodesically complete and isometric to a product $(\mathbb{R} \times H, -dt^2 \oplus h)$ where (H,h) is a complete Riemannian manifold. The factor $(\mathbb{R}, -dt^2)$ is represented by γ and (H,h) is represented by a level set of a Busemann function associated to γ.

With the signature convention $(-, +, \ldots, +)$ for a Lorentzian metric, everywhere nonpositive timelike sectional curvatures corresponds to nonnegative Riemannian sectional curvatures and also implies $\mathrm{Ric}(v,v) \geq 0$ for all timelike v. Thus Theorem 4.2 is the complete analogue to the Toponogov splitting theorem of Riemannian geometry mentioned above. Note that Theorem 4.2 provides an affirmative answer to the question raised by Yau for globally hyperbolic space-time with nonpositive timelike sectional curvatures.

We would like to thank J.-H. Eschenburg and E. Heintze for providing us with a copy of [8].

A Toponogov Splitting Theorem for Lorentzian Manifolds

John K. Beem*, Paul E. Ehrlich*, Steen Markvorsen+, Gregory J. Galloway

1. Introduction

The Toponogov Splitting Theorem [5], [7], [14], [15] states that a complete Riemannian manifold (H,h) of nonnegative sectional curvature which contains a line $\gamma:\mathbb{R} \to H$ (i.e., a complete absolutely minimizing geodesic) must be isometric to a product $\mathbb{R} \times H'$. In [5] Cheeger and Gromoll gave a proof of this theorem stemming from their soul construction. Subsequently, Cheeger and Gromoll [6] were able to generalize this Riemannian splitting theorem to the case of nonnegative Ricci curvature using the Busemann functions associated to the line γ.

In [18, p. 696], S. T. Yau raised the question of showing that a geodesically complete Lorentzian 4-manifold of nonnegative timelike Ricci curvature which contains a timelike line (i.e., a complete absolutely maximizing timelike geodesic) is isometrically the Cartesian product of that geodesic and a spacelike hypersurface.

Galloway [10] has recently considered this question for space-times which are spatially closed, i.e., which admit a smooth time function whose level sets are compact (smooth) Cauchy surfaces. Let (M,g) be such a globally hyperbolic space-time which satisfies the strong energy condition $Ric(v,v) \geq 0$ for all timelike vectors v in TM. Suppose further (M,g) contains a timelike curve which is

* This material is based upon work supported by the National Science Foundation under grant no. DMS-840 1599. This work was also supported in part by a grant from the Weldon Spring fund of the University of Missouri.

+ Supported in part by the Danish Natural Science Research Council.

2. Busemann Functions

In this paper (M,g) will always be a connected, time oriented Lorentzian manifold which is globally hyperbolic with metric g of signature $(-, +, \cdots, +)$. This means that (M,g) is a strongly causal space-time and for each pair $p, q \in M$ the intersection of the causal future $J^+(p)$ of p with the causal past $J^-(q)$ of q is compact, cf. [2, p. 30]. A point q is in the chronological future $I^+(p)$ if there is a future directed timelike curve from p to q. The chronological past $I^-(p)$ is defined dually. We write $p \leq q$ if $q \in J^+(p)$ and $p \ll q$ if $q \in I^+(p)$.

Given $p, q \in M$, set $d(p,q) = 0$ if $q \notin J^+(p)$ and let $d(p,q)$ be the supremum of lengths of future directed causal curves from p to q if $q \in J^+(p)$. The Lorentzian distance function d satisfies the reverse triangle inequality $d(p,q) \geq d(p,r) + d(r,q)$ whenever $p \leq r \leq q$. Since (M,g) is globally hyperbolic, the Lorentzian distance function is both finite valued and continuous, cf. [12], [2, p. 86].

If $\gamma: (-\infty, \infty) \to M$ is a future directed timelike line, then for each fixed $r \geq 0$ we define $b_r^+: M \to \mathbb{R}$ by $b_r^+(x) = r - d(x,\gamma(r))$, cf. [3, p. 131], [6, p. 119]. These functions are continuous functions of both x and r because (M,g) is globally hyperbolic.

If $\gamma(r) \notin I^+(x)$, then $d(x,\gamma(r)) = 0$ and $b_r^+(x) = r$. It follows that $b_r^+(x)$ is an increasing function of r for fixed x as long as $\gamma(r) \notin I^+(x)$. On the other hand, if $x \ll \gamma(r)$ for some $r \geq 0$, then there is a smallest $r_0 \geq 0$ such that $x \ll \gamma(r)$ for all $r_0 < r < \infty$ and $x \in J^-(\gamma(r_0)) - I^-(\gamma(r_0))$. Assuming such an r_0 exists, it follows easily from the reverse triangle inequality that $b_r^+(x)$ is a monotone decreasing function of r for all $r > r_0$. If we then allow the possible values of $-\infty$ and $+\infty$, the Buseman function

$$b^+(x) = \lim_{r\to\infty} b_r^+(x)$$

exists for all $x \in M$. In the case $x \notin I^-(\gamma)$ we have $b^+(x) = +\infty$. In general, b^+ need not be a continuous function of x for globally hyperbolic space-times. In fact, examples conformal to a subset of the Minkowski plane L^2 may be constructed with b^+ discontinuous and with $b^+(x) = +\infty$ for some points x of M.

If $x \ll \gamma(r)$ for all $r_0 < r < \infty$, given a sequence of points $\{x_n\}$ converging to x and a sequence of numbers $\{r_n\}$ diverging to $+\infty$ we will have $x_n \ll \gamma(r_n)$ for all sufficiently large n by the openness of chronological sets. We will implicitly use the properties of limit curves (cf. [2]) to define the notion of co-ray to γ as follows. A (future) <u>co-ray</u> to γ from x will be a causal curve starting at x which is future inextendible and is the limit curve of a sequence of maximal length timelike geodesic segments from x_n to $\gamma(r_n)$ for two sequences $\{x_n\}$, $\{r_n\}$ with $x_n \to x$ and $r_n \to \infty$, cf. [2, pp. 33-45]. Past co-rays are defined dually. This definition of co-ray corresponds to that used by Busemann for G-spaces in [3, p. 130]. A (future) co-ray is always a maximal length causal geodesic starting at a point x. The reverse triangle inequality implies that any timelike co-ray must have infinite length and thus be (future) complete. The maximality of co-rays follows from a slight variation of the proof of Theorem 7.10 in [2, p. 206]. However, there may be more than one co-ray to γ from x. Furthermore, a co-ray to γ from x may be a null geodesic rather than a timelike geodesic. To rule out this possibility, we impose the following condition on (M,g).

<u>Definition 2.1</u>: The globally hyperbolic space-time (M,g) satisfies the <u>timelike co-ray condition</u> for the timelike line $\gamma: (-\infty, \infty) \to M$ if for each $x \in I^+(\gamma) \cup I^-(\gamma)$ all future and past co-rays to γ

from x are timelike.

Using the reverse triangle inequality, one may show that the Busemann function b^+ is continuous on the set $I^-(\gamma)$ provided (M,g) satisfies the timelike co-ray condition for γ.

Since $\gamma: (-\infty,\infty) \to M$ is a timelike <u>line</u> rather than just a ray, one may also define a Busemann function b^- on $I^+(\gamma)$ by

$$b_r^-(x) = r - d(\gamma(-r), x) \quad , \; r \geq 0 \; ,$$

and

$$b^-(x) = \lim_{r\to+\infty} b_r^-(x) \; .$$

Provided (M,g) satisfies the timelike co-ray condition for γ, both b^+ and b^- are continuous on $I(\gamma) = I^+(\gamma) \cap I^-(\gamma)$. Also, the reverse triangle inequality implies that

$$B(x) := b^+(x) + b^-(x) \geq 0 \tag{2.1}$$

on $I(\gamma)$ and $B(x) = 0$ if x is on γ.

At this juncture, we do <u>not</u> know that B, b^+ or b^- are differentiable functions. Thus as in [8], we now need to define smooth local support functions at each $p \in I(\gamma)$ for b^+ and b^- in order to prove that b^+ and b^- are in fact smooth on $I(\gamma)$ under the hypotheses of Theorem 4.2. To this end fix $p \in I(\gamma)$ and a sequence of real numbers $\{r_n\}$ with $r_n \to +\infty$. Set $\alpha_n = d(p, \gamma(r_n))$. Then α_n is positive for all sufficiently large n, $\alpha_n \to \infty$ by the reverse triangle inequality, and

$$b^+(p) = \lim_{n\to\infty} (r_n - \alpha_n) \; . \tag{2.2}$$

Furthermore, for each sufficiently large n there is some unit future directed timelike vector $v_n^+ \in T_pM$ such that $\exp_p(\alpha_n v_n^+) = \gamma(r_n)$. By taking a subsequence if necessary and by using the timelike co-ray condition we may assume w.l.g. that $v_n^+ \to v^+$ where v^+ is a unit future directed timelike vector at p. The future inextendible timelike geodesic with initial velocity vector v^+ is a co-ray to γ from p.

Define a family of functions $b_{p,a}^+$ as follows (cf. [8, p. 2]):

$$b_{p,a}^+(x) = b^+(p) + a - d(x, \exp_p av^+) \ . \tag{2.3}$$

For sufficiently large a these functions are continuous <u>local super support functions</u> for the Busemann function b^+; that is, $b_{p,a}^+(p) = b^+(p)$ and $b_{p,a}^+(x) \geq b^+(x)$ for all x sufficiently close to p.

Now given $p \in I(\gamma)$, construct a unit past directed timelike vector $v^- \in T_pM$ using the same technique as above for the future case. Then

$$b_{p,a}^-(x) = b^-(p) + a - d(\exp_p(av^-), x) \tag{2.4}$$

provides a family of local super support functions for b^-.

For any fixed parameter value a, the nonspacelike cut locus of $\exp_p(av^{\pm})$ is closed (cf. [2, p. 242]) and since $s \to \exp_p(sv^{\pm})$, $s \in [0,a]$ is maximal we get that there is a neighborhood of p in which $b_{p,a}^+$ (resp., $b_{p,a}^-$) is a <u>smooth</u> super support function for the continuous Busemann function b^+ (resp., b^-). Hence $B_{p,a} = b_{p,a}^+ + b_{p,a}^-$ is also a smooth super support function for $B = b^+ + b^-$ near p.

3. The Significance of Nonpositive Timelike Sectional Curvature

Let u,v be tangent vectors at some point $x \in M$ which span a nondegenerate two-plane E. Then the sectional curvature $K(E)$ of E is given by

$$K(E) = \frac{\langle R(u,v)v, u\rangle}{\langle u,u\rangle\langle v,v\rangle - \langle u,v\rangle^2}$$

cf. [2, p. 409]. The plane E is <u>timelike</u> if the metric induced on E is Lorentzian. In this case, $\langle u,u\rangle \langle v,v\rangle - \langle u,v\rangle^2$ is negative.

A first important consequence of nonpositive timelike sectional curvature is

<u>Proposition 3.1</u>: Let (M,g) be a globally hyperbolic space-time of dimension ≥ 2 with everywhere nonpositive timelike sectional curvatures $K \leq 0$ which contains a complete timelike line γ. Then the timelike co-ray condition holds on $I(\gamma) = I^+(\gamma) \cap I^-(\gamma)$. Thus b^+ and b^- are both continuous on $I(\gamma)$.

This result is proved using S. Harris' [11, p. 303] Lorentzian version of the Toponogov triangle comparison theorem.

A second consequence of the sectional curvature hypothesis is that we may obtain the following estimate on the Hessians of the (smooth) support functions $b^+_{p,a}$ and $b^-_{p,a}$ for the Busemann functions b^+ , b^- , cf. [8], [6].

<u>Proposition 3.2</u>: Let (M,g) be a globally hyperbolic space-time with $K \leq 0$ and suppose that (M,g) contains a timelike line $\gamma: (-\infty, \infty) \to (M,g)$. Then for any $p \in I(\gamma)$ and $a > 0$, we have

$$\text{Hess}(b^+_{p,a})(w,w) \le \frac{\alpha_+^2(w)}{a} \tag{3.1}$$

and

$$\text{Hess}(b^-_{p,a})(w,w) \le \frac{\alpha_-^2(w)}{a} \tag{3.2}$$

for any $w \in T_pM$ where $\alpha_+^2(w) = \langle w,w\rangle + (\langle w,v^+\rangle)^2$ and $\alpha_-^2(w) = \langle w,w\rangle + (\langle w,v^-\rangle)^2$.

Recall that the extended real valued function B is defined on M by

$$B(x) = b^+(x) + b^-(x) .$$

We have already shown that $B(x) \ge 0$ on $I(\gamma)$, that B is continuous on $I(\gamma)$, and that B vanishes on γ. Using the timelike co-ray condition, Proposition 3.2 and a one-dimensional Calabi-type maximum principle argument, one may now show that B vanishes on the connected set $I(\gamma)$, cf.[8, Lemma 3]. Hence $b^+(x) = -b^-(x)$ for $x \in I(\gamma)$. Furthermore, the Busemann functions b^+ and b^- are once differentiable on $I(\gamma)$ and the vector field $V = \text{grad } b^+ = -\text{grad } b^-$ is a continuous unit past directed timelike vector field defined on $I(\gamma)$. At each point $p \in I(\gamma)$ there is a unique future directed co-ray $c^+(t) = \exp_p(-tV)$ and a unique past directed co-ray $c^-(t) = \exp_p(tV)$ to γ. These co-rays to γ at p fit together to form a (distance realizing and complete) timelike line.

4. Splitting $I(\gamma)$ and M .

We are now ready to show that $I(\gamma)$ is a metric product.

<u>Lemma 4.1</u>. The set $I(\gamma)$ is isometric to a Lorentzian product $\mathbb{R} \times H$ where (H,h) is a spacelike hypersurface of $I(\gamma)$. Furthermore, each spacelike slice $\{t_0\} \times H$ corresponds to the intersection of $I(\gamma)$ with a level set of b^+ (resp., b^-).

<u>Proof</u>: Fix $p \in I(\gamma)$ and let c be a geodesic with $c(0) = p$. Using $b^+_{p,a}(x) \geq b^+(x) = -b^-(x) \geq -b^-_{p,a}(x)$ for all x near p we find that $b^+ \circ c$ has both super support functions $b^+_{p,a} \circ c$ and subsupport functions $-b^-_{p,a} \circ c$. Proposition 3.2 shows that these support functions have arbitrarily small second derivatives for all t near 0. If $L:\mathbb{R} \to \mathbb{R}$ is any affine function, then the same is true of $b^+ \circ c - L$. It follows that $b^+ \circ c$ is an affine function near $t = 0$ and this implies $b^+ \circ c$ is an affine function for any geodesic c with image in $I(\gamma)$. Let $H(t_0) = \{q \in I(\gamma) \,|\, b^+(q) = t_0\}$ be the t_0 level set of b^+ in $I(\gamma)$ and let c be any geodesic segment $c\colon [0,a] \to I(\gamma)$ with endpoints in $H(t_0)$. Then $b^+ \circ c(0) = b^+ \circ c(a)$ implies $b^+ \circ c(t) = b^+ \circ c(0)$ for all $0 \leq t \leq a$. Thus $c([0,a]) \subset H(t_0)$ which shows $H(t_0)$ is totally geodesic.

Fixing $p \in I(\gamma)$ we let $e_1 = -\operatorname{grad} b^+(p), e_2, \cdots, e_n$ be an orthonormal basis of T_pM and use this basis to obtain normal coordinates $x_1, x_2, \cdots, x_n$ near p. The affine function $b^+ \circ c$ is given in these local coordinates by $b^+(x) = x_1 + b^+(p)$. Thus b^+ is smooth.

The vector field $\operatorname{grad} b^+$ is everywhere orthogonal to the totally geodesic level surfaces $H(t_0)$ and $X = \operatorname{grad} b^+$ is a unit normal field to $H(t_0)$. The second fundamental form S_X must vanish on $H(t_0)$ because this surface is totally geodesic. Thus if v,w are tangent to $H(t_0)$, $S_X(v,w) = \langle -\nabla_v X, w\rangle = 0$. On the other hand, $\langle X,X\rangle \equiv -1$ yields that $\nabla_v X$ is orthogonal to X and hence tangential to $H(t_0)$. We conclude that $\nabla_v X = 0$ if v is tangential to $H(t_0)$. Furthermore, X is the unit tangent to the (geodesic) co-ray to γ through each $p \in I(\gamma)$ and hence $\nabla_X X \equiv 0$. Thus $X = \text{grad } b^+$ is a parallel timelike vector field on $I(\gamma)$. Hence $I(\gamma)$ splits locally isometrically by Wu's version of the Lorentzian de Rham Theorem, cf. [17, p. 299].

The vector field $\text{grad } b^+$ is complete since all co-rays to γ are complete geodesics which are contained in $I(\gamma)$. Consequently, the map $I(\gamma) \to I(\gamma)$ given by $p \to \exp_p(t \text{ grad } b^+(p))$ is an isometry of $I(\gamma)$ onto $I(\gamma)$ for each fixed $t \in \mathbb{R}$. This isometry takes level sets of b^+ to level sets of b^+. Using the induced metric h on $H(0)$ and the product metric $-dt^2 \oplus h$ on $\mathbb{R} \times H(0)$, we find that $F: \mathbb{R} \times H(0) \to I(\gamma)$ given by $f(t,p_0) = \exp_{p_0}(t \text{ grad } b^+(p_0))$ is an isometry onto $I(\gamma)$. This establishes the result. □

We are finally ready to prove the main theorem.

Theorem 4.2. Let (M,g) be a globally hyperbolic space-time of dimension ≥ 2 with everywhere nonpositive timelike sectional curvatures $K \leq 0$ which contains a complete timelike line $\gamma: (-\infty, \infty) \to (M,g)$. Then (M,g) is isometric to a product $(\mathbb{R} \times H, -dt^2 \oplus h)$ where (H,h) is a complete Riemannian manifold. The factor $(\mathbb{R}, -dt^2)$ is represented by γ and (H,h) is represented by a level set of a Buseman function associated to γ.

Proof: The set $I(\gamma)$ must be strongly causal because it is an open subset of the globally hyperbolic space-time (M,g). Furthermore, $p, q \in I(\gamma)$ implies the compact set $J^+(p) \cap J^-(q)$ also lies in $I(\gamma)$. Thus $I(\gamma)$ is globally hyperbolic. Lemma 4.1 shows that $I(\gamma)$ is isometric to $\mathbb{R} \times H$, but $\mathbb{R} \times H$ is globally hyperbolic iff H is complete (cf. [2, p. 66]). Furthermore, the geodesics γ of $\mathbb{R} \times H$ are of the form $\gamma(t) = (at, c(t))$ where

$a > 0$ and c is a geodesic of H. Hence H is complete iff $\mathbb{R} \times H$ is geodesically complete. Thus the global hyperbolicity of $I(\gamma)$ together with the splitting of $I(\gamma)$ yield that $I(\gamma)$ is geodesically complete. Consequently, $I(\gamma)$ is inextendible (cf. [2, p. 160]) and must therefore be all of M. □

Corollary 4.3: (M,g) is geodesically complete and the level surfaces of the Busemann functions b^+ and b^- are complete (spacelike) Cauchy hypersurfaces of (M,g).

We conclude by stating two related results whose proofs will appear elsewhere.

Proposition 4.4. Let (M,g) be a globally hyperbolic space-time with $Ric(v,v) \geq 0$ on all timelike vectors v. Assume that (M,g) obtains a complete timelike line γ such that every co-ray to γ is timelike and without focal points. Then (M,g) is isometric to a product $(\mathbb{R} \times H, -dt^2 \oplus h)$.

Theorem 4.5. Let (M,g) be a space-time which contains a compact Cauchy surface and has everywhere nonpositive timelike sectional curvatures. Then either (M,g) is timelike geodesically incomplete or (M,g) splits isometrically as $(\mathbb{R} \times H, -dt^2 \oplus h)$ where (H,h) is compact.

References

1. J. K. Beem and P. E. Ehrlich, "Constructing maximal geodesics in strongly causal space-times", Math. Proc. Camb. Phil. Soc. 90, 183-190(1981; Zbl. 483.53056).

2. J. K. Beem and P. E. Ehrlich, Global Lorentzian Geometry, Marcel Dekker Pure and Applied Mathematics, Vol. 67, New York (1981; Zbl. 462.53001).

3. H. Busemann, The Geometry of Geodesics, Academic Press, New York (1955; Zbl. 112,370).

4. E. Calabi, "An extension of E. Hopf's maximum principle with an application to Riemannian geometry", Duke Math. J. 25, 45-56 (1957; Zbl. 79,118).

5. J. Cheeger and D. Gromoll, "On the structure of complete manifolds of nonnegative curvature", Ann. of Math. 96, 413-443 (1972; Zbl. 246.53049).

6. J. Cheeger and D. Gromoll, "The splitting theorem for manifolds of nonnegative Ricci curvature", J. Diff. Geo. 6, 119-128 (1971; Zbl. 223.53033).

7. S. Cohn-Vossen, "Totalkrümmung und geodätische Linien auf einfach zusammenhängenden offenen vollständigen Flachenstücken", Mat. Sb. (N.S.) 1, 43, 139-163 (1936; Zbl. 14,276).

8. J.-H. Eschenburg and E. Heintze, "An elementary proof of the Cheeger-Gromoll splitting theorem", (to appear in Ann. of Global Anal. Geom.)

9. J.-H. Eschenburg and J. J. O'Sullivan, "Jacobi tensors and Ricci curvature", Math. Annalen 252, 1-26 (1980; Zbl. 433.53029).

10. G. J. Galloway, "Splitting theorems for spatially closed space-times", preprint, 1984.

11. S. G. Harris, A triangle comparison theorem for Lorentz manifolds, Indiana Univ. Math. J. 31, 289-308 (1982; Zbl. 496.53042).

12. S. W. Hawking and G.F.R. Ellis, The Large Scale Structure of Space-time, Cambridge University Press, Cambridge (1973; Zbl. 265.53054).

13. B. O'Neill, Semi-Riemannian Geometry, Academic Press, New York, (1983; Zbl. 531.53051).

14. V. A. Toponogov, "Riemannian space containing straight lines", Dokl. Akad. Nauk, SSSR 127 (1959) 976-979 (Russian). - Amer. Math. Soc. Transl. (2) 37 (1964) 287-290 (Zbl. 94,347).

15. V. A. Toponogov, "The metric structure of Riemannian spaces with nonnegative curvature which contain straight lines", Sibirsk. Mat. Z. 5 (1964) 1358-1369 (Russian). - Amer. Math. Soc. Transl. (2) 70 (1968) 225-239 (Zbl. 145,185).

16. H. Wu, "An elementary method in the study of nonnegative curvature", Acta Mathematica 142, 57-78 (1979; Zbl. 403.53022).

17. H. Wu, "On the de Rham decomposition theorem", Illinois J. Math. 8, 291-311(1964; Zbl. 122,400).

18. Ṡ. T. Yau, ed., Seminar on Differential Geometry, Annals of Math. Studies, Princeton Univ. Press, Vol. 102, (1982; Zbl. 479.53001).

J. Beem, P. Ehrlich, S. Markvorsen
Mathematics Department
University of Missouri
Columbia, Mo. 65211

G. Galloway
Mathematics Department
University of Miami
Coral Gables, Fl. 33124

A SURVEY ON CR - SUBMANIFOLDS OF KAEHLERIAN MANIFOLDS

BY

AUREL BEJANCU

§ 1. Introduction.

The theory of submanifolds in Kaehlerian manifolds is one of the most interesting topics in differential geometry. According to the behaviour of the tangent bundle of a submanifold with respect to the action of the almost complex structure of the ambient manifold we have three typical classes of submanifolds : holomorphic submanifolds (see Ogiue [21]) , totally real submanifolds (see Yano - Kon [29]) and CR - submanifolds. The notion of CR - submanifold has been introduced by the author in [3] as follows:

Let N be an almost Hermitian manifold and J be the almost complex structure on N. A real submanifold M of N is called a CR - submanifold if there exists a differentiable distribution D on M satisfying

$$\text{(i)}\quad J(D_x) = D_x \quad \text{and} \quad \text{(ii)}\quad J(D_x^{\perp}) \subset T_xM^{\perp}, \text{ for each } x \in M,$$

where $D^{\perp}$ is the complementary orthogonal distribution to D and $T_xM^{\perp}$ is the normal space to M at x. If we have $\dim . D_x^{\perp} = \dim . T_xM^{\perp}$ then we say that M is an anti - holomorphic submanifold. Thus, holomorphic submanifolds and totally real submanifolds are particular cases of CR - submanifolds. Moreover, each real hypersurface of N is a CR - submanifold which is neither a holomorphic submanifold nor a totally real submanifold.

The purpose of the present lecture is to discuss shortly some of the main problems in geometry of CR - submanifolds. First, we study the integrability of both distributions D and $D^{\perp}$ on M. Then we give results on the geometry of some special classes of CR - submanifolds. Finally, we show that the theory of CR - submanifolds stimulated the work on new structures on submanifolds of various classes of manifolds.

§ 2. Integrability and geometry of leaves

Let M be a CR - submanifold of a Kaehlerian manifold N. We shall study the integrability of D and $D^{\perp}$ and the immersions of their leaves either in M or in N.

First we state

Theorem 1. Let M be a CR - submanifold of a Kaehlerian manifold N. Then we have

(i) the distribution $D^{\perp}$ is integrable,

(ii) the distribution D is integrable if and only if the second fundamental form h of M satisfies

$$h(X,JY) = h(Y,JX) \quad , \text{ for any } X,Y \in D.$$

The assertion (i) is due to Blair - Chen [14] and the assertion (ii) has been obtained by the author in [3]. Also, Blair and Chen constructed in [14] an example of a CR - submanifold of a Hermitian manifold on which $D^{\perp}$ is not integrable. The integrability of D and $D^{\perp}$ on CR - submanifolds in almost Hermitian manifolds and nearly Kaehlerian manifolds has been studied by the author in [4] and respectively by Urbano [28] and Sato [24].

Next, we denote by P and Q the projection morphisms of TM to D and respectively $D^{\perp}$. Then we have a tensor field φ of type (1,1) on M given by $\varphi X = JPX$ and a normal bundle - valued 1 - form ω on M given by $\omega X = JQX$, for any vector field X tangent to M. We say that a linear connection ∇ on M is a φ - connection if we have $\nabla_X \varphi = 0$, for each X tangent to M. We obtained in [9] that all φ - connections on a CR -

submanifold M of an almost Hermitian manifold N are given by

$$\nabla_X Y = P(\overset{o}{\nabla}_X PY) + Q(\overset{o}{\nabla}_X QY) + \frac{1}{2}\left\{(\overset{o}{\nabla}_X \varphi)\varphi Y + PK(X,PY) - \varphi K(X,\varphi Y)\right\} + QS(X,QY) \;,$$

for all X,Y tangent to M, where $\overset{o}{\nabla}$ is a linear connection with respect to which both distributions D and $D^{\perp}$ are parallel and K,S are arbitrary tensor fields of type (1,2) on M.

We say that a CR - submanifold M of an almost Hermitian manifold N is a <u>CR - product</u> if both distributions D and $D^{\perp}$ are integrable and M is locally a Riemannian product $M_1 \times M_2$, where M_1 is a leaf of D and M_2 is a leaf of $D^{\perp}$. If M is a CR - product with $D \neq \{0\}$ and $D^{\perp} \neq \{0\}$ we say that M is a <u>proper CR - product</u>.

<u>Theorem 2</u>. Let M be a CR - submanifold of an almost Hermitian manifold N. If the Levi - Civita connection on M is a φ - connection then M is a CR - product.

Chen proved in [16] that a CR - submanifold of a Kaehlerian manifold is a CR - product if and only if the Levi - Civita connection is a φ - connection. With respect to the existence of CR - products we give the following important result

<u>Theorem 3</u> (<u>Sekigawa</u> [25]). There exist no proper CR - products in the sphere S^6.

A CR - submanifold M is called <u>D - geodesic</u> (resp. <u>$D^{\perp}$ - geodesic</u>) if its second fundamental form satisfies

$$h(D,D) = \{0\} \qquad (\text{resp. } h(D^{\perp},D^{\perp}) = \{0\}).$$

Denote by $\tilde{D}$ the complementary orthogonal subbundle to $JD^{\perp}$ in $TM^{\perp}$. Then with respect to the immersions of leaves of both distributions we have

Theorem 4 (Chen [16]). Let M be a CR - submanifold of a Kaehlerian manifold N. Then we have

(i) the distribution D is integrable and its leaves are totally geodesic immersed in M if and only if $h(D,D) \subset \widetilde{D}$,

(ii) the distribution D is integrable and its leaves are totally geodesic immersed in N if and only if M is D - geodesic.

Theorem 5. Let M be a CR - submanifold of a Kaehlerian manifold N. Then we have

(i) each leaf of $D^{\perp}$ is totally geodesic immersed in M if and only if $h(D,D^{\perp}) \subset \widetilde{D}$,

(ii) each leaf of $D^{\perp}$ is totally geodesic immersed in N if and only if M is $D^{\perp}$ - geodesic and $h(D,D^{\perp}) \subset \widetilde{D}$.

The first assertion was proven by Bejancu - Kon - Yano in [12] and the second by Chen in [16]. Other results on the immersions of leaves either in M or in N were obtained by the author in [6].

§ 3. Special classes of CR - submanifolds

First, for totally umbilical CR - submanifolds we have the following important result

Theorem 6 (Chen [15] and Bejancu [8]). Let M be a CR - submanifold of a Kaehlerian manifold N. If M is totally umbilical then we have

(i) M is totally geodesic, or

(ii) M is totally real, or

(iii) $\dim.D_x^{\perp} = 1$, for each $x \in M$.

On the other hand, each totally geodesic CR - submanifold is a CR - product. Thus by Theorem 6 each totally umbilical CR - submanifold with $\dim.D_x^{\perp} > 1$ is a CR - product. The case (iii) is not studied yet generally. We have only some results on totally umbilical real hypersurfaces (see Tashiro - Tachibana [27], Okumura [22]).

Now, by using the tensor fields φ and ω we define a tensor field S of type (1,1) on each CR - submanifold M by

$$S(X,Y) = [\varphi,\varphi](X,Y) - 2\left\{ Jd\ (X,Y)\right\}^T \quad , X,Y \in TM,$$

where $[\varphi,\varphi]$ is the Nijenhuis tensor of φ and $\{\cdot\}^T$ is the tangent component of $\{\cdot\}$. If S vanishes identically on M then we say that M is a normal CR - submanifold. Of course, each normal real hypersurface (see Blair [13], Okumura [23]) is an example of a normal CR - submanifold. We denote by A_V the fundamental tensor of Weingarten with respect to the normal section V. Then we have

Theorem 7. A CR - submanifold M of a Kaehlerian manifold N is normal if and only if we have

$$A_{\omega Y}(\varphi X) = \varphi(A_{\omega Y}X) \quad , \quad \text{for any } X \in D \text{ and } Y \in D^{\perp}.$$

For normal real hypersurfaces in an Euclidean space we have a theorem of classification (see Okumura [22]). It is an open problem whether such a theorem holds for normal CR - submanifolds.

We say that a CR - submanifold is cosymplectic if it is normal and the differential form ω is closed. Then it is proven (see [10]) that an anti - holomorphic submanifold of a Kaehlerian manifold is cosymplectic if and only if it is a CR - product.

Concerning CR - products in complex space forms it was proven the non - existence of CR - products in a complex space form N(c) with $c < 0$ (see [16]) and of anti - holomorphic CR - products in N(c) with $c > 0$ (see [12]). Examples of proper CR - products in a complex projective space have been constructed by Chen in [16]. Also, we have to say that due to Shimizu (see [26]) we have many examples of proper CR - submanifolds which are neither CR - products nor real hypersurfaces.

§ 4. Some extensions to the quaternion case

Let N be a 4n - dimensional manifold and g be a Riemannian metric on N. Then N is said to be a quaternion Kaehlerian manifold if there exists a 3 - dimensional vector bundle V of tensors of type (1,1) with local basis of almost Hermitian structures J_1, J_2, J_3 satisfying

$$J_1 \circ J_2 = -J_2 \circ J_1 = J_3 \quad \text{and} \quad \tilde{\nabla}_X J_a = \sum_{b=1}^{3} Q_{ab}(X) J_b, \quad a = 1,2,3,$$

for all vector fields X tangent to N, where $\tilde{\nabla}$ is the Levi - Civita connection determined by g on N and Q_{ab} are certain 1 - forms locally defined on N such that $Q_{ab} + Q_{ba} = 0$.

Suppose $\{J_1, J_2, J_3\}$ is a local basis for the vector bundle V on a coordinate neighborhood U on N. We take another coordinate neighborhood $\tilde{U}$ with the corresponding local basis $\{\tilde{J}_1, \tilde{J}_2, \tilde{J}_3\}$ and $U \cap \tilde{U} \neq \emptyset$. Then we have

$$\tilde{J}_a = \sum_{b=1}^{3} A_{ab} J_b \quad , \qquad a = 1,2,3,$$

where $[A_{ab}]$ is an element of the special orthogonal group SO(3).

Now, let M be an m - dimensional Riemannian manifold isometrically immersed in N. We say that M is a quaternion CR - submanifold (see [1]), if there exists a differentiable distribution D on M satisfying

$$J_a(D) = D \quad \text{and} \quad J_a(D^{\perp}) \subset TM^{\perp}, \quad \text{for all } a = 1,2,3,$$

where $D^{\perp}$ is the complementary orthogonal distribution to D.

Of course, a quaternion CR - submanifold is the natural extension of the notion of CR - submanifold to the quaternionic case. However, in this case a real hypersurface of N is not a quaternion CR -

submanifold.

Further, we note that a CR - structure, $(D,D^{\perp})$ on the tangent bundle to a submanifold in a Kaehlerian manifold induces a CR - structure $(\tilde{D},JD^{\perp})$ on the normal bundle $TM^{\perp}$ and viceversa. But it is easily seen that this is not the case for quaternion CR - submanifolds. For this reason we say that M is a QR - (quaternion - real) submanifold if there exists a vector subbundle ν of the normal bundle $TM^{\perp}$ such that

$$J_a(\nu) = \nu \quad \text{and} \quad J_a(\nu^{\perp}) \subset TM \ , \text{ for all } a = 1,2,3,$$

where $\nu^{\perp}$ is the complementary orthogonal subbundle to ν. Then each real hypersurface of N is a non-trivial example of a QR - submanifold. The geometry of both, quaternion CR - submanifolds and QR - submanifolds is still under research and many results are expected to be obtained especially in quaternion space forms.

Finally, we note that the notion of CR - submanifolds has been considered also in Sasakian manifolds, locally conformal manifolds, locally product manifolds, etc. (see [11], [17] - [20], [31]). A parallel study of CR - submanifolds in Kaehlerian and Sasakian manifolds is given by Yano - Kon in [32] via the method of Riemannian fibre bundles.

REFERENCES

1. Barros M., Chen B.Y. and Urbano F., Quaternion CR - submanifolds of quaternion manifolds, Kodai Math.J., 4, 399-417 (1981; Zbl. 484.53030).

2. Barros M. and Urbano F., CR - submanifolds of generalized complex space forms , An.St.Univ.Al.I.Cuza Iasi 25, 355-363 (1979; Zbl. 418.53022).

3. Bejancu A., CR - submanifolds of a Kaehler manifold I., Proc. of A.M.S. 69, 134-142 (1978; Zbl. 368.53040).

4. ———— , On integrability conditions on a CR - submanifold, An.St.Univ.Al.I.Cuza Iasi 24, 21-24 (1978; Zbl. 409.53038).

5. ———— , CR - submanifolds of a Kaehler manifold II, Trans. of A.M.S. 250, 333-345 (1979; Zbl. 368.53041).

6. ———— , On the geometry of leaves on a CR - submanifold, An.St.Univ.Al.I.Cuza Iasi 25, 393-398 (1979; Zbl. 425.53030).

7. ———— , Normal CR - submanifolds of a Kaehler manifold, An.St.Univ.Al.I.Cuza Iasi 26, 123-132 (1980; Zbl. 441.53051).

8. ———— , Umbilical CR - submanifolds of a Kaehler manifold, Rendiconti di Mat. 13, 431-446 (1980; Zbl. 455.53040).

9. ———— , F - connections on a CR - submanifold, Bul. Inst.P.Iasi 27, 33-40 (1981; Zbl. 498.53043).

10. ———— , Anti - holomorphic submanifolds of a Kaehler manifold, Math.Rep.Toyama Univ. 6, 179-196 (1983; Zbl. 519.53034).

11. ———— , Semi - invariant submanifolds of locally product Riemannian manifolds, An.Univ.Timisoara , 22(1984), 3 - 11.

12. Bejancu A., Kon.M. and Yano K., CR - submanifolds of a complex space form, J. Differential Geometry 16, 137-145 (1981; Zbl. 469.53048).

13. Blair D.E., Contact manifolds in Riemannian geometry, Lecture Notes in Math. 509, Springer - Verlag, Berlin, (1976; Zbl. 319.53026).

14. Blair D.E. and Chen B.Y., On CR - submanifolds of Hermitian manifolds, Israel J.Math. 34, 353-363 (1979; Zbl. 453.53018).

15. Chen B.Y., Totally umbilical submanifolds of Kaehler manifolds, Archiv der Math. 36, 83-91 (1981; Zbl. 459.53044).

16. ———— , CR - submanifolds of a Kaehler manifold I, II, J. Differential Geometry 16, 305-323 (1981; Zbl. 431.53048); 493-509 (1981; Zbl. 485.53051).

17. Ianus S. and Mihai I., Semi - invariant submanifolds of an almost paracontact manifold, Tensor N.S. 39, 195-200 (1982; Zbl. 514.53015).

18. Kobayashi M., CR - submanifolds of a Sasakian manifold, Tensor N.S. 35, 297-307 (1981; Zbl. 484.53048).

19. Matsumoto K., On submanifolds of locally product Riemannian manifolds, TRU Math. 18-2, 145-157 (1982; Zbl. 532.53019).

20. ———— , On CR - submanifolds of locally conformal Kaehler manifolds, J. Korean Math.Soc. 21(1984), 49-61.

21. Ogiue K., Differential geometry of Kaehler submanifolds, Advances in Math. 13, 73-114 (1974; Zbl. 275.53035).

22. Okumura M., Certain almost contact hypersurfaces in Euclidean spaces, Kodai Math.Sem.Rep. 16, 44-54 (1964; Zbl. 116, 389).

23. ———— , On some real hypersurfaces of a complex projective space, Trans. of A.M.S. 212, 355-364 (1975; Zbl. 288.53043).

24. Sato N., Certain CR - submanifolds of almost Hermitian manifolds, to appear.

25. Sekigawa K., Some CR - submanifolds in a 6 - dimensional sphere, Tensor N.S.,41 (1984), 13-20.

26. Shimizu Y., On a construction of homogeneous CR - submanifolds in a complex projective space, Commentarii Math. Univ. Sancti Pauli 32, 203-207 (1982; Zbl. 518.53055).

27. Tashiro Y. and Tachibana S., On Fubinian and C - Fubinian manifolds, Kodai Math.Sem.Rep. 15, 176-183 (1963; Zbl. 116, 390).

28. Urbano F., CR - submanifolds. Totally real submanifolds of quaternionic manifolds. Thesis at Granada University (1979), (in Spanish).

29. Yano K. and Kon.M., Anti - invariant submanifolds, M. Dekker, New York (1976; Zbl. 349.53055).

30. ——— , Differential geometry of CR - submanifolds, Geometriae Dedicata 10, 369-391 (1981; Zbl. 464.53044).

31. ——— , Contact CR - submanifolds, Kodai Math.J., 5, 238-252 (1982; Zbl. 496.53038).

32. ——— , CR - submanifolds of Kaehlerian and Sasakian manifolds, Birkhäuser, Boston (1983; Zbl. 496.53037).

POLITECHNIC INSTITUTE OF IASI
DEPARTMENT OF MATHEMATICS
R.S.ROMANIA.

Address for correspondence:

C.P.606, IASI 1,
6600 IASI, R.S.ROMANIA.

ISOPERIMETRIC INEQUALITIES, HEAT EQUATION AND GEOMETRIC APPLICATIONS

Pierre H. Bérard
Université de Savoie
73011 Chambery Cedex / France

0. This text is a summary of a lecture given at the Conference on Global Differential Geometry and Global Analysis, T.U. Berlin 1984. For detailed proofs we refer the reader to [B-G] and [B-B-G].

All Riemannian manifolds (M,g) are assumed to be C^∞, compact, connected, without boundary.

1. The main question we shall deal with is the following *"Can one give bounds on geometric or topological invariants of a Riemannian manifold in terms of local estimates on curvature ?"*

2. EXAMPLES. (i) in dimension 2, the Gauss-Bonnet theorem answers the above question as follows: if the curvature of (M,g), K(M,g), is bounded from below by $-k (k \geq 0)$ and if the volume of (M,g), Vol(M,g), is bounded from above by v then the first Betti number $b_1(M)$ is bounded from above by $2 + \frac{kv}{2\pi}$

$$\begin{matrix} K(M,g) \geq -k \\ Vol(M,g) \leq v \end{matrix} \quad \Longrightarrow \quad b_1(M) \leq 2 + \frac{kv}{2\pi}$$

(ii) When the Ricci curvature of (M,g) is nonnegative, Bochner's theorem gives a partial answer to our question.

$$\text{Ricci }(M,g) > 0 \qquad b_1(M) = 0$$
$$\text{Ricci }(M,g) \geq 0 \qquad b_1(M) \leq \dim M .$$

A natural question here is "what happens when Ricci (M,g) is allowed to take negative values ?"

3. REMARKS Let us point out two facts

(i) curvature assumptions alone are not sufficient (one can always scale the metric).

(ii) curvature and volume assumptions are not sufficient, see [B-G], page XV.2.

In what follows, taking into account the above remarks, we shall use assumptions on CURVATURE AND DIAMETER. In general, we shall make assumptions on the Ricci curvature or on the sectional curvature because scalar curvature does not carry enough information.

4. FRAME WORK. We shall work in the following framework, involving

WEITZENBÖCK formulas.

Let $E \to M$ be a fiber bundle of rank l, equipped with a Riemannian metric $<.,.>$, a compatible connexion D and a natural Laplacian $\tilde{\Delta}$. We assume that $\tilde{\Delta}$ can be written as $\tilde{\Delta} = D^*D + R_x$ where $D^*D = \bar{\Delta}$ is called the <u>rough Laplacian</u> and where R_x is a <u>symmetric endomorphism</u> of E which can be estimated in terms of the curvature of M [this is a <u>WEITZENBÖCK formula</u>].

We are interested in upper bounds of dim Ker$\tilde{\Delta}$ and lower bounds on the eigenvalues of $\tilde{\Delta}$.

5. <u>EXAMPLES</u>. The above situation (which can be generalized) is quite common in geometry

(i) $E = T^*M$, $\tilde{\Delta}$ = Laplacian on 1-forms, $\tilde{\Delta} = \bar{\Delta}$ + Ricci : we look for upper bounds on $b_1(M)$ in terms of Ricci and Diameter.

(ii) $E = \Lambda^p T^*M$, $\tilde{\Delta}$ = Laplacian on p-forms: we look for upper bounds on $b_p(M)$ in terms of Sectional Curvature and Diameter.

(iii) $E = M \times \mathbb{R}$, $\tilde{\Delta} = \Delta - \lambda$, where Δ is the Laplacian acting on $C^\infty(M)$: we look for upper bounds on $N(\lambda) = \#\{\lambda_j(M,g) \leq \lambda\}$ i.e. for lower bounds on the eigenvalues of Δ, in terms of Ricci and Diameter.

(iv) other examples involve the $\hat{A}$-genus, dimensions of moduli spaces (e.g. for Einstein metrics), lower bounds on the eigenvalues of the Laplacian in a domain of a closed Riemannian manifold, with Dirichlet boundary condition, ... see [B-G], page XV.19 ff or [B-B-G], §III.

6. <u>HISTORY</u> Let us briefly sketch the history of this type of estimates

1953 Bochner: $b_1(M)$ (Ricci ≥ 0)

1968 Lichnerowicz : $\hat{A}$-genus (Scal ≥ 0)

1980 Li : $b_i(M)$ (curvature allowed to take negative values; uses Sobolev inequalities and Croke isoperimetric constant)

1980 Gromov : $b_i(M)$ (purely geometric methods)

1981 Gallot : $b_i(M)$, $\hat{A}$-genus, $N(\lambda)$,... (curvature allowed to take negative values; uses Sobolev inequalities and isoperimetric constants improving Li's results).

1982 Maurey-Meyer : improve some of the preceding results.

7. <u>NEW PROOF</u> Integrating the Weitzenböck formula of n°4, we can write, for a section s of the fiber bundle E,

$$\int_M <\tilde{\Delta}s,s> = \int_M |Ds|^2 + \int_M <Rs,s>.$$

If we now define

$$R(x) = \inf \{\frac{<R_x s,s>}{<s,s>_x} \; ; \; s \in E_x\}$$

and $R_{min} = \inf\{R(x): x \in M\}$, the mini-max gives

$\dim \operatorname{Ker} \tilde{\Delta} < \tilde{Z}(t) =: \sum_{i=0}^{\infty} \exp(-\tilde{\lambda}_i t) \leqslant \exp(-R_{min} t)\bar{Z}(t)$ where

$\bar{Z}(t) =: \sum_{i=0}^{\infty} \exp(-\bar{\lambda}_i t)$; here $\tilde{\lambda}_i$ (resp. $\bar{\lambda}_i$) is the ith eigenvalue of $\tilde{\Delta}$ (resp. $\bar{\Delta}$).

Now, Kato's inequality ([H-S-U]) says that

$$\bar{Z}(t) \leqslant Z_M(t) = l \sum_{i=0}^{\infty} \exp(-\lambda_i t)$$

where λ_i is the ith eigenvalue of Δ acting on $C^{\infty}(M)$.

Finally, we can write

(8) $\dim \operatorname{Ker} \tilde{\Delta} \leqslant \tilde{Z}(t) \leqslant l \exp(-R_{min} t)\, Z_M(t)$

9. <u>BOCHNER REVISITED</u> If we want to give upper bounds on $b_1(M)$, we can use (8) with $E = T^*M$ and

$R_{min} = r_{min} = \inf \{Ricci(u,u)/(u,u) : u \in UM\}$. We then have

$$b_1(M) \leqslant n \exp(-r_{min} t)\, Z_M(t) =: a_M(t) \text{ for all } t > o.$$

Recalling that $\lim_{t\to\infty} Z_M(t) = 1$, we obtain Bochner's results as a by-product:

(i) if $r_{min} > 0$ $b_1(M) \leqslant \lim_{t\to\infty} a_M(t) = 0$

(ii) if $r_{min} \geqslant 0$ $b_1(M) \leqslant \lim_{t\to\infty} a_M(t) \leqslant n$.

On the other hand, we can also write, for any value of r_{min}

$$b_1(M) \leqslant n \inf \{\exp(-r_{min} t)\, Z_M(t) : t > 0\}$$

which gives upper-bounds for $b_1(M)$ provided one can control $Z_M(t)$. We shall now describe how that can be done.

10. <u>ISOPERIMETRIC INEQUALITY FOR THE HEAT KERNEL</u>

<u>Assumptions</u> : there is an isoperimetric function, i.e. a function $L : [0,1] \to \mathbb{R}_+$ (which depends on M, we shall see examples later on) such that for all regular domain $\Omega \subset M$

$$(*) \quad \frac{Vol(\partial\Omega)}{Vol(M)} \geqslant L\left(\frac{Vol(\Omega)}{Vol(M)}\right) .$$

Given such a function L, on can construct a model space, M^*, namely a manifold with revolution symmetry, such that for geodesic balls Ω^* centered at O

$$\frac{Vol(\partial\Omega^*)}{Vol(M^*)} = L\left(\frac{Vol(\Omega^*)}{Vol(M^*)}\right) .$$

<u>Conclusions</u> Using the classical idea of symmetrization, one can prove the following theorem which we have obtained in a joint work with S. Gallot (see [B-G], page XV.13).

11. <u>THEOREM</u> Under the assumptions 10., one has the following inequalities

$$Z_M(t) \leqslant \mathrm{Vol}(M) \sup_x k_M(t,x,x) \leqslant \mathrm{Vol}(M^*) k_{M^*}(t,0,0)$$

where $k_M(t,x,y)$ stands for the heat kernel of (M,g) .

12. <u>REMARKS</u> (i) Theorem 11 is non empty if and only if (*) is non trivial. Here are two instances of such inequalities

(a) When $\mathrm{Ricci}(M,g) \geqslant (n-1)g$, we can take $M^* = (S^n, \mathrm{can})$ and use Gromov's isoperimetric inequality for (*) .

(b) When $\mathrm{Ricci}(M,g) \geqslant -(n-1)kg$, $k \in \mathbb{R}_+$, we can use $M^* = G^n$ a double-ball see [B-G] page XV.17.

(ii) Theorem 11 also gives estimates on the L^∞-norm of eigenfunctions.

We shall now describe a new isoperimetric inequality which can be used for heat kernel inequalities.

13. <u>A GENERALIZED GROMOV TYPE INEQUALITY</u>

For a Riemannian manifold (M,g) we introduce the following isoperimetric function

$$h(\beta) = \inf \{\frac{\mathrm{Vol}(\partial\Omega)}{\mathrm{Vol}(M)} \mid \Omega \subset M \text{ and } \frac{\mathrm{Vol}(\Omega)}{\mathrm{Vol}(M)} = \beta\},\ 0 \leqslant \beta \leqslant 1.$$

We call $\mathrm{Is}(\beta)$ the same function for the canonical sphère (S^n,can) of radius 1 in $\mathbb{R}^{n+1}$.

The following theorem has been obtained in a joint work with G. Besson and S. Gallot (see [B-B-G],p.3).

14. <u>THEOREM</u>. Let (M,g) be a smooth, compact, n-dimensional Riemannian manifold without boundary, satisfying

$$r_{min}(M) d^2(M) \geqslant \varepsilon(n-1)\alpha^2 \quad \text{where } \varepsilon \in \{-1,0,1\},\ \alpha \in \mathbb{R}_+,$$

$r_{min}(M)$ is defined in n° 9 and $d(M)$ = Diameter of (M,g).

Then, for all $\beta \in [0,1]$

$$d(M) h(\beta) \geqslant a(n,\varepsilon,\alpha)\, \mathrm{Is}(\beta) \quad \text{where}$$

$$a(n,\varepsilon,\alpha) = \begin{cases} \alpha[\omega_n/2 \int_0^{\alpha/2} (\cos t)^{n-1} dt]^{1/n} & \text{if } \varepsilon = +1 \\ (1+n\omega_n)^{1/n} - 1 & \text{if } \varepsilon = 0 \\ \alpha c(\alpha) & \text{if } \varepsilon = -1 \end{cases}$$

where $\omega_n = \int_0^\pi (\sin t)^{n-1} dt$ and $c(\alpha)$ is the unique positive root of the equation $x \int_0^{\alpha} (\mathrm{ch}t + \mathrm{sh}t)^{n-1} dt = \omega_n$.

15. HEAT KERNEL ESTIMATES

From theorems 11 and 14 we can deduce the following results, under the assumptions of Theorem 14.

(C1) $Z_M(t) \leqslant Z_{S^n}(a^2(n,\varepsilon,\alpha)t/d^2(M))$

(C2) $\lambda_1(M) \geqslant n\, a^2(n,\varepsilon,\alpha)/d^2(M)$ ($>n$ if $r_{min}(M) \geqslant (n-1)$ and (M,g) is not (S^n, can); see also [B-B-G] Corollaire 17)

(C3) $b_1(M) \leqslant n \inf_t [\exp(-r_{min}(M)t) Z_{S^n} (a^2(n,\varepsilon,\alpha)t/d^2(M))]$

(C4) $N_M(\lambda) \leqslant eZ_M(1/\lambda) \leqslant eZ_{S^n} (a^2(n,\varepsilon,\alpha)/\lambda d^2(M))$

which implies that

$$\lambda_i(M)\, d^2(M) \geqslant C(n, r_{min} d^2)\, i^{2/n}.$$

Let us point out that (C3) and (C4) are qualitatively sharp : see [B-G], page XV.21 ff.

For other applications (including a gap theorem for λ_{n+2} and a pinching theorem for λ_1 in the case $r_{min}(M) \geqslant n-1$) we refer to [B-G] , page XV.24 and [B-B-G], Corollaire 18.

16. FINAL REMARKS

(i) Although one shall in most cases use Theorem 14, the general formulation of Theorem 12 might turn out to be useful in some instances. Given an isoperimetric function L as in (*) n° 10, the manifold M^* is given by solving an ordinary differential equation satisfied by the meridian curve.

(ii) Other comparison theorems can be deduced from (*) n° 10. One can for example study an inequality of the following type $\Delta u \leqslant \lambda u$ a.e on M, $u \geqslant 0$ on M. From this one can also derive estimates for various geometric or topological invariants : see [GA].

(iii) Let us finally point out that <u>philosophical interpretations</u> of our results are provided by Gromov compactness and pre-compactness theorem as explained in [B-G].

REFERENCES

[B-B-G] Bérard, P. - Besson, G. - Gallot, S.: Sur une inégalité isopérimétrique qui généralise celle de Paul Lévy-Gromov, Preprint Université de Grenoble, Institut Fourier 1984.

[B-G] Bérard, P. - Gallot, S.: Inégalités isopérimétriques pour l'équation de la chaleur et application à l'estimation de quelques invariants, Séminaire Goulaouic-Meyer-Schwartz 1983-84 Exposé n° XV, Ecole Polytechnique Palaiseau 1984.

[GA] Gamara, N. : Thèse de 3ième cycle, Université de Savoie, 1984.

[H-S-U] Hess, H. - Schrader, R. - Uhlenbrock D.A. : Kato's inequality and the spectral distribution of Laplacians on compact Riemannian manifolds, J. Diff. Geom. 15, 27 - 37 (1980; Zbl. 442.58032).

The editors add the following list of references to § 6 (History).

[B-Y] Bochner, S. - Yano, K.: Curvature and Betti numbers. Annals Math. Studies 32. Princeton University Press 1953.

[L] Lichnerowicz, A.: Géométrie des groupes de transformations. Dunod 1958

[Li] Li, P.: On the Sobolev constant and the p-spectrum of a compact Riemannian manifold. Ann. Sc. Ec. Norm. Sup. 13, 451 - 469 (1980).

[G-1] Gromov, M.: Structures métriques pour les variétés riemanniennes. Redigé par J. Lafontaine et P. Pansu. Cédic F. Nathan 1981

[G-2] Gromov, M.: Curvature, diameter and Betti numbers. Commentarii Math. Helv. 56, 179 - 195 (1981).

[GAL-1] Gallot, S.: Estimées de Sobolev quantitatives sur les variétés et applications. C. R. Acad. Sc. 292, 375 (1981).

[GAL-2] Gallot, S.: A Sobolev inequality and some geometric applications. Actes du Séminaire Franco-Japonais Kyoto 1981. Kaigai Publ. to appear.

[GAL-3] Gallot, S.: Inégalités isopérimétriques, courbure de Ricci et invariants géométriques. Part. I. C.R. Acad. Sci. 296, 333 (1983). Part. II. C.R. Acad. Sci. 296, 365 (1983).

[M-M] Maurey, B. - Meyer, D.: Sur un lemme de géométrie hilbertienne. Pretirage U. Paris VII (1982).

SYMMETRIC IMMERSIONS IN PSEUDO-RIEMANNIAN SPACE FORMS

Carol Blomstrom
Department of Mathematics
Wellesley College
Wellesley, MA 02181

1. Introduction

In the papers [2] and [3], Ferus classified parallel submanifolds of Euclidean space, showing that if M^n is a complete, connected Riemannian manifold and $f : M \longrightarrow \mathbb{E}^N$ is an irreducible, parallel isometric immersion, then $f(M)$ is congruent to a standard imbedded symmetric R-space. Symmetric R-spaces had been studied by Kobayashi and Takeuchi in [5] and [15] and classified by Kobayashi and Nagano in [6]. Takeuchi, in [14], extended this result to the other Riemannian space forms. In the present work this problem is considered for pseudo-Riemannian manifolds. In addition, the extrinsic local symmetry for submanifolds of space forms which was studied by Ferus [3] and Strübing [12] is extended to the pseudo-Riemannian case.

We consider isometric immersions $f : M_r^n \longrightarrow \tilde{M}_s^N(\tilde{c})$, where $\tilde{M}$ is the complete, connected pseudo-Riemannian space form of signature $(s,N-s)$ and constant sectional curvature $\tilde{c}$. We say f is <u>parallel</u> if its second fundamental form α is covariantly constant: $\nabla^*\alpha = 0$.

For $p \in M$, let σ_p be the rigid motion of $\tilde{M}$ defined by

$$\sigma_p(f(p)) = f(p),$$
$$(\sigma_p)_* f_* X = -f_* X \quad \text{for} \quad X \in T_pM,$$
$$(\sigma_p)_* \xi = \xi \quad \text{for} \quad \xi \in N_{f(p)}M,$$

i.e. σ_p is reflection in $N_{f(p)}M$. Then f is <u>locally symmetric</u> if for each $p \in M$, $\sigma_p(f(U)) = f(U)$ for some neighborhood U of p in M, and f is <u>globally symmetric</u> if $\sigma_p(f(M)) = f(M)$ for all $p \in M$. Section 2 is devoted to the proof of

<u>Theorem 1</u>. An isometric immersion $f : M_r^n \longrightarrow \tilde{M}_s^N(\tilde{c})$ is locally symmetric if and only if it is parallel. Furthermore, if M is complete then f is globally symmetric.

Symmetric indefinite R-spaces are introduced in Section 3. As in the Riemannian case, these are constructed from semisimple Lie algebras $\mathfrak{g}$ with decompositions $\mathfrak{g} = \mathfrak{k} \oplus \mathfrak{p}$, but here the Killing form of $\mathfrak{g}$ is only required to be nondegenerate on each of $\mathfrak{k}$ and $\mathfrak{p}$, instead of negative-definite on $\mathfrak{k}$ and positive-definite on $\mathfrak{p}$. Properties of the standard imbedding are discussed (e.g. it is parallel and has zero mean curvature in a hypersphere), and a number of examples are described.

In Section 4 the following classification results are proved. We say that $f : M_r^n \longrightarrow \tilde{M}_s^N$ is <u>full</u> if $f(M)$ is contained in no totally geodesic hypersurface of $\tilde{M}$.

<u>Theorem 2</u>. Let $f : M_r^n \longrightarrow \mathbb{R}_s^N$ be a full, parallel isometric immersion of a complete pseudo-Riemannian manifold M such that $f(M)$ is a submanifold with zero mean curvature in a hypersphere S_s^{N-1} or H_{s-1}^{N-1}. Then there exist a symmetric indefinite R-space $\bar{M}_r^n$ with standard imbedding $\bar{f} : \bar{M} \longrightarrow \mathbb{R}_s^N$ and an isometric covering $q : M \longrightarrow \bar{M}$ such that $f = \bar{f} \circ q$ up to a rigid motion of $\mathbb{R}_s^N$.

This is the analogue of Ferus's classification. The next theorem extends this result to the other pseudo-Riemannian space forms, which we generally identify with their images under inclusion in pseudo-Euclidean space, so that for $\tilde{c} > 0$,

$$\tilde{M}_s^N(\tilde{c}) = S_s^N(\tilde{c}) = \left\{ x \in \mathbb{R}_s^{N+1} : \langle x,x \rangle = \frac{1}{\tilde{c}} \right\},$$

and for $\tilde{c} < 0$,

$$\tilde{M}_s^N(\tilde{c}) = H_s^N(\tilde{c}) = \left\{ x \in \mathbb{R}_{s+1}^{N+1} : \langle x,x \rangle = \frac{1}{\tilde{c}} \right\}.$$

<u>Theorem 3</u>. Let $f : M_r^n \longrightarrow \tilde{M}_s^N(\tilde{c})$, $\tilde{c} \neq 0$, be a full, parallel isometric immersion with zero mean curvature of a complete pseudo-Riemannian manifold M. Then there exist a symmetric indefinite R-space $\bar{M}_r^n$ with standard imbedding $\bar{f} : \bar{M} \longrightarrow \mathbb{R}_j^{N+1}$ ($j = s$ if $\tilde{c} > 0$, $j = s+1$ if $\tilde{c} < 0$), and an isometric covering $q : M \longrightarrow \bar{M}$ such that if $i : \tilde{M} \longrightarrow \mathbb{R}_j^{N+1}$ is the usual inclusion, then $i \circ f = \bar{f} \circ q$ up to rigid motion of $\mathbb{R}_s^N$.

After this work was completed, it was learned that Naitoh, in [10] and [11], has also defined symmetric R-spaces for the pseudo-Riemannian case. In particular, Theorem 2 is proved and the infinitesimal classification of symmetric indefinite R-spaces is given in [11].

This paper is taken from a part of the author's thesis. I would like to thank my adviser, Professor Katsumi Nomizu, for all of his help and guidance.

2. Proof of Theorem 1.

We see as follows that a locally symmetric immersion $f : M_r^n \longrightarrow \tilde{M}_s^N(\tilde{c})$ is parallel: Let $X \in T_pM$ and extend X along the geodesic $\exp(tX)$ of M by parallel translation in TM. Then

$$(\overset{*}{\nabla}_X\alpha)(X,X) = \nabla_X^{\perp}\alpha(X,X) - 2\alpha(\nabla_X X,X) = \nabla_X^{\perp}\alpha(X,X) \in N_{f(p)}M ,$$

so $(\sigma_p)_*((\overset{*}{\nabla}_X\alpha)(X,X)) = (\overset{*}{\nabla}_X\alpha)(X,X)$.
On the other hand, since σ_p is affine,

$$\begin{aligned}(\sigma_p)_*((\overset{*}{\nabla}_X\alpha)(X,X)) &= (\overset{*}{\nabla}_{(\sigma_p)_*X}\alpha)((\sigma_p)_*X,(\sigma_p)_*X) \\ &= (\overset{*}{\nabla}_{-X}\alpha)(-X,-X) = -(\overset{*}{\nabla}_X\alpha)(X,X) .\end{aligned}$$

Hence $(\overset{*}{\nabla}_X\alpha)(X,X) = 0$, so by the Codazzi equation $\overset{*}{\nabla}\alpha = 0$ and f is parallel.

The remainder of this section is devoted to proving the other statements of Theorem 1, following the method of Strübing in [12]. We first develop a Frenet theory for curves in a pseudo-Riemannian manifold and then use this to complete the proof of the theorem.

<u>Definition</u>: Let V^k be a k-dimensional vector space with a (possibly degenerate) inner product $\langle\ ,\ \rangle$. Then $\{V_1,\ldots,V_k\}$ is a <u>pseudo-orthonormal</u> <u>basis</u> for V if

1) $\{V_1,\ldots,V_k\}$ is a linearly independent set in V.
2) If $\langle V_m,V_m\rangle \neq 0$, then $|\langle V_m,V_m\rangle| = 1$ and $\langle V_m,V_j\rangle = 0$ for all V_j such that $\langle V_j,V_j\rangle \neq 0$.

To pseudo-orthonormalize a given basis $\{E_1,\ldots,E_k\}$ of V, we use a generalization of the Gram-Schmidt process, defining V_m by

$$(*)\qquad V_m = \begin{cases} E_m\ , \text{ if } \ \langle E_m, E_m\rangle = \sum_{j=1}^{m-1} \langle V_j,V_j\rangle\langle E_m,V_j\rangle^2 = 0\ , \\[2ex] \dfrac{E_m - \sum_{j=1}^{m-1}\langle V_j,V_j\rangle\langle E_m,V_j\rangle V_j}{\left\| E_m - \sum_{j=1}^{m-1}\langle V_j,V_j\rangle\langle E_m,V_j\rangle V_j\right\|}\ , \text{ otherwise,} \end{cases}$$

where $\|V\| = |\langle V,V\rangle|^{1/2}$.
A straightforward calculation shows that the set $\{V_1,\ldots,V_k\}$ defined by (*) is a pseudo-orthonormal basis of V .

Let J be an open interval of the real line, and let $c : J \longrightarrow \tilde{M}$ be a C^∞ curve in a pseudo-Riemannian manifold $\tilde{M}$ with Levi-Civita connection $\tilde{\nabla}$. As in [12], we say c is a <u>Frenet curve of osculating rank</u> k if for all $t \in J$, the set

$\{\dot{c}(t), (\tilde{\nabla}_t\dot{c})(t), \ldots, (\tilde{\nabla}_t^{k-1}\dot{c})(t)\}$

is linearly independent while the set

$\{\dot{c}(t), (\tilde{\nabla}_t\dot{c})(t), \ldots, (\tilde{\nabla}_t^{k}\dot{c})(t)\}$

is linearly dependent in $T_{c(t)}\tilde{M}$. We pseudo-orthonormalize the set $\{\dot{c}(t), \ldots, (\tilde{\nabla}_t^{k-1}\dot{c})(t)\}$ for each $t \in J$ to obtain $\vec{V}(t) = (V_1(t), \ldots, V_k(t))$, the <u>Frenet frame</u> of c , and we differentiate $\vec{V}(t)$ to obtain the <u>Frenet curvature matrix</u> $\kappa(t) = [\kappa_{ij}(t)]$, $1 \le i,j \le k$, of c , defined by

$$(\tilde{\nabla}_t\vec{V})(t) = \kappa(t)\vec{V}(t)\ .$$

<u>Lemma 1</u>. Let $f : M_r^n \longrightarrow \tilde{M}_s^N$ be a parallel isometric immersion, and let $c(t)$, $t \in J$, be a C^∞ curve of the form

a) $f(x(t))$ for a geodesic $x(t)$ of M , or

b) $(f(x(t)), Y_t)$ in f_*TM (resp. N_fM) , where $x(t)$ is a geodesic in M and Y_t is a ∇-parallel (resp. $\nabla^\perp$-parallel) vector field along $x(t)$.

(We require t to be an affine parameter for $x(t)$.) Then:

(1) There exists k such that c is a Frenet curve of osculating rank k in (a) $\tilde{M}$ or (b) $T\tilde{M}$, where we identify $T_{f(x)}\tilde{M}$ with $f_*T_xM \oplus N_{f(x)}M$.

(2) If c is in $\tilde{M}$ or N_fM , then the Frenet frame $\vec{V}(t)$ of c satisfies

$$V_i(t) \in \begin{cases} f_*T_{x(t)}M, & i \text{ odd}, \\ N_{f(x(t))}M, & i \text{ even}, \end{cases}$$

while if c is in TM,

$$V_i(t) \in \begin{cases} N_{f(x(t))}M, & i \text{ odd}, \\ f_*T_{x(t)}M, & i \text{ even}. \end{cases}$$

(3) The Frenet curvature matrix $\kappa(t)$ of c is constant.

<u>Proof</u>: Let $X_t = \dot{x}(t)$. In case (a) we compute

$$\begin{aligned}
(\tilde{\nabla}_t\dot{x})(t) &= f_*\nabla_t X_t + \alpha(X_t,X_t) \\
&= \alpha(X_t,X_t) \quad \text{since } x(t) \text{ is a geodesic;} \\
(\tilde{\nabla}_t^2\dot{x})(t) &= -f_*A_{\alpha(X_t,X_t)}X_t + \nabla_t^\perp\alpha(X_t,X_t) \\
&= -f_*A_{\alpha(X_t,X_t)} \quad \text{since } \overset{*}{\nabla}_t\alpha = 0\,; \\
(\tilde{\nabla}_t^3\dot{x})(t) &= -f_*\nabla_t A_{\alpha(X_t,X_t)}X_t - \alpha(A_{\alpha(X_t,X_t)}X_t,X_t) \\
&= -\alpha(A_{\alpha(X_t,X_t)}X_t,X_t) \quad \text{since } \nabla_t A = 0\,, \text{ et cetera.}
\end{aligned}$$

In particular, $(\tilde{\nabla}_t^j\dot{x})(t)$ is ∇-parallel for j even and $\nabla^\perp$-parallel for j odd. This implies (1), and (2) follows by induction from the formula (*) for V_m. For (3) we compute

$$(**)\quad (\tilde{\nabla}_t V_m)(t) = \begin{cases} (\tilde{\nabla}_t^m\dot{x})(t), \text{ if } \langle\tilde{\nabla}_t^{m-1}\dot{x},\tilde{\nabla}_t^{m-1}\dot{x}\rangle = \sum\limits_{j=1}^{m-1}\langle V_j,V_j\rangle\langle\tilde{\nabla}_t^{m-1}\dot{x},V_j\rangle^2 = 0\,, \\[2ex] \dfrac{(\tilde{\nabla}_t^m\dot{x})(t) - \sum\limits_{j=1}^{m-1}\langle V_j,V_j\rangle\langle\tilde{\nabla}_t^{m-1}\dot{x},V_j\rangle(\tilde{\nabla}_t V_j)(t)}{\|(\tilde{\nabla}_t^{m-1}\dot{x})(t) - \sum\limits_{j=1}^{m-1}\langle V_j,V_j\rangle\langle\tilde{\nabla}_t^{m-1}\dot{x},V_j\rangle V_j(t)\|}\,, \text{ otherwise.} \end{cases}$$

Every scalar function in this formula is constant because all of the vector fields involved are ∇- or $\nabla^\perp$-parallel. From (*) and (**), $(\tilde{\nabla}_t V_m)(t)$ is a linear combination with constant coefficients of $V_j(t)$, $1 \le j \le m+1$, and $(\tilde{\nabla}_t V_j)(t)$, $1 \le j \le m-1$, and then it follows by induction on m that $\tilde{\nabla}_t V_m$ is a constant linear combination of $V_j(t)$, $1 \le j \le m+1$, which proves (3).

The proof for case (b) is very similar; for the curve $c(t) = (f(x(t)),Y_t)$ in $T\tilde{M}$ we use the connection in $T\tilde{M}$ to identify the horizontal subspace of $T_{c(t)}(T\tilde{M})$ with $T_{f(x(t))}\tilde{M}$, so that the

k^{th} derivative of $\dot{c}(t)$ is $(\tilde{\nabla}_t^k f_* X_t, \tilde{\nabla}_t^{k+1} Y_t)$. The proofs of (1), (2), and (3) follow as in case (a). ∎

We are now ready to show that if $f : M_r^n \longrightarrow \tilde{M}_s^N(\tilde{c})$ is parallel then f is locally symmetric.

Let $x(t)$, $|t| < \varepsilon$, be a geodesic in M with $x_0 = p$, and let $c(t) = f(x(t))$. Write σ for σ_p and define $\tilde{c}(t) = \sigma(c(-t))$; we must show that $\tilde{c}(t) = c(t)$, $|t| < \varepsilon$. By Lemma 2, the Frenet frame $\vec{V}(t) = (V_1(t),\ldots,V_k(t))$ of c satisfies

$$V_i(t) \in \begin{cases} f_* T_{x(t)} M \ , & i \text{ odd} \ , \\ N_{c(t)} M \ , & i \text{ even} \ , \end{cases}$$

so that $\sigma_* V_i(0) = (-1)^i V_i(0)$.

For $\tilde{c}(t)$ we have

$$\dot{\tilde{c}}(t) = -\sigma_* \dot{c}(-t)$$
$$(\tilde{\nabla}_t^i \dot{\tilde{c}})(t) = (-1)^{i+1} \sigma_* (\tilde{\nabla}_t^i \dot{c})(-t) \ ,$$

so the Frenet frame $\tilde{\vec{V}}(t) = (\tilde{V}_1(t),\ldots,\tilde{V}_{\tilde{k}}(t))$ of $\tilde{c}$ satisfies $\tilde{k} = k$ and $\tilde{V}_i(t) = (-1)^i \sigma_* V_i(-t)$. In particular, $\tilde{V}_i(0) = (-1)^i \sigma_* V_i(0) = V_i(0)$, $1 \le i \le k$. Also, since the Frenet curvature matrix $\kappa(t)$ of c is constant by Lemma 2, and

$$(\tilde{\nabla}_t \tilde{V}_i)(t) = (-1)^{i+1} \sigma_* (\tilde{\nabla}_t V_i)(-t) \ ,$$

the matrix $\tilde{\kappa}(t)$ of $\tilde{c}$ must be constant. Now

$$(\tilde{\nabla}_t \tilde{V}_i)(0) = \sum_{j=1}^{k} \tilde{\kappa}_{ij} \tilde{V}_j(0) \ ,$$

but also $(\tilde{\nabla}_t \tilde{V}_i)(0) = (-1)^{i+1} \sigma_* (\tilde{\nabla}_t V_i)(0)$

$$= (-1)^{i+1} \sigma_* \sum_{j=1}^{k} \kappa_{ij} V_j(0) \ ,$$

so that $\tilde{V}_j(0) = V_j(0)$ implies $\tilde{\kappa}_{ij} = \kappa_{ij}$, $1 \le i,j \le k$. Hence $\vec{V}(t)$ and $\tilde{\vec{V}}(t)$ both satisfy the first-order linear system

$$(\tilde{\nabla}_t \vec{U})(t) = \kappa \vec{U}(t)$$

with the same initial condition, so $\vec{V}(t) = \tilde{\vec{V}}(t)$. Since

$\tilde{c}(0) = f(p) = c(0)$, $\tilde{c}(t) = c(t)$ for $|t| < \varepsilon$, proving that f is locally symmetric.

To finish the proof of Theorem 1, we show that if M is (geodesically) complete, then f is symmetric, i.e. $\sigma_p(f(M)) = f(M)$ for all $p \in M$. Given two points $p,q \in M$, join p to q by a sequence of geodesic segments

$$x_1(t) \ , \ 0 \le t \le t_1 \ , \ x_1(0) = p \ ,$$
$$x_2(t) \ , \ 0 \le t \le t_2 \ , \ x_2(0) = x_1(t_1) \ ,$$
$$\vdots$$
$$x_m(t) \ , \ 0 \le t \le t_m \ , \ x_m(0) = x_{m-1}(t_{m-1}) \ , \ x_m(t_m) = q \ .$$

Let $c_i(t) = f(x_i(t))$ and again write σ for σ_p .

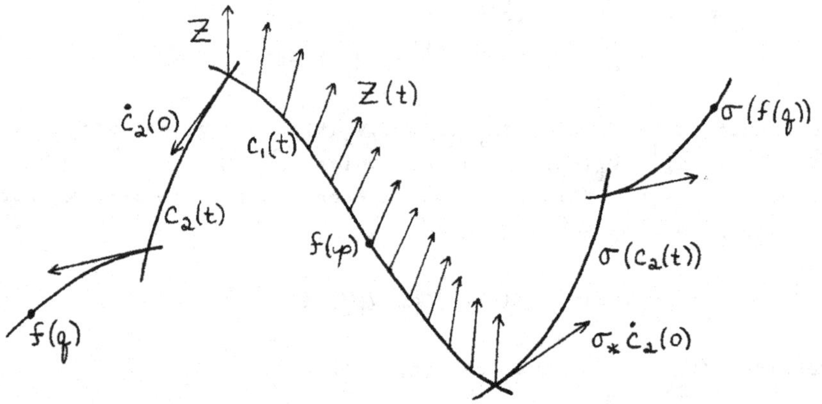

We have seen that $\sigma(c_1(t)) = c_1(-t)$ for all t . As suggested by the diagram, the proof will be completed by application of the following arguments, in the order $A, B_1, C_1, B_2, \ldots, C_{m-2}, B_{m-1}$.

A. Let $Z \in T_{c_2(0)}M$ (resp. $N_{c_2(0)}M$) and define $Z(t)$ by ∇-parallel (resp. $\nabla^{\perp}$-parallel) translation of Z along $c_1(t)$, (so that $Z(t_1) = Z$) . Then $\sigma_* Z(t) = -Z(-t)$ (resp. $\sigma_* Z(t) = Z(-t)$) .

B_i. For $Y_i = \dot{c}_{i+1}(0)$ let $y(t)$ be the geodesic of $f(M)$ starting at $\sigma(c_{i+1}(0))$ with initial tangent vector $\sigma_* Y_i$. Then

$y(t) = \sigma(c_{i+1}(t))$.

C_i. Let $Z \in T_{c_{i+2}(0)}M$ (resp. $N_{c_{i+2}(0)}M$) and define $Z(t)$ by ∇-parallel (resp. $\nabla^{\perp}$-parallel) translation of Z along $c_{i+1}(t)$, (so that $Z(t_{i+1}) = Z$) . Define $\tilde{Z}(t)$ by ∇-parallel (resp. $\nabla^{\perp}$-parallel) translation of $\sigma_*(Z(0))$ along $\sigma(c_{i+1}(t))$, (so $\tilde{Z}(0) = \sigma_* Z(0))$. Then $\tilde{Z}(t) = \sigma_* Z(t)$.

We only prove A, as the proofs of B_i and C_i follow a similar pattern.

Suppose $Y \in T_{c_2(0)}M$. (The case of $N_{c_2(0)}M$ is essentially the same.) For the curve $(c_1(t), Y(t))$ in f_*TM , Lemma 1 shows that

$$V_i(t) \in \begin{cases} N_{c_1(t)}M , & i \text{ odd} , \\ f_*T_{c_1(t)}M , & i \text{ even} , \end{cases}$$

so that $\sigma_* V_i(0) = (-1)^{i+1} V_i(0)$.

Let $\tilde{Y}(t) = -\sigma_* Y(-t)$; we must show that $\tilde{Y}(t) = Y(t)$. We see that

$$(\tilde{\nabla}_t^i \tilde{Y})(t) = (-1)^{i+1} \sigma_* (\tilde{\nabla}_t^i Y)(-t) ,$$

which implies $\tilde{V}_i(t) = (-1)^{i+1} \sigma_* V_i(-t)$, where $\vec{\tilde{V}}$ is the Frenet frame for $\tilde{Y}(t)$. For the same reason as in the previous argument, $\tilde{\kappa}_{ij} = \kappa_{ij}$, $1 \le i,j \le k$, and

$$\tilde{V}_i(0) = (-1)^{i+1} \sigma_* V_i(0) = V_i(0) , \quad 1 \le i \le k .$$

From this we get $\vec{V}(t) = \vec{\tilde{V}}(t)$, and, since $\tilde{Y}(0) = -\sigma_* Y(0) = Y(0)$, it follows that $\tilde{Y}(t) = Y(t)$. ∎

3. Symmetric Indefinite R-spaces

Let $\mathfrak{g}$ be a real, semisimple Lie algebra with a decomposition $\mathfrak{g} = \mathfrak{k} \oplus \mathfrak{p}$, where $\mathfrak{k}$ is a subalgebra and $\mathfrak{p}$ a vector subspace, such that

$$[\mathfrak{k},\mathfrak{k}] \subset \mathfrak{k} , \quad [\mathfrak{p},\mathfrak{p}] \subset \mathfrak{k} , \quad \text{and} \quad [\mathfrak{k},\mathfrak{p}] \subset \mathfrak{p} .$$

From Koecher [7; Ch. II §4] we have the following:

Lemma 2. For $X_1, X_2 \in \mathfrak{k}$ and $Y_1, Y_2 \in \mathfrak{p}$,

$$B(X_1+Y_1, X_2+Y_2) = \mathrm{tr}_{\mathfrak{g}}\,\mathrm{ad}X_1\mathrm{ad}X_2 + 2\mathrm{tr}_{\mathfrak{p}}\,\mathrm{ad}Y_1\mathrm{ad}Y_2 .$$

In particular, $B(\mathfrak{k},\mathfrak{p}) = 0$ so that B is nondegenerate when restricted to each of $\mathfrak{k}$ and $\mathfrak{p}$. (This is a generalization of the definition of a Cartan decomposition of $\mathfrak{g}$; see [4].)

Suppose there exists an element $E \in \mathfrak{p}$ such that ad E has eigenvalues $0, 1, -1$ and $\mathfrak{g}$ admits a decomposition into eigenspaces: $\mathfrak{g} = \mathfrak{g}_0 \oplus \mathfrak{g}_1 \oplus \mathfrak{g}_{-1}$. Denote by $\mathfrak{g}^C$ the complexification of $\mathfrak{g}$ and let G^C be the adjoint group of $\mathfrak{g}^C$; that is,

$$G^C = \mathrm{Ad}(\mathfrak{g}^C) = \exp(\mathrm{ad}(\mathfrak{g}^C)) \subset GL(\mathfrak{g}^C) .$$

Let K be the connected subgroup of G^C generated by $\mathfrak{k}$; we define

$$K_0 = \{k \in K : \mathrm{Ad}(k)E = E\}$$

and form the differentiable manifold $M = K/K_0$.

Let $\mathfrak{g}_+ = \mathfrak{g}_1 \oplus \mathfrak{g}_{-1}$. The automorphism $\mathrm{Ad}(e^{i\pi E})$ of $\mathfrak{g}^C$ leaves $\mathfrak{k}$ invariant, so conjugation in G^C by $e^{i\pi E}$ leaves K invariant. Denote by τ the restriction to K of conjugation by $e^{i\pi E}$, as well as the restriction to $\mathfrak{k}$ of $\mathrm{Ad}(e^{i\pi E})$. Decompose $\mathfrak{k}$ into $\mathfrak{k}_0 \oplus \mathfrak{m}$, the $+1$ and -1 eigenspaces of τ; in fact, $\mathfrak{k}_0 = \mathfrak{k} \cap \mathfrak{g}_0$ and $\mathfrak{m} = \mathfrak{k} \cap \mathfrak{g}_+$. Then $\mathfrak{k}_0$ is the Lie algebra of K_0 and $\mathfrak{m}$ is identified with $T_{[e]}M$, where e is the identity of G^C. Since τ leaves K_0 pointwise-fixed, M is a symmetric space.

We define an imbedding $f : M \longrightarrow \mathfrak{p}$ by $f([k]) = \mathrm{Ad}(k)E$. Using either $B|_{\mathfrak{p}}$ or $-B|_{\mathfrak{p}}$ for a metric makes $\mathfrak{p}$ into a pseudo-Euclidean space, and we give M the induced pseudo-Riemannian metric. To see that this is indeed nondegenerate, we note that the differential of f at $[e]$ is given by $f_*X = \mathrm{ad}(X)E$ for $X \in \mathfrak{m}$; if we define $\mathfrak{p}_+ = \mathfrak{p} \cap \mathfrak{g}_+$ and $\mathfrak{p}_0 = \mathfrak{p} \cap \mathfrak{g}_0$, then

$$f_*(T_{[e]}M) = \mathrm{ad}(E)\mathfrak{m} = \mathfrak{p}_+ ,$$

on which B is nondegenerate.

We say that M is a <u>symmetric indefinite R-space</u> and f is its <u>standard imbedding</u>.

Proposition. A standard imbedded symmetric indefinite R-space $f : M \longrightarrow \mathfrak{p}$ has the following properties:

(1) f is locally symmetric, and hence parallel by Theorem 1.

(2) $f(M)$ is contained in a hypersphere of $\mathfrak{p}$ as a submanifold with zero mean curvature.

Proof. (1) See [3].

(2) $\langle f([k]), f([k])\rangle = B(\mathrm{Ad}(k)E, \mathrm{Ad}(k)E)$

$= B(E,E) = \mathrm{tr}\,(\mathrm{ad}E)^2 = 2\dim M$,

so that if $\mathfrak{p} = \mathbb{R}^N_s$, then $f(M) \subset S^{N-1}_s(\frac{1}{2n})$, $n = \dim M$. (If $-B|_{\mathfrak{p}}$ is used, then $\mathfrak{p} = \mathbb{R}^N_{N-s}$ and $f(M) \subset H^{N-1}_{N-s-1}(-\frac{1}{2n})$.)

Let $\tilde{f} : M \longrightarrow S^{N-1}_s$ be defined by $f = i \circ \tilde{f}$, where $i : S^{N-1}_s \longrightarrow \mathfrak{p}$ is the usual inclusion. (The case of H^{N-1}_{N-s-1} is the same.) If $\tilde{A}$ is the shape operator and $\tilde{\eta}$ the mean curvature normal for $\tilde{f}$, then $\mathrm{tr}\tilde{A}_\xi = \langle\tilde{\eta},\xi\rangle$, so to show $\tilde{\eta} = 0$ it suffices to show $\mathrm{tr}\tilde{A}_\xi = 0$ for all $\xi \in N_{\tilde{f}[e]}M$. We have seen that $N_{f[e]}M = \mathfrak{p}_0$, and since $\tilde{f}([e]) = E$, we have $N_{\tilde{f}[e]}M = \mathfrak{p}_0 \cap \{E\}^\perp$.

As in the Riemannian case,

$$\tilde{A}_\xi X = -\mathrm{ad}\,E\,\mathrm{ad}\,\xi(X) \quad \text{for} \quad \xi \in \mathfrak{p}_0 \cap \{E\}^\perp \quad \text{and} \quad X \in \mathfrak{m}.$$

Using Lemma 2, $0 = B(E,\xi) = 2\,\mathrm{tr}_{\mathfrak{p}}\mathrm{ad}\,E\,\mathrm{ad}\xi$, so

$$0 = \mathrm{tr}_{\mathfrak{k}}\mathrm{ad}\,E\,\mathrm{ad}\,\xi = \mathrm{tr}_{\mathfrak{k}_0}\mathrm{ad}\,E\,\mathrm{ad}\,\xi + \mathrm{tr}_{\mathfrak{m}}\mathrm{ad}\,E\,\mathrm{ad}\,\xi \ .$$

But for $Y \in \mathfrak{k}_0$, $\mathrm{ad}\,E\,\mathrm{ad}\,\xi(Y) = [\xi,[E,Y]] + [[E,\xi],Y] = 0$.

Hence $\mathrm{tr}_{\mathfrak{m}}\tilde{A}_\xi = 0$ and $\tilde{f}(M)$ has zero mean curvature. ∎

Remark. If the imbedding $f : M \longrightarrow \mathfrak{p}$ is followed by a homothetic transformation of $\mathfrak{p}$, this composition is also called a standard imbedding of M .

Examples. We start with some analogues of (positive-definite) symmetric R-spaces; see [5] and [6] for details. Notation for the classical Lie algebras is that of [4].

(1) For $\mathfrak{g}$ we use:

(a) $\mathfrak{sl}(n;\mathbb{R})$, (b) $\mathfrak{sl}(n;\mathbb{C})$, (c) $\mathfrak{su}^*(2n)$

$\mathfrak{k}$ is: (a) $\mathfrak{so}(p,n-p)$, (b) $\mathfrak{su}(p,n-p)$, (c) $\mathfrak{sp}(p,n-p)$ for any p between 0 and n .

$\mathfrak{p}$ is: (a) $\mathbb{R}^{\frac{1}{2}(n-1)(n+2)}_{p(n-p)}$, (b) $\mathbb{R}^{n^2-1}_{2p(n-p)}$, (c) $\mathbb{R}^{n(2n+1)}_{4p(n-p)}$.

$$E = \begin{bmatrix} -\frac{n-k}{n} I_s & & & \\ & \frac{k}{n} I_{p-s} & & \\ & & -\frac{n-k}{n} I_{k-s} & \\ & & & \frac{k}{n} I_{n-k-p+s} \end{bmatrix}$$

for any k and s, $1 \le k \le n-1$, $\max(0,k+p-n) \le s \le \min(k,p)$ in cases (a) and (b); for (c) we use $\begin{bmatrix} E & \\ & E \end{bmatrix}$.

M is: (a) $O(p,n-p)/O(s,k-s) \times O(p-s,n-k-p+s)$,

(b) $U(p,n-p)/U(s,k-s) \times U(p-s,n-k-p+s)$,

(c) $Sp(p,n-p)/Sp(s,k-s) \times Sp(p-s,n-k-p+s)$,

the Grassmann manifold of k-planes of signature $(s,k-s)$ in, respectively, real, complex or quaternionic n-space of signature $(p,n-p)$. The standard imbedding $f : M \longrightarrow \mathfrak{p}$ is the natural generalization of the Veronese imbedding of $\mathbb{R}P^{n-1}$ in $\mathbb{R}^{\frac{1}{2}(n-1)(n+2)}$.

(2) $\mathfrak{g} = \mathfrak{so}(p,q)$, $p,q \ge 1$

$\mathfrak{k} = \mathfrak{so}(p-r) \oplus \mathfrak{so}(r,q)$, $0 \le r \le p-1$.

$\mathfrak{p} = \mathbb{R}^{(r+q)(p-r)}_{r(p-r)}$.

$$E = \begin{array}{c} \\ p \\ \\ q \end{array}\left[\begin{array}{c|c} 0 & \begin{matrix} 1 & 0 \cdots 0 \\ 0 & \\ \vdots & 0 \\ 0 & \end{matrix} \\ \hline \begin{matrix} 1 & 0 \cdots 0 \\ 0 & \\ \vdots & 0 \\ 0 & \end{matrix} & \end{array}\right]$$

(columns: p, q)

$M = SO(p-r) \times SO(r,q)/S(O(p-r-1) \times O(r,q-1)) = S^{p-r-1} \times S^{r+q-1}_r/\pm I$.

In case $r = p-1$, $M = S^{p+q-2}_{p-1}$ (or H^{p+q-2}_{q-1}) and f is the usual imbedding into $\mathbb{R}^{p+q-1}_{p-1}$ (or $\mathbb{R}^{p+q-1}_{q}$).

(3) $\mathfrak{g}$: (a) $\mathfrak{so}(n,n)$, (b) $\mathfrak{su}(n,n)$, (c) $\mathfrak{sp}(n,n)$.

$\mathfrak{k}$: (a) $\mathfrak{so}(p,n-p) \oplus \mathfrak{so}(p,n-p)$

(b) $\mathfrak{su}(p,n-p) \oplus \mathfrak{su}(p,n-p) \oplus \mathbb{R}$

(c) $\mathfrak{sp}(p,n-p) \oplus \mathfrak{sp}(p,n-p)$, $0 \le p \le n$.

$\mathfrak{p}$: (a) $\mathbb{R}^{n^2}_{2p(n-p)}$, (b) $\mathbb{R}^{2n^2}_{4p(n-p)}$, (c) $\mathbb{R}^{4n^2}_{8p(n-p)}$.

E : (a) and (b) $\frac{1}{2}\begin{bmatrix} & I_n \\ I_n & \end{bmatrix}$; (c) $\frac{1}{2}\begin{bmatrix} & & & I_n \\ & & I_n & \\ & I_n & & \\ I_n & & & \end{bmatrix}$

M : (a) SO(p,n-p) , (b) U(p,n-p) , (c) Sp(p,n-p) .
The imbedding f is the inclusion of M into the space of $n \times n$ (a) real, (b) complex, (c) quaternionic matrices.
The remaining examples have no positive-definite form; examples (4) and (5) are complex manifolds considered in [8].

(4) $\mathfrak{g} = \mathfrak{sl}(n;\mathbb{C})$

$\mathfrak{k} = \mathfrak{so}(n;\mathbb{C})$ $\qquad \mathfrak{p} = \mathbb{R}^{(n-1)(n+2)}_{\frac{1}{2}(n-1)(n+2)} \cong \mathbb{C}^{\frac{1}{2}(n-1)(n+2)}$

$$E = \begin{bmatrix} -\frac{n-k}{n} I_k & \\ & \frac{k}{n} I_{n-k} \end{bmatrix} \quad 1 \le k \le n-1 .$$

$M = O(n;\mathbb{C}) / O(k;\mathbb{C}) \times O(n-k;\mathbb{C})$,

the Grassmann manifold of nondegenerate complex k-planes in $\mathbb{C}^n$.
The imbedding f is the complex analogue of the Veronese imbedding.

(5) $\mathfrak{g} = \mathfrak{so}(1,n+1;\mathbb{C})$

$\mathfrak{k} = \mathfrak{so}(n+1;\mathbb{C})$ $\qquad \mathfrak{p} = \mathbb{R}^{2(n+1)}_{n+1} \cong \mathbb{C}^{n+1}$

$$E = \begin{bmatrix} 0 & 1 & 0 \cdots 0 \\ 1 & & \\ 0 & & \\ \vdots & & 0 \\ 0 & & \end{bmatrix} \qquad M = SO(n+1;\mathbb{C}) / SO(n;\mathbb{C}) \ ;$$

this is the complex n-sphere and f is its natural imbedding in $\mathbb{C}^{n+1}$.

(6) $\mathfrak{g} = \mathfrak{sl}(2n;\mathbb{R})$

$\mathfrak{k} = \mathfrak{sl}(n;\mathbb{C}) \oplus \mathbb{R}$ $\qquad \mathfrak{p} = \mathbb{R}^{2n^2}_{n(n-1)}$ $\qquad E = \frac{1}{2}\begin{bmatrix} -I_n & \\ & I_n \end{bmatrix}$

$M = SL(n;\mathbb{C}) \times T / SL(n;\mathbb{R})$, where T is the circle group.

(7) $\mathfrak{g} = \mathfrak{sl}(2n;\mathbb{C})$

$\mathfrak{k} = \mathfrak{su}^*(2n)$; $\qquad \mathfrak{p} = \mathbb{R}^{4n^2-1}_{2n^2-n-1}$

$$E = \frac{1}{2}\begin{bmatrix} -I_n & \\ & I_n \end{bmatrix} , \ M = SU^*(2n) / SL(n;\mathbb{C}) \times T .$$

4. Classification of Parallel Submanifolds.

In this section we prove Theorems 2 and 3, but first we mention some examples of parallel submanifolds of pseudo-Euclidean space which do not satisfy the hypotheses of the theorems and hence cannot be realized as symmetric indefinite R-spaces.

One group of such examples is given by the totally umbilic hypersurfaces of $\tilde{M}^N_s(\tilde{c})$ defined by $Q^{N-1}_{j-1} = \tilde{M} \cap P^N$, where P^N is a degenerate hyperplane in $\mathbb{R}^{N+1}_j$ not through the origin, and $j = s$ if

$\tilde{c} > 0$, $j = s+1$ if $\tilde{c} < 0$. Such a submanifold is parallel in $\mathbb{R}_j^{N+1}$ but does not have zero mean curvature in $\tilde{M}$. Also, it is full in the degenerate subspace P^N.

Another set of examples consists of the parallel B-scroll immersion of $\mathbb{R}_1^2$ in $\mathbb{R}_1^3$ and its higher-dimensional analogues studied by Magid in [8] and [9]. Such an immersion is full in $\mathbb{R}_s^N$ but contained in no hypersphere.

A third example is Example 1.13 of [8], a parallel isometric immersion of $\mathbb{R}_1^2$ in $\mathbb{R}_2^4$ which is full and contained in a hypersphere but not with zero mean curvature.

That no such "non-examples" arise in the positive-definite case is a consequence of a result of Ferus [1]: any irreducible, parallel isometric immersion $f : M^n \longrightarrow \mathbb{R}^N$ is minimal in a hypersphere.

<u>Proof of Theorem 2</u>. Our proof follows that of Ferus [2] and [3] very closely - as does Naitoh's in [11] - so we refer the reader to these papers for details. The method of proof is first to construct from the immersion f a standard imbedded symmetric indefinite R-space $\bar{f} : \bar{M}_r^n \longrightarrow \mathbb{R}_s^N$; this construction proceeds exactly as in [3] and its result is a parallel submanifold of $\mathbb{R}_s^N$ with, at one point, the same tangent plane and second fundamental form as $f(M)$. The second step uses the Gauss maps of f and $\bar{f}$, just as in [2]. (For $f : M_r^n \longrightarrow \mathbb{R}_s^N$, the Gauss map $\gamma_f : M \longrightarrow G$ is defined by $\gamma_f(x) = f_*(T_xM)$, where $G = G_r^n(\mathbb{R}_s^N)$ denotes the Grassman manifold of n-planes of signature $(r,n-r)$ in $\mathbb{R}_s^N$.) It is shown that γ_f and $\gamma_{\bar{f}}$ are affine covering maps onto the same totally geodesic submanifold of G, and hence there exists an isometric covering map $q: M \longrightarrow \bar{M}$ such that $\gamma_f = \gamma_{\bar{f}} \circ q$.

This map q induces a connection-preserving bundle map $\bar{q} : N_fM \longrightarrow N_{\bar{f}}\bar{M}$ such that $\bar{q}\big|_{N_{f(x)}M} = Id_{p_0}$ and

$$\bar{q}(\alpha(X,Y)) = \bar{\alpha}(q_*X, q_*Y) \quad \text{for all} \quad X,Y \in TM ,$$

so that we now have two isometric immersions f and $f' = \bar{f} \circ q$ of M into $\mathbb{R}_s^N$ with $f'(x) = f(x)$, $(f'_*)_x = (f_*)_x$, and

$$\alpha'(X,Y) = \bar{\alpha}(q_*X, q_*Y) = \bar{q}(\alpha(X,Y)) = \alpha(X,Y) \quad \text{for} \quad X,Y \in T_xM .$$

Define vector bundles $E = TM \oplus N_fM$ and $E' = TM \oplus N_{f'}M$ (Whitney sums) over M and give them the direct sum metrics. If e (resp. e') denotes the inclusion of TM into E (resp. E') and g (resp. g') denotes the inclusion of N_fM into E (resp. $N_{f'}M$ into E'), then E has a natural metric connection D given by

$$D_X(eY) = e\nabla_X Y + g\alpha(X,Y) ,$$

$$D_X(g\xi) = -eA_\xi X + g\nabla^\perp_X \xi,$$

for X,Y C^∞ sections of TM and ξ a C^∞ section of N_fM, and similarly E' has D'; computations using the Gauss, Codazzi, and Ricci equations for f and f' show that D and D' are flat connections. In addition,

$$D_X(eY)-D_Y(eX)-e[X,Y] = e\nabla_X Y+g\alpha(X,Y)-e\nabla_Y X-g\alpha(Y,X)-e[X,Y] = 0 ,$$

and the same for D'.

We want to apply the following Fundamental Theorem for mappings: Let M^n be a simply connected differentiable manifold and E a vector bundle over M with a fiber metric of signature (s,N-s), a flat metric connection D, and a bundle inclusion $e : TM \longrightarrow E$ satisfying $D_X(eY)-D_Y(eX)-e[X,Y] = 0$. Then there exist a differentiable map $f : M \longrightarrow \mathbb{R}^N_s$ and a bundle isometry $F : E \longrightarrow f^{-1}T\mathbb{R}^N_s$ such that $f_* = F\circ e$, and f is unique up to rigid motion of $\mathbb{R}^N_s$.

This version of the Fundamental Theorem for mappings and its application to isometric immersions with prescribed second fundamental form are due to Nomizu, and constitute a new proof of the results of Szczarba [13].

We will now show that the bundles E and E' are isometrically isomorphic with D and D' corresponding, whence the Fundamental Theorem proves that f = f'.

Define a map $\tilde{q} : E \longrightarrow E'$ by

$$\tilde{q}\circ e = e' \quad \text{on TM} ,$$

$$\tilde{q}\circ g(\alpha(X,Y)) = g'(\alpha'(X,Y)) \quad \text{on } N_fM .$$

This is clearly a bundle isomorphism which preserves the fiber metric, since $\alpha'(X,Y) = \bar{q}\alpha(X,Y)$ and $\bar{q}$ is isometric. Also,

$$\tilde{q}D_X(eY) = D'_{\tilde{q}X}\tilde{q}(eY) ,$$

$$\tilde{q}D_X(g\alpha(Y,Z)) = D'_{\tilde{q}X}\tilde{q}(g\alpha(Y,Z)) ,$$

so that $\tilde{q}$ preserves the connection and is therefore the required bundle isometry.

This completes the proof of Theorem 2. ∎

Proof of Theorem 3. Theorem 3 will follow from Theorem 2 once it is shown that $i \circ f$ is full whenever f is full. As in the Riemannian case, we have:

Lemma 3: If $\tilde{M}^N_s(\tilde{c})$ is isometrically imbedded in $\mathbb{R}^{N+1}_j$ in the usual way, centered at the origin, then its complete, connected, totally geodesic hypersurfaces are connected components of its intersections with hyperplanes through the origin.

To prove Theorem 3, let $\tilde{f} = i \circ f$ and suppose $\tilde{f}(M) \subset P^N$, a hyperplane of $\mathbb{R}^{N+1}_j$. Let $\hat{P}^N$ be the hyperplane through the origin such that $P^N = \tilde{f}(x) + \hat{P}^N$ for all $x \in M$; if $\tilde{\eta}$ is the mean curvature normal of $\tilde{f}$, then $\tilde{\eta} \in \hat{P}^N$. But $\eta = 0$ implies that $\tilde{\eta}$ is normal to $\tilde{M}$, so $\tilde{\eta}_x$ is parallel to $\tilde{f}(x)$ and thus $\tilde{f}(x) \in \hat{P}^N$ for all $x \in M$. Then $P^N = \hat{P}^N$ so that $\tilde{M} \cap P^N$ is a totally geodesic hypersurface of $\tilde{M}$, and hence f is not full. ∎

Bibliography

[1] Ferus, D.: Produkt-Zerlegung von Immersionen mit paralleler zweiter Fundamentalform. Math. Ann. 211, 1-5 (1974; Zbl. 273. 53044)

[2] Ferus, D.: Immersions with parallel second fundamental form. Math. Z. 140, 87-93 (1974; Zbl. 279.53048)

[3] Ferus, D.: Symmetric submanifolds of Euclidean space. Math. Ann. 247, 81-93 (1980; Zbl. 446.53041)

[4] Helgason, S.: Differential Geometry, Lie Groups, and Symmetric Spaces. Academic Press, New York (1978; Zbl. 451.53038)

[5] Kobayashi, S.: Isometric imbeddings of compact symmetric spaces. Tôhoku Math. J. 20, 21-25 (1968; Zbl. 175,483)

[6] Kobayashi, S. and Nagano, T.: On filtered Lie algebras and geometric structures I, II. J. Math. Mech. 13 (1964), 875-908, 14 (1965), 513-522. (Zbl. 142,195 and 163,281)

[7] Koecher, M. : An Elementary Approach to Bounded Symmetric Domains. Rice University, Houston, 1969.

[8] Magid, M.: Isometric immersions between indefinite flat spaces with parallel second fundamental forms. Thesis, Brown University, 1978.

[9] Magid, M.: Shape operators of Einstein hypersurfaces in indefinite space forms. Proc. Am. Math. Soc. 84, 237-242 (1982; Zbl. 485.53023)

[10] Naitoh, H.: Parallel submanifolds of complex space forms I, II, Nagoya J. Math. 90 (1983), 85-117, 91 (1983), 110-149 (Zbl. 502.53044 and 502.53045)

[11] Naitoh, H.: Pseudo-Riemannian symmetric R-spaces. Osaka J. Math. 21, 733-764 (1984).

[12] Strübing, W.: Symmetric submanifolds of Riemannian manifolds. Math. Ann. 245, 37-44 (Zbl. 424.53025)

[13] Szczarba, R.: On existence and rigidity of isometric immersions. Bull. Am. Math. Soc. 75,783-787 (1969; Zbl. 177,246)

[14] Takeuchi, M.: Parallel submanifolds of space forms. Manifolds and Lie Groups: Papers in Honor of Y. Matsushima, Birkhäuser, Basel, 429-447 (1981; Zbl. 481.53047)

[15] Takeuchi, M. and Kobayashi, S.: Minimal imbeddings of R-spaces. J. Differ. Geom. 2, 203-215 (1968; Zbl. 165,249)

Immersions of Surfaces into Space Forms

by

J. Bolton, T.J.Willmore and L.M.Woodward

§1. Introduction

This reports on some research work in progress at Durham University. Our interest in the subject arose from the work of Jensen and Rigoli [2], [3], [5], although most of the ideas can be found in earlier works of Calabi and Chern. (See for example [1]). These authors consider an immersion $\psi : M \to W$, where M is an oriented 2-dimensional manifold and W is an n-dimensional Riemannian manifold, usually restricted to be of constant sectional curvature. Then M has a complex structure defined by its orientation and the metric induced on it by ψ. By means of the method of moving frames Jensen and Rigoli obtain from ψ, when $n = 3$ and W is oriented, a globally defined symmetric quadratic differential form Ψ with complex values. They prove

Theorem 1.1. Let W be an oriented 3-dimensional space form and let $\psi : M \to W$ be an immersion of the oriented surface M. If the mean curvature is constant, then the quadratic differential form Ψ is holomorphic.

More generally, for arbitrary n, they produce a globally defined $\mathbb{C}$-valued symmetric 4-form on M which we will denote by Λ_1. They prove

Theorem 1.2. Let W be an n-dimensional space form and let $\psi : M \to W$ be an immersion of the oriented surface M. If the mean curvature vector H of M is parallel in the normal bundle, then the quartic form Λ_1 is holomorphic.

The above theorems are interesting but they do not claim to characterize the holomorphicity of the symmetric forms. In fact the converse of Theorem 1.1 is true but the converse of Theorem 1.2 is probably false. Moreover, the forms Ψ and Λ_1 are constructed locally and only subsequently shown to be globally defined.

It follows from the Riemann-Roch Theorem that any holomorphic form on S^2 is in fact identically zero. This observation motivated Jensen and Rigoli to consider minimal immersions of surfaces into space forms which have Λ_1 identically zero. Such an immersion they call isotropic of order 1. When ψ is not totally geodesic they construct locally a $\mathbb{C}$-valued symmetric 6-form, which we will denote by Λ_2. They then prove that Λ_2 is globally defined and holomorphic. The immersion ψ is said to be isotropic of order 2 if it is isotropic of order 1, and if Λ_2 is identically zero. In that case the authors construct a $\mathbb{C}$-valued symmetric 8-form, which we call Λ_3. Clearly this procedure leads to the concept of isotropy of order k and the local construction of a globally defined symmetric $(2k + 2)$-form Λ_k.

Again, we stress that the forms Ψ, Λ_k are constructed locally and then shown to exist globally.

The object of the present paper is to give an alternative treatment which, we feel, has several advantages. Firstly, we give global definitions of the forms Ψ and Λ_i, and these use other globally defined bundle-valued forms which appear to be geometrically significant. Moreover, we obtain an analogue of Theorem 1.2 which gives a characterization of those immersions into space forms which have parallel mean curvature vector.

§2. Holomorphic bundle-valued forms

Let M be a 2-dimensional oriented Riemannian manifold. Then M has a canonically defined complex structure. Our alternative treatment depends upon an examination of the holomorphicity of a tensor field on M, covariant of degree p, with values in a complex vector bundle equipped with a connection. Since M is a complex manifold the tangent bundle TM is a complex line bundle over M, which possesses a connection ∇ given by the Levi-Cività connection of M.

Let E be a complex vector bundle over M equipped with a connection $\tilde{\nabla}$, and let F be an E-valued p-covariant tensor field on M.

The <u>covariant differential</u> $*\nabla F$ is the E-valued (p+1)-covariant tensor field on M given by

$$\begin{aligned} *\nabla F(X_1,X_2,\ldots,X_{p+1}) &= \tilde{\nabla}_{X_{p+1}}(F(X_1,\ldots,X_p)) \\ &\quad - F(\nabla_{X_{p+1}}X_1,X_2,\ldots,X_p) \\ &\quad - F(X_1,\nabla_{X_{p+1}}X_2,\ldots,X_p) \\ &\quad \cdots\cdots \\ &\quad - F(X_1,X_2,\ldots,\nabla_{X_{p+1}}X_p). \end{aligned} \tag{2.1}$$

If F has holomorphy type (p,0), we say that F is <u>holomorphic</u> if $*\nabla F$ has holomorphy type (p+1,0).

We now obtain a description of the above using a local complex coordinate (z,U) on M. On U a tensor F of holomorphy type (p,0) has the form

$$F = f\,dz^p \tag{2.2}$$

where $f : U \to E$ is given by

$$f = F\left(\frac{\partial}{\partial z},\ldots,\frac{\partial}{\partial z}\right). \tag{2.3}$$

We now prove

<u>Lemma 2.1.</u> <u>F is holomorphic if and only if</u>

$$\tilde{\nabla}_{\partial/\partial\bar{z}}f = 0.$$

To prove the lemma we note that it is sufficient to check $\mathbb{C}$-linearity in the last slot. Now

$$
\begin{aligned}
{}^*\nabla_{\partial/\partial\bar{z}} F &= {}^*\nabla_{\partial/\partial\bar{z}} (f\, dz^p) \\
&= (\tilde{\nabla}_{\partial/\partial\bar{z}} f)\, dz^p + p.f dz^{p-1}\, \nabla_{\partial/\partial\bar{z}} dz
\end{aligned}
$$

However, $\nabla_{\partial/\partial\bar{z}}\, dz = 0$, so the lemma now follows.

We illustrate the above theory by re-casting a theorem due to Chern [1] into our framework. This theorem is fundamental to the definition of the higher order forms mentioned in §1.

Theorem 2.2 (Chern). Let $\omega_\alpha(z)$ be complex-valued functions which satisfy the differential system

$$\frac{\partial\omega_\alpha}{\partial\bar{z}} = \sum_{\beta=1}^{n} a_{\alpha\beta}\,\omega_\beta, \qquad 1 \leq \alpha,\beta \leq n,$$

in a neighbourhood of $z = 0$, where $a_{\alpha\beta}$ are complex-valued C^1 functions. Suppose that the ω_α do not all vanish identically. Then the ω_α are of the form

$$\omega_\alpha(z) = z^m\,\omega_\alpha^*(z),$$

where m is a positive integer and $\omega_\alpha^*(0)$ are not all zero.

The above theorem will be used to prove the following result, in which we use notation from the first part of this section.

Theorem 2.3. Assume that the E-valued tensor field F on M is holomorphic, and let $x \in M$ be such that $F(x) = 0$. Let (z,U) be a local complex coordinate on M with $z(x) = 0$. Then either $F \equiv 0$ on U; or $F = z^m F^*$, where m is a positive integer and $F^*(x) \neq 0$.

To prove the theorem we let

$$F = f\,dz^p$$

on U and let $\{e_1, e_2, \ldots, e_n\}$ be a local framing of E as a complex vector bundle. Since F is holomorphic, we have from Lemma 2.1,

$$0 = \tilde{\nabla}_{\partial/\partial\bar{z}} f = \tilde{\nabla}_{\partial/\partial\bar{z}} \sum_{i=1}^{n} (f^i e_i) = \sum_{i=1}^{n} \frac{\partial f^i}{\partial \bar{z}} e_i + \sum_{i=1}^{n} f^i \tilde{\nabla}_{\partial/\partial\bar{z}} e_i \,.$$

Hence, if $\{\omega^i\}$ is the dual basis of $\{e_i\}$ we have

$$\frac{\partial f^i}{\partial \bar{z}} = \sum_{j=1}^{n} \alpha_j^i f^j ,$$

where $\alpha_j^i = -\omega^i(\tilde{\nabla}_{\partial/\partial\bar{z}} e_j)$.

We now apply Theorem 2.2 to see that if F doe. not vanish identically on U, then there are complex functions f^{*i}, $(i = 1,2,\ldots,n)$, and a positive integer m such that

$$f^i = z^m f^{*i} ,$$

and $f^{*i}(0)$ are not all zero.

Then

$$F^* = \sum_{i=1}^{n} f^{*i} e_i \, dz^p$$

is the tensor required to complete the proof of Theorem 2.3.

The general theory of holomorphic bundle-valued forms will be applied in the following situation. Let E be a <u>real</u> vector bundle equipped with a connection $\tilde{\nabla}$ and let F be an E-valued tensor on M covariant of degree p. Let $\tilde{F}$ denote that part of the complexification $^{\mathbb{C}}F$ of F which is of holomorphy type (p,0). Thus if (z,U) is a local complex coordinate of M, then

$$\tilde{F} = \tilde{f}\,dz^p$$

where

$$\tilde{f} = {}^{c}F\left(\frac{\partial}{\partial z}, \dots, \frac{\partial}{\partial z}\right).$$

We see that $\tilde{F}$ is an ${}^{c}E$-valued covariant tensor on M of holomorphy type (p,0). Since ${}^{c}E$ has a natural connection (also denoted $\tilde{\nabla}$) induced by the connection on E it is possible to discuss the homomorphicity of $\tilde{F}$.

§3. Application to immersions

Let (W^n,g) be a Riemannian manifold, not necessarily of constant sectional curvature. Let $\psi : M \to W$ be an immersion. Then the derivative ψ_* is a tensor covariant of degree 1 with values in the pull-back ψ^*TW of the tangent bundle of W. This bundle has a connection $\tilde{\nabla}$ naturally induced by the Levi-Cività connection on TW, and we give M the metric induced from W by the immersion ψ.

Lemma 3.1. The immersion $\psi : M \to W$ is minimal if and only if $\tilde{\psi}_*$ is holomorphic.

To prove the lemma, we let $z = x + iy$ be a local complex coordinate on M. Then

$$\tilde{\psi}_* = \tilde{f}\,dz ,$$

where

$$\tilde{f} = \tfrac{1}{2}\left(\psi_* \frac{\partial}{\partial x} - i\,\psi_* \frac{\partial}{\partial y}\right).$$

Hence

$$4\,\tilde{\nabla}_{\partial/\partial\bar{z}}\tilde{f} = \tilde{\nabla}_{\partial/\partial x}\psi_*\frac{\partial}{\partial x} + \tilde{\nabla}_{\partial/\partial y}\psi_*\frac{\partial}{\partial y} + i\left(\tilde{\nabla}_{\partial/\partial y}\psi_*\frac{\partial}{\partial x} - \tilde{\nabla}_{\partial/\partial x}\psi_*\frac{\partial}{\partial y}\right).$$

We see from this that the real part of $4\,\tilde{\nabla}_{\partial/\partial\bar{z}}\tilde{f}$ is equal to $g\left(\frac{\partial}{\partial x}, \frac{\partial}{\partial x}\right)H$, where H is the mean curvature of the immersion. The imaginary part is equal to the Lie bracket of $\psi_* \frac{\partial}{\partial x}$ and $\psi_* \frac{\partial}{\partial y}$, and this is equal to zero. This completes the proof of the lemma.

We may regard the second fundamental form II_1 of the immersion ψ as an NM-valued tensor field on M of covariant degree 2, where NM is the normal bundle of the immersion. Hence we may use II_1 to construct an ${}^{\mathbb{C}}$NM-valued tensor field $\tilde{II}_1$ of holomorphy type (2,0) as detailed at the end of §2. Further, NM has a connection $\nabla^{\perp}$ induced by the Levi-Cività connection of W. The main theorem of this secion is

Theorem 3.2. Let W be a space form. Then the mean curvature vector H of ψ is parallel in NM if and only if $\tilde{II}_1$ is holomorphic.

We prove the theorem by using a local positively oriented orthonormal frame field $\{e_1, e_2\}$ on M. If we put $Z = e_1 - ie_2$, and if ${}^*\nabla \tilde{II}_1$ is the covariant differential of $\tilde{II}_1$, then $\tilde{II}_1$ is holomorphic if and only if

$$ {}^*\nabla({}^{\mathbb{C}}II_1)(Z,Z,\bar{Z}) = 0. \tag{3.1} $$

However it follows from the Codazzi equation for submanifolds of a space form (see e.g. [4] page 25) that

$$ \begin{aligned} {}^*\nabla({}^{\mathbb{C}}II_1)(Z,Z,\bar{Z}) &= {}^*\nabla({}^{\mathbb{C}}II_1)(Z,\bar{Z},Z) \\ &= \nabla^{\perp}_Z({}^{\mathbb{C}}II_1(Z,\bar{Z})) - {}^{\mathbb{C}}II_1(\nabla_Z Z,\bar{Z}) - {}^{\mathbb{C}}II_1(Z,\nabla_Z\bar{Z}). \end{aligned} \tag{3.2} $$

Let $\nabla_Z\bar{Z} = \alpha\bar{Z}$ and $\nabla_Z Z = \beta Z$. We differentiate the equation

$$ g(Z,\bar{Z}) = 2 $$

with respect to Z, and deduce that $\alpha + \beta = 0$.

Hence the final two terms on the right hand side of (3.2) cancel, so that

$$ {}^*\nabla({}^{\mathbb{C}}II_1)(Z,Z,\bar{Z}) = \nabla^{\perp}_Z H. \tag{3.3} $$

The proof of the theorem now follows immediately.

We will now obtain a local expression for $\tilde{II}_1$ using moving frames. This will enable us to compare our tensor fields with those constructed by Jensen and Rigoli. Let $\{e_1, e_2, \ldots, e_n\}$ be a local orthonormal framing of ψ^*TW such that $\{e_1, e_2\}$ is a positively oriented basis of the tangent space to M. Jensen and Rigoli call this a "first order frame field". Let $\{\omega^1, \omega^2\}$ be the basis dual to $\{e_1, e_2\}$ and let $\phi = \omega^1 + i\omega^2$. Then

$$\tilde{II}_1 = \tilde{f}\,\phi^2,$$

where $\tilde{f}$ is the ${}^{\mathbb{C}}NM$-valued function given by

$$\begin{aligned} \tilde{f} &= \frac{1}{4}\tilde{II}_1(e_1 - ie_2, e_1 - ie_2) \\ &= \frac{1}{4}(II_1(e_1,e_1) - II_1(e_2,e_2)) - \frac{i}{2}(II_1(e_1,e_2)). \end{aligned} \tag{3.4}$$

In the special case in which n = 3 and W is oriented, NM may be canonically identified with $M \times \mathbb{C}$ and $\tilde{II}_1$ becomes a $\mathbb{C}$-valued 2-form on M. The expression above for $\tilde{f}$ shows that $\tilde{II}_1$ is the form Ψ introduced by Jensen and Rigoli in [2], so Theorem 1.1 (and, indeed, its converse) becomes a special case of our Theorem 3.2.

In the case of arbitrary n, the above expression for $\tilde{f}$ shows that the symmetric 4-form Λ_1 of Jensen and Rigoli is given by

$$\Lambda_1(X_1, X_2, X_3, X_4) = g(\tilde{II}_1(X_1, X_2), \tilde{II}_1(X_3, X_4)),$$

where g denotes the extension of the Riemannian metric of W to a complex-valued complex bilinear map.

Using the fact that $\tilde{\nabla}$ is a metric connection, it is clear that if $\tilde{II}_1$ is holomorphic, then Λ_1 is a holomorphic $\mathbb{C}$-valued symmetric

4-form. However it seems unlikely that the converse is true. We have now shown that Theorem 1.2 is also a special case of Theorem 3.2, and we also have a characterization of immersions with parallel mean curvature vector in terms of the symmetric 2-form $\tilde{II}_1$.

§4. Higher order Forms

In this section we will outline the way in which the higher order forms of Jensen and Rigoli may be constructed using our methods. We begin by assuming that (W,g) is a space form and $\psi : M \to W$ is a minimal immersion of an oriented surface M which is not totally umbilical.

In order to define the next form of higher order we must assume that every vector in the image of $\tilde{II}_1$ is isotropic, i.e. if $X + iY$ is in the image of $\tilde{II}_1$, then $g(X + iY, X + iY) = 0$. Since TM is a complex line bundle, this is equivalent locally to assuming that every vector in the image of the ${}^{\mathbb{C}}NM$-valued function f of equation (3.4) is isotropic. This assumption is the same as that of Jensen and Rigoli that "ψ is isotropic of order 1". From now on we shall make this assumption.

Let $x \in M$ be such that $\tilde{II}_1(x) \neq 0$ (i.e. x is not an umbilical point), and let $a + ib$ be any non-zero vector in the image of $\tilde{II}_1(x)$, where a, b are normal vectors. Since ψ is isotropic of order 1, we have $g(a + ib,\ a + ib) = 0$, and hence

$$\left.\begin{aligned} g(a,a) &= g(b,b) \\ g(a,b) &= 0 \end{aligned}\right\} .$$

The subspace of the normal space $N(x)$ spanned by $\{a,b\}$ is independent of the choice of $a + ib$ and is clearly 2-dimensional. We shall call it the <u>first principal normal space</u> $P_1(x)$ at x.

Suppose now that $x \in M$ is an umbilical point. Then, although $\tilde{II}_1(x) = 0$, we shall show that the first principal normal space at x can still be defined. From Theorem 2.3, it follows that x is an isolated umbilical point. Moreover, if z is a local complex coordinate of M such that $z(x) = 0$, then there is a positive integer m such that

$$\tilde{II}_1 = z^m \tilde{II}_1^*, \qquad \text{with } \tilde{II}_1^*(x) \neq 0.$$

If $a + ib$ is a non-zero vector in the image of $\tilde{II}_1^*(x)$, we define $P_1(x)$ to be that subspace of $N(x)$ spanned by $\{a,b\}$.

We see that the set of first principal normal spaces forms a vector bundle. We call this the <u>first principal normal bundle</u> and denote it by P_1M. Clearly it is a subbundle of the normal bundle NM - in fact it is a smooth 2-plane subbundle of NM. Let N_2M be the subbundle of NM which is the orthogonal complement of P_1M. Then if $X \in NM$ we shall use $(X)^{N_2M}$ to denote the orthogonal projection of X onto N_2M. We now define an N_2M-valued tensor on M of covariant degree 3 as follows

$$II_2(X,Y,Z) = (\tilde{\nabla}_X(II_1(Y,Z)))^{N_2M} - (II_1(\nabla_X Y,Z))^{N_2M} - (II_1(Y,\nabla_X Z))^{N_2M}.$$

Incidentally, we have written the final two terms on the right hand side in order to make the tensorial nature of II_2 clear. In fact, minimality of ψ implies that the image of II_1 is contained in P_1M, so that both these last two terms are zero.

Then $\tilde{II}_2$ is an ${}^{\mathbb{C}}N_2M$-valued tensor on M of holomorphy type (3,0), and it is the next form in the sequence after $\tilde{II}_1$. Following our analysis in Sections 2 and 3 it can be shown that $\tilde{II}_2$ is a holomorphic ${}^{\mathbb{C}}N_2M$-valued tensor, where N_2M is given the connection induced on it as a subbundle of NM. Hence, unless $\tilde{II}_2$ is identically zero, we can use

$\tilde{II}_2$ to define the second principal normal bundle P_2M, and the process may be continued to define a sequence $\tilde{II}_1, \tilde{II}_2, \tilde{II}_3, \ldots,$ of holomorphic higher order bundle-valued forms.

§5. Relation with forms of Jensen and Rigoli

We shall now obtain a local expression for $\tilde{II}_2$ using moving frames in a neighbourhood of any non-umbilical point. So, we assume that $\psi : M \to W$ is a minimal immersion of an oriented surface into a space form (W,g), and we assume that ψ is isotropic of order 1. Let $\{e_1, e_2, \ldots, e_n\}$ be a local orthonormal framing of ψ^*TW such that $\{e_1, e_2\}$ is a positively oriented basis of the tangent space to M and such that $\tilde{II}_1(e_1 - ie_2, e_1 - ie_2) = k(e_3 + ie_4)$ for some positive real number k. Then $\{e_1, e_2, \ldots, e_n\}$ is a second order frame field in the terminology of Jensen and Rigoli. Let $\{\omega^1, \omega^2\}$ be the basis dual to $\{e_1, e_2\}$ and let $\phi = \omega^1 + i\omega^2$.

Then

$$\tilde{II}_1(e_1 - ie_2, e_1 - ie_2) = 2II_1(e_1, e_1) - 2iII_1(e_1, e_2),$$

so that

$$\begin{aligned} ke_3 &= 2II_1(e_1, e_1) \\ ke_4 &= -2II_1(e_1, e_2) \, . \end{aligned} \tag{5.1}$$

Then

$$\tilde{II}_2 = L\phi^3 \, ,$$

where

$$8L = \tilde{II}_2(e_1 - ie_2, e_1 - ie_2, e_1 - ie_2).$$

Recall that $\tilde{\nabla}$ is the connection on ψ^*TW induced by the Levi-Cività connection on TW. We compute the real part Re(L) of L as follows.

$$8\mathrm{Re}(L) = \left[\mathrm{Re}\,\tilde{\nabla}_{e_1 - ie_2}(\tilde{II}_1(e_1 - ie_2, e_1 - ie_2))\right]^{N_2M}$$

$$= \left[\mathrm{Re}\,\tilde{\nabla}_{e_1 - ie_2}(k(e_3 + ie_4))\right]^{N_2M}$$

$$= k\left[\tilde{\nabla}_{e_1} e_3 + \tilde{\nabla}_{e_2} e_4\right]^{N_2M}.$$

The imaginary part of L can be found in a similar manner. In fact we find that

$$8L = k\left[\tilde{\nabla}_{e_1} e_3 + \tilde{\nabla}_{e_2} e_4\right]^{N_2M} + ik\left[\tilde{\nabla}_{e_1} e_4 - \tilde{\nabla}_{e_2} e_3\right]^{N_2M}.$$

By comparing this with the local expressions for Λ_2 in [2], it follows that Λ_2 is constructed from $\tilde{II}_2$ in the same way that Λ_1 was constructed from $\tilde{II}_1$. In fact, modulo a real scalar multiple, we find that Λ_2 is the holomorphic $\mathbb{C}$-valued 6-form given by

$$\Lambda_2(X_1,\ldots,X_6) = g(\tilde{II}_2(X_1,X_2,X_3),\ \tilde{II}_2(X_4,X_5,X_6)).$$

As noted by Rigoli and Jensen, the local expression for L may be simplified because one can show that

$$(\tilde{\nabla}_{e_1} e_3 - \tilde{\nabla}_{e_2} e_4)^{N_2M} = 0$$

and

$$(\tilde{\nabla}_{e_1} e_4 + \tilde{\nabla}_{e_2} e_3)^{N_2M} = 0.$$

It *is* clear that the forms $\Lambda_3, \Lambda_4, \ldots,$ of Jensen and Rigoli will be obtained from $\tilde{II}_3, \tilde{II}_4, \ldots,$ by the same construction.

References

1. S.S.Chern, "On the minimal immersions of the two sphere in a space of constant curvature", Problems in Analysis, Princeton, N.J., 1970, 27-40.

2. G. Jensen and M. Rigoli, "Minimal surfaces in spheres by the method of moving frames", Seminaire de l'institute É. Cartan (1983), University de Nancy I, Nancy.

3. G. Jensen and M. Rigoli, "Minimal surfaces in spheres", Preprint.

4. S. Kobayashi and K. Nomizu, Foundations of Differential Geometry, Interscience, Vol. II (1969).

5. M. Rigoli, "Surfaces with parallel mean curvature vector in a 4-space form", Preprint.

Durham University
Department of Mathematical Sciences
Science Laboratories
South Road
Durham DH1 3LE, England.

Examples of 1-Codimensional non Totally Geodesic Isometric Immersions of Pseudo-riemannian Space Forms with the Same Positive Constant Curvature and the Same Space-like Rank.

by

Marcos Dajczer
IMPA
Estrada Dona Castorina 110
22460 Rio de Janeiro (Brazil)

Peter Dombrowski
Math. Institut
Universität zu Köln
Weyertal 86 - 90
5000 Köln 41 (Germany)

1. Introduction.

If $n \geqq 2$ and $f: M^n(c) \longrightarrow \tilde{M}^{n+p}(c)$ is an isometric immersion of complete riemannian manifolds of the same constant sectional curvature $c > 0$ and of the dimensions n and $n+p$ respectively, D. Ferus ([2], p. 155) has proved, that for $p < n$ such an isometric immersion f must be totally geodesic. - For $p = n$ it is still unknown, whether a similar result is true. For $p > n \geqq 2$ however there exist simple examples of non totally geodesic isometric immersions $f: S^n(c) \longrightarrow S^{n+p}(c)$, e.g.

$$f(a) := \alpha^{-1}(\cos(\alpha a_0), \sin(\alpha a_0), \ldots, \cos(\alpha a_n), \sin(\alpha a_n), \underbrace{0, \ldots, 0}_{p-(n+1)}) \tag{1}$$

for $a = (a_0, \ldots, a_n)$ in

$$S^n(c) := \{a \in \mathbb{R}^{n+1} \mid a_0^2 + \ldots + a_n^2 = c^{-1}\} \tag{2}$$

and with $\alpha := \sqrt{(n+1)c}$. - In other words, for "high codimension" p (namely for $p > n$) the standard totally geodesic isometric imbedding

$$S^n(c) \hookrightarrow S^{n+p}(c), \quad ((a_0, \ldots, a_n) \longmapsto (0, \ldots, 0, a_0, \ldots, a_n)) \tag{3}$$

is no more rigid. [For $n = 1$ and $p \geqq 1$ this is trivial, since for every $k \in \mathbb{N}$ with $k \geqq 2$ one has a non totally geodesic isom. immersion $z_k: S^1(1) \longrightarrow S^{1+p}(1)$ induced by the (2π)-periodic map

$$\mathbb{R} \longrightarrow S^{1+p}(1), \; (t \longmapsto (k^{-1}\cos(kt), k^{-1}\sin(kt), \sqrt{1-k^{-2}}, \underbrace{0, \ldots, 0}_{p-1})) \text{ .]} \tag{4}$$

The main purpose of this article is to present examples of non totally geodesic isometric immersions $f: S^n_s(c) \longrightarrow S^{n+p}_t(c)$ of pseudo--riemannian manifolds of dimensions n and $n+p$ and time-like ranks

s and t respectively for all $n \in \mathbb{N}_+$ already with codimension $p = 1$, in which examples however the space-like rank $n-s$ of the original manifold $S^n_s(c)$ will be 1 too. [It was conjectured and proved in special cases (see section 7) by the first author, that, if the codimension p is less or at least sufficiently small compared (not with the dimension n but) with the space-like rank $n-s$ of the original manifold $S^n_s(c)$, such an isometric immersion must be always totally geodesic (similar to the above-mentioned result in the riemannian case). This conjecture is somehow backed by a special result of L. Graves and K. Nomizu ([4], Theorem, p. 129).]

2. Preliminaries.

2.0. Let V be a real vector space of finite dimension with a nondegenerate, possibly indefinite, inner product $\langle\cdot,\cdot\rangle$. A vector $v \in V$ is called space-like, time-like resp. light-like (or null) according to whether $\langle v,v\rangle$ is positive, negative or zero. If $|\langle v,v\rangle| = 1$, then v is called a unit vector (of its type). For any vector subspace W of V we define its orthogonal vector subspace

$$W^{\perp} := \{v \in V \mid \langle v,w\rangle = 0 \text{ for all } w \in W\} . \tag{5}$$

and we call W degenerate resp. nondegenerate (w.r.t. $\langle\cdot,\cdot\rangle$) if the restriction of $\langle\cdot,\cdot\rangle$ to W is degenerate resp. not (i.e. if $W \cap W^{\perp}$ contains a non-vanishing vector resp. not). Non-degeneracy of W (w.r.t. $\langle\cdot,\cdot\rangle$), is therefore equivalent to the fact, that W induces the orthogonal splitting of V:

$$V = W + W^{\perp} \quad \text{and} \quad W \cap W^{\perp} = \{o\}. \tag{6}$$

2.1. Suppose $n,s \in \mathbb{N}$ and $s \leqq n$. Then $\mathbb{R}^n_s$ denotes the standard space form of signature $(s,n-s)$ [1] and of constant curvature 0 with the vector space $\mathbb{R}^n$ as its manifold and the (pseudo-)riemannian metric induced from the canonical inner product on $\mathbb{R}^n$ of signature $(s,n-s)$ [1] given by

$$\langle v,w\rangle^n_s := -\sum_{i=1}^{s} v_i w_i + \sum_{j=s+1}^{n} v_j w_j \quad \text{for } v,w \in \mathbb{R}^n \;, \;^{2)} \tag{7}$$

and we write for brevity $\langle v,w\rangle := \langle v,w\rangle^n_o$ for $v,w \in \mathbb{R}^n$.

---

1) i.e. in other words: Of time-like resp. space-like rank s resp. $n-s$.

2) we write more simply $\langle .,.\rangle_s := \langle .,.\rangle^n_s$, if there is no doubt about the dimension in question.

For $c \in \mathbb{R}_+$ the standard space form $S^n_s(c)$ of signature $(s,n-s)$ [1] and of constant curvature c is given as the hypersurface

$$S^n_s(c) \; := \; \{x \in \mathbb{R}^{n+1} \mid \langle x,x\rangle_s \; = c^{-1}\} \tag{8}$$

of $\mathbb{R}^{n+1}_s$ with the induced riemannian metric (which proves to be of signature $(s,n-s)$.

For $c \in \mathbb{R}_-$ the standard space form $H^n_s(c)$ of signature $(s,n-s)$ [1] and of constant curvature c is given as the hypersurface

$$H^n_s(c) \; := \; \{x \in \mathbb{R}^{n+1} \mid \langle x,x\rangle_{s+1} \; = \; c^{-1}\} \tag{9}$$

of $\mathbb{R}^{n+1}_{s+1}$ with the induced riemannian metric.

In case $s = 0$, i.e. in the positive definite case, we write $S^n(c)$ for $S^n_o(c)$ and $H^n(c)$ for the connected component of $H^n_o(c)$ containing $(1/\sqrt{|c|}, 0,\dots,0) \in \mathbb{R}^{n+1}$. Evidently $S^n := S^n(1)$ resp. $H^n := H^n(-1)$ is the standard unit sphere resp. the hyperbolic space (of curvature 1 resp. -1).

2.2. Some elementary properties of the standard space forms of signature $(s,n-s)$ and of constant curvature $c \in \mathbb{R}_+$:

(i) $$\begin{array}{ccc} \mathbb{R}^s \times S^{n-s} & \rightarrow & S^n_s(c) \\ (x,y) & \longmapsto & (1/\sqrt{c})(x, \sqrt{1+\langle x,x\rangle}\, y) \end{array} \quad \text{is a } C^\infty \text{diffeomorphism} \tag{10}$$

(see e.g. [6],2.4.6), in particular, $S^n_s(c)$ for $0 < s < n$, resp. $S^n_n(c)$, is - considered as a quadric hypersurface of $\mathbb{R}^{n+1}$ - a hyperboloid of one resp. of two sheets and we define

$$\left.\begin{array}{l} \tilde{S}^n_s(c) := \text{connected component of } (0,\dots,0,1/\sqrt{c}) \text{ in } S^n_s(c), \\ \qquad \text{i.e. } \tilde{S}^n_s(c) = S^n_s(c) \text{ for } s < n \text{ and} \\ \qquad \tilde{S}^n_n(c) = \{x \in S^n_n(c) \mid x_{n+1} > 0\} \text{ (= "upper" sheet of } S^n_n(c)). \end{array}\right\} \tag{11}$$

(ii) For all $x \in S^n_s(c)$ the tangent space to $S^n_s(c)$ at x can be (and will be in the following) identified with the following n-dim vector subspace of $\mathbb{R}^{n+1}$:

$$T_x S^n_s(c) \; = \; \{v \in \mathbb{R}^{n+1} \mid \langle v,x\rangle_s \; = 0\} \; . \tag{12}$$

<u>(iii)</u> For all $k,p\in\mathbb{N}$ with $k\leq p$ the map (see (11))

$$\left.\begin{array}{rcl} g\,(=g_{n,p,k}):\ \tilde{S}^n_n(c)\times S^p_k(1) & \longrightarrow & S^{n+p}_{n+k}(c) \\ (x,y) & \longmapsto & (x_1,\ldots,x_n,x_{n+1}y_1,\ldots,x_{n+1}y_{p+1}) \end{array}\right\} \quad (13)$$

is an isometry of the warped product $\tilde{S}^n_n(c)\times_{(x_{n+1})}S^p_k(1)$ onto $S^{n+p}_{n+k}(c)$, i.e. for all $(x,y)\in\tilde{S}^n_n(c)\times S^p_k(1)$ one has:

$$\left.\begin{array}{l} \text{If } (v,w),\ (\tilde{v},\tilde{w})\in T_x\tilde{S}^n_n(c)\oplus T_yS^p_k(1)\ \cong\ T_{(x,y)}(\tilde{S}^n_n(c)\times S^p_k(1)),\ \text{then} \\ \langle g_*(v,w),g_*(\tilde{v},\tilde{w})\rangle^{n+p+1}_{n+k} = \langle v,\tilde{v}\rangle^{n+1}_n + x^2_{n+1}\langle w,\tilde{w}\rangle^{p+1}_k . \end{array}\right\} \quad (14)$$

<u>(iv)</u> The group of isometries of $S^n_s(c)$ consists of the restrictions of all the linear isometries of $\mathbb{R}^{n+1}_s$ to $S^n_s(c)$ (see e.g. [6], 2.4.4) and this group acts transitively on $S^n_s(c)$, i.e. $S^n_s(c)$ is a homogeneous submanifold of $\mathbb{R}^{n+1}_s$.

<u>(v)</u> For every $x\in S^n_s(c)$ and every m-dimensional $(m\in\mathbb{N})$ vector subspace W of $T_xS^n_s(c)$ (see (12)) there is exactly one m-dimensional auto-parallel (see e.g. [5], p. 53), geodesically complete connected submanifold L of $S^n_s(c)$ with $x\in L$ and $T_xL = W$ and this L is the connected component of x in the intersection of $S^n_s(c)$ with the (m+1)--dimensional vector subspace V of $\mathbb{R}^{n+1}$, which is characterized by $x\in V$ and $W\subset T_xV$. If W is non-degenerate, this L is even totally geodesic in $S^n_s(c)$ in the sense that the second fundamental form of L in $S^n_s(c)$ vanishes, (observe, that the second fundamental form of a submanifold of a pseudo-riemannian manifold is only defined, if the tangent spaces of the submanifold give orthogonal splittings of the tangent spaces of the ambient space analogous to (6), which for an autoparallel submanifold is true iff for at least one point the tangent space is nondegenerate).
[Since $S^n_s(c)$ intersects every vector subspace V of $\mathbb{R}^{n+1}$ "ortho-transversal" w.r.t. $\langle\cdot,\cdot\rangle^{n+1}_s$, i.e. for all $x\in S^n_s(c)\cap V$ one has $\mathbb{R}x = (T_xS^n_s(c))^{\perp}\subset T_xV$, the fact, that $S^n_s(c)\cap V$ is totally geodesic, is a well-known result in the definite, riemannian case (see e.g. [1], p. 180,

Corollary 2(i)) and remains true here (in the nondegenerate case without any change in the proof) for the possibly indefinite case.

(vi) For all $n,p,s,t \in \mathbb{N}$ with $s \leqq n$ and $s \leqq t \leqq s+p$ the map

$$\left.\begin{array}{rcl} i\,(=i_{n,p,s,t}):S^n_s(c) & \hookrightarrow & S^{n+p}_t(c) \\ x & \longmapsto & (x_1,\ldots,x_s,\underbrace{0,\ldots,0}_{p},x_{s+1},\ldots,x_{n+1}) \end{array}\right\} \quad (15)$$

is a ("the canonical") totally geodesic isometric imbedding of $S^n_s(c)$ into $S^{n+p}_t(c)$ and every other totally geodesic isometric immersion of $S^n_s(c)$ (see (11)) into $S^{n+p}_t(c)$ is obtained by composing the restriction of the map (15) to $S^n_s(c)$ with a linear isometry of $\mathbb{R}^{n+p+1}_t$, (as follows easily from (v)). In particular:

If $f:S^n_s(c) \to S^{n+p}_t(c)$ is any isometric immersion and there exist $n+2$ points $z_1,\ldots,z_{n+2} \in S^n_s(c)$ such that $f(z_1),\ldots,f(z_{n+2})$ are linearly independent vectors in $\mathbb{R}^{n+p+1}$, then f is not totally geodesic.

(vii) The linear "anti-isometry"

$$a^n_s: \mathbb{R}^{n+1}_s \longrightarrow \mathbb{R}^{n+1}_{n+1-s} \qquad (x \longmapsto (x_{s+1},\ldots,x_{n+1},x_1,\ldots,x_s)) \qquad (16)$$

(i.e. $\langle a^n_s(v),a^n_s(w)\rangle_{n+1-s} = -\langle v,w\rangle_s$ for all $v,w \in \mathbb{R}^{n+1}$) maps $S^n_s(c)$ anti-isometrically and hence affinely onto $H^n_{n-s}(-c)$ (see [6],2.4.6(iii)), thus mapping the auto-parallel resp. totally geodesic submanifolds of $S^n_s(c)$ onto the ones of $H^n_{n-s}(-c)$. Moreover every isometric immersion $f:S^n_s(c) \to S^{\tilde{n}}_{\tilde{s}}(\tilde{c})$ gives rise (in a functorial way) to an isometric immersion

$$\bar{f} := a^{\tilde{n}}_{\tilde{s}} \circ f \circ (a^n_s)^{-1}:H^n_{n-s}(-c) \to H^{\tilde{n}}_{\tilde{n}-\tilde{s}}(-\tilde{c}). \qquad (17)$$

3. Lemma [3)]: Let n,p,q,k,l denote natural integers (with $k \leqq \min\{l,p\}$, $\max\{l,p\} \leqq q$), $f:S^p_k(1) \to S^q_l(1)$ an isometric immersion and α^f the

3) This Lemma holds m.m. for an isometric immersion $f:N \to \tilde{N}$ and the corresponding immersion $id \times f: M \times_\varphi N \to M \times_\varphi \tilde{N}$ of warped products, if M, N, $\tilde{N}$ are arbitrary pseudo-riemannian manifolds and $\varphi:M \to \mathbb{R}_+$ a C^∞ function.

second fundamental form of f (see [5],p. 10), i.e. for all vector fields $X,Y \in \mathfrak{X}(S^p_k(1))$:

$$\alpha^f(X,Y) = (\nabla_X f_* Y)^\perp , \tag{18}$$

where ∇ is the Levi-Civita covariant derivative of $S^q_l(1)$ and $(\dots)^\perp$ means the f- normal component in $TS^q_l(1)$. The latter is defined, since for all $x \in S^p_k(1)$ the subspace $f_* T_x S^p_k(1)$ is nondegenerate in $T_{f(x)}S^q_l(1)$, hence (compare (6)):

$$T_{f(x)}S^q_l(1) = f_*(T_x S^p_k(1)) \oplus (f_* T_x S^p_k(1))^\perp .$$

Then: (i) The map

$$\begin{array}{ccc} (\mathrm{id}\times f): \tilde{S}^n_n(c)\times S^p_k(1) & \longrightarrow & \tilde{S}^n_n(c)\times S^q_l(1) \\ (x,y) & \longmapsto & (x,f(y)) \end{array} \tag{19}$$

is an isometric immersion of the warped products (see 2.2.iii)

$$\tilde{S}^n_n(c)\times_{(x_{n+1})}S^p_k(1) \quad \text{resp.} \quad \tilde{S}^n_n(c)\times_{(x_{n+1})}S^q_l(1). \tag{20}$$

(ii) For all $(x,y) \in \tilde{S}^n_n(c)\times S^p_k(1)$ and all $(v,w),(\tilde{v},\tilde{w}) \in T_x\tilde{S}^n_n(c) \oplus T_y S^p_k(1) \cong T_{(x,y)}(\tilde{S}^n_n(c)\times S^p_k(1))$ the second fundamental form $\alpha^{(\mathrm{id}\times f)}$ of $(\mathrm{id}\times f)$ (see(19)) satisfies (see(18)):

$$\alpha^{(\mathrm{id}\times f)}((v,w),(\tilde{v},\tilde{w})) = (o, \alpha^f(w,\tilde{w})) . \tag{21}$$

and therefore (see(19)):

$$f \text{ totally geodesic} \iff \begin{cases} (\mathrm{id}\times f) \text{ is a totally geodesic map} \\ \text{of the warped products (20)} \end{cases} . \tag{22}$$

Proof: For (i): Straight-forward (using (14)).
For (ii): Using the standard computational techniques in warped product manifolds, one finds for all $(x,y) \in \tilde{S}^n_n(c)\times S^p_k(1)$ and all $X,\tilde{X} \in \mathfrak{X}(\tilde{S}^n_n(c))$, $Y,\tilde{Y} \in \mathfrak{X}(S^p_k(1))$, if $\nabla, \overset{1}{\nabla}, \overset{2}{\nabla}$ denotes the Levi-Civita covariant derivative of the second warped product manifold in (20) resp. of its first and second factor:

$$(\nabla_{(X,Y)}(\mathrm{id}\times f)_*(\tilde{X},\tilde{Y}))_{(x,y)} =$$

$$= (\mathrm{id}\times f)_*\Big((\overset{1}{\nabla}_X\tilde{X})_x \,,\, (X\cdot\log x_{n+1})(x)Y_y + (X\cdot\log x_{n+1})(x)Y_y + (\overset{2}{\nabla}_Y\tilde{Y})_y\Big) +$$
$$+ (o,\alpha^f(Y,\tilde{Y})_y) .$$

(21) is then an easy consequence of the last equation, (22) an immediate Corollary of (21).

4. Example of a real analytic isometric imbedding $f: S^1_o(1) \to S^2_1(1)$, which is not totally geodesic. [4)]

For all $\alpha \in \mathbb{R}$ consider the

$$(2\pi)\text{-periodic, real analytic map } c_\alpha : \mathbb{R} \to S^2_1(1) \tag{23}$$

defined by: For all $t \in \mathbb{R}$

$$c_\alpha(t) := \sinh(\tfrac{1}{5} + \alpha\cos 4t)(1,0,0) + \cosh(\tfrac{1}{5} + \alpha\cos 4t)(0,\cos t,\sin t). \tag{24}$$

One verifies immediately (see(8),(7)) $c_\alpha(\mathbb{R}) \subset S^2_1(1)$,

$$c_\alpha | [0,2\pi[\quad \text{is injective} \tag{25}$$

and for all $t \in \mathbb{R}$ (see(7))

$$\langle \dot c_\alpha(t), \dot c_\alpha(t)\rangle_1 = \cosh^2(\tfrac{1}{5} + \alpha\cos 4t) - (4\alpha)^2\sin^2(4t) . \tag{26}$$

From (26) follows:

$$c_\alpha : \mathbb{R} \to S^2_1(1) \quad \text{is a space-like immersion for } \alpha \in [0,\tfrac{1}{4}] \tag{27}$$

and evidently

$$\text{Length}(c_0 | [0,2\pi]) = \int_0^{2\pi} \sqrt{\langle \dot c_0(t), \dot c_0(t)\rangle_1}\, dt = \cosh(\tfrac{1}{5})2\pi > 2\pi . \tag{28}$$

Moreover one finds for $\alpha = (1/4)$:

$$\text{Length}(c_{(1/4)} | [0,2\pi]) < 2\pi . \tag{29}$$

Proof for (29): First we have the following numerical estimates:

$$1 \leqq \cosh(\tfrac{1}{5} + \tfrac{1}{4}\cos 4t) \leqq \cosh(\tfrac{1}{5} + \tfrac{1}{4}) < \tfrac{9}{8} \quad \text{for all } t \in \mathbb{R} \tag{30}$$

and

$$\frac{\sqrt{3}}{2} \leqq \sin t, \text{ hence } \sqrt{(\tfrac{9}{8})^2 - \sin^2 t} \leqq \frac{\sqrt{81-48}}{8} < \frac{6}{8} \quad \text{for } t \in [\tfrac{\pi}{3}, 2\tfrac{\pi}{3}] \tag{31}$$

Using (26), (30), (31), the substitution formula for integrals and the π-periodicity of $\sin^2(x)$ we get:

$$\text{Length}(c_{(1/4)} | [0,2\pi]) = \int_0^{2\pi} \sqrt{\langle \dot c_{(1/4)}(t), \dot c_{(1/4)}(t)\rangle_1}\, dt =$$

4) At first sight this example seems to be only a special case of our example 8 below. However, the latter one will be only C^∞, where as here we stress the r e a l a n a l y t i c i t y of the example 4 !

$$= \int_0^{2\pi} \sqrt{\cosh^2(\tfrac{1}{5} + \tfrac{1}{4}\cos 4t) - \sin^2 4t}\, dt \quad <$$

$$< \int_0^{2\pi} \sqrt{(9/8)^2 - \sin^2 4t}\, dt = \tfrac{1}{4}\int_0^{8\pi} \sqrt{(9/8)^2 - \sin^2 x}\, dx \quad =$$

$$= 2\int_0^{\pi} \sqrt{(9/8)^2 - \sin^2 x}\, dx$$

$$< 2\left[\int_0^{(\pi/3)} (9/8)dx + \int_{(\pi/3)}^{(2\pi/3)} (6/8)dx + \int_{(2\pi/3)}^{\pi} (9/8)dx\right] \quad = \quad 2\pi\,,$$

which proves (29).

According to (27) the function $\alpha \longmapsto \text{Length}(c_\alpha | [0,2\pi])$ is defined and due to (26) continuous on $[0,(1/4)]$. Moreover this function attains on this interval (see (28), (29)) values greater and smaller than 2π. Therefore there must exist

$$\alpha \in\,]0,(1/4)[\text{ such that } \quad \text{Length}(c_\alpha | [0,2\pi]) = 2\pi\ . \tag{32}$$

Re-parametrization of c_α with respect to arclength gives therefore a (2π)-periodic (see(32)!) real analytic space-like unit speed path $\gamma: \mathbb{R} \to S_1^2(1)$ and, due to (25), $\gamma | [0,2\pi[$ is injective. γ induces therefore, since $S_0^1(1) = S^1$, a

$$\left.\begin{array}{l} \text{real analytic isometric imbedding } \; z: S_0^1(1) \to S_1^2(1) \\ \text{and this imbedding is not totally geodesic,} \end{array}\right\} \tag{33}$$

because (see above 2.2.vi) the following three points of $z(S_0^1(1)) = c_\alpha(\mathbb{R})$:

$$c_\alpha(0),\ c_\alpha(\pi/2),\ c_\alpha(\pi) \quad \text{are linearly independent vectors of } \mathbb{R}^3,$$

since $\det\big(c_\alpha(0), c_\alpha(\pi/2), c_\alpha(\pi)\big) \;=\; 2\sinh(\tfrac{1}{5}+\alpha)\cosh^2(\tfrac{1}{5}+\alpha) \;\neq 0$.

Finally we observe, that the map (33) represents a generator of the fundamental group of $S_1^2(1)$, which is diffeomorphic (see (10)) to $\mathbb{R}\times S^1$, and therefore

$$\pi_1(S_1^2(1)) \;\cong\; \mathbb{Z}\ . \tag{34}$$

[This generator property of (33) is clear, because by (24) every

$c_\alpha | [0,2\pi]$ winds around exactly once about the axis of rotational symmetry of the hyperboloid $S_1^2(1)$ of one sheet (see 2.2.i).

5. Proposition: For every $n \in \mathbb{N}$ and $c \in \mathbb{R}_+$ there exists a real analytic isometric imbedding $f: S_n^{n+1}(c) \to S_{n+1}^{n+2}(c)$, which is not totally geodesic.

Proof: Let $z: S_o^1(1) \to S_1^2(1)$ denote the real analytic isometric imbedding, which is not totally geodesic, given by (33). Then due to Lemma 3 :

$$(id \times z): \tilde{S}_n^n(c) \times_{(x_{n+1})} S_o^1(1) \longrightarrow \tilde{S}_n^n(c) \times_{(x_{n+1})} S_1^2(1) \tag{35}$$

is a real analytic isometric imbedding, which is not totally geodesic. Then the map (see (13), (35)):

$$f := (g_{n,2,1}) \circ (id \times z) \circ (g_{n,1,0})^{-1} : S_n^{n+1}(c) \to S_{n+1}^{n+2}(c) \tag{36}$$

is - since the mappings $g_{n,p,k}$ of (13) are real analytic isometries - a real analytic isometric imbedding, which is not totally geodesic, q.e.d.

Remark: Using the translation recipe of 2.2.vi for passing to the pseudo-riemannian space forms of constant negative curvature, Proposition 5 translates as follows:

For every $n \in \mathbb{N}$ and $c \in \mathbb{R}_+$ there exists a real analytic isometric imbedding $f: H_1^{n+1}(-c) \to H_1^{n+2}(-c)$, which is not totally geodesic.

Similar to Proposition 5 one can obtain the even simpler:

6. Proposition (For $n = 1$ a stronger version of this can already be found in L. Graves and K. Nomizu [4] , p. 128):
For every $n \in \mathbb{N}$ and $c \in \mathbb{R}_+$ there exists a real analytic isometric immersion $f: S_n^{n+1}(c) \to S_n^{n+2}(c)$, which is not totally geodesic.

Proof.: Let $z_k: S_o^1(1) \to S_o^2(1)$ denote (for $k \in \mathbb{N}$ and $k \geq 2$) the real analytic isometric immersion induced by (4), which is not totally geodesic. The rest of the proof goes exactly like the one of Proposition 5, only substituting there "z" by "z_k" and "$g_{n,2,1}$" by "$g_{n,2,0}$".

Remark: Again applying 2.2.vi we get from Proposition 6:

For every $n \in \mathbb{N}$ and $c \in \mathbb{R}_+$ there exists a real analytic isometric immersion $f: H_1^{n+1}(-c) \to H_2^{n+2}(-c)$, which is not totally geodesic.

7. Can assumptions on $n,p,s \in \mathbb{N}$ imply, that an isometric immersion $f: S_s^n(c) \to S_s^{n+p}(c)$ is totally geodesic ?

(i) Definition: Let $f: M \to \tilde{M}$ be an isometric immersion of pseudo-riemannian manifolds, let α^f denote the second fundamental form of f and for all $x \in M$ we define the null space of α^f as usual by

$$N^f(x) := \{ v \in T_xM \mid \alpha^f(v,w) = 0 \text{ for all } w \in T_xM \}, \tag{37}$$

sometimes called the "relative nullity space of f at x". Let G denote the non-void, open subset of M, where the $\mathbb{N}$-valued function $x \mapsto \dim_{\mathbb{R}} N^f(x)$ attains its minimum value on M. We say, that f is "of nondegenerate minimal relative nullity", if for all $x \in G$ the nullspace $N^f(x)$ of α^f (see (37)) is a nondegenerate subspace of T_xM (see 2.0).

Remarks: a) If $f: M \to \tilde{M}$ is a totally geodesic isometric immersion of pseudo-riemannian manifolds, then $\alpha^f = 0$ and therefore (see (37)) $N^f(x) = T_xM$ and T_xM is a nondegenerate subspace of itself for all $x \in M$. Therefore f is then "of nondegenerate minimal relative nullity" in the sense of the preceding definition and consequently this last property of f is a necessary condition for f being totally geodesic.

b) Thus it doesn't surprise to find this necessary condition as a hypothesis on f in the following criteria (ii) and (iii) for f to be totally geodesic. These criteria were conjectured by the first author and the following valid version of which [incorporating (unfortunately, see below (iv)) the concept of Definition (i) as a hypothesis] was brought up by the second author:

Suppose, that n,p,s are natural integers. Then:

(ii) If $f: S_s^n(c) \to S_s^{n+p}(c)$ is an isometric immersion which is of nondegenerate minimal relative nullity (see (i)) and if $p = 0$ or $p < (n-s)$ (= space-like rank of $S^n_s(c)$), then f is totally geodesic.

Remarks: a) For $s = 0$ this is the Theorem of D. Ferus [2].
b) The result (ii) is sharp, since e.g. for $n = 2$, $p = s = 1$, and consequently $p = n-s$, the isometric immersion $f:S^2_1(1) \to S^3_1(1)$ constructed in Proposition 6, and which is not totally geodesic, is - according to L. Graves and K. Nomizu (see [4], p. 128, line 11+) - of nondegenerate minimal relative nullity [more precisely, $N^f(x)$ is here for all $x \in S^2_1(1)$ a time-like 1-dimensional vector subspace of $T_x S^2_1(1)$ and the relative nullity foliation is given by the meridian (= profile) hyperbolas of $S^2_1(1)$, considered as a 2-dimensional surface of revolution in $\mathbb{R}^3$ around the x_1-axis, (see (8), (7)).]

(iii) If $f:S^n_s(c) \to S^{n+p}_{s+1}(c)$ is an isometric immersion which is of nondegenerate minimal relative nullity (see (i)) and if $2p \leqq n-s$ (= space-like rank of $S^n_s(c)$), then f is totally geodesic.

Remarks: a) The hypotheses of (iii) imply $p \geqq 1$ and $2 \leqq 2p+s \leqq n$.
b) The result (iii) is sharp, since e.g. for $n = p = 1$, $s = 0$, and consequently $2p > n-s$, the map (33) of section 4 provides an isometric imbedding $f:S^1_0(1) \to S^2_1(1)$, which is not totally geodesic, which however is space-like (see (27)) and therefore of nondegenerate minimal relative nullity.

(iv) We omit proofs for (ii) and (iii) [which run - due to the nondegeneracy assumption for the relative nullity spaces of f - essentially along similar lines as in the well-known riemannian (positive definite) case] because of the following reason:

If the hypothesis of (ii) or (iii), that f is of nondegenerate minimal relative nullity, has to be verified by going through the definition (37) of the relative nullity spaces of f, then one has to know α^f already so well, that on that basis one can decide directly (without any further "criterion"), whether α^f vanishes everywhere or not, i.e. whether f is totally geodesic or not. Therefore, the statements (ii) and (iii) seem to us to be useful (for deciding whether f is totally geodesic or not) only, if one possesses additional criteria, which guarantee a priori (e.g. for dimensional or space-like resp. time-like rank reasons on the original or the target manifold of f), that f is of nondegenerate minimal relative nullity. Unfortunately, we don't know of such criteria (except for trivial ones). That one faces here a problem, is indicated by the fact, that - at least in the case of the flat space forms $\mathbb{R}^n_s$ with sectional curvature $c = 0$ (see 2.1) - **L. Graves** has given in his Ph. D. Thesis ([3] , p. 66 and 68) examples of isometric immersions $f: \mathbb{R}^n_1 \to \mathbb{R}^{n+1}_1$ for all $n \geq 2$, for which the spaces of relative nullity are degenerate, and he classified them all. Of course (see (i), Remark a)) these immersions are not totally geodesic.

We conclude the article by another example, refining the examples of section 4 and Proposition 5 respectively.

8. Example of an isometric C^∞ immersion $f: S^1_o(1) \to S^2_1(1)$, which is not totally geodesic and generates a prescribed (non-zero) homotopy class in the fundamental group $\pi_1(S^2_1(1)) \cong \mathbb{Z}$. - More precisely:

Proposition: For every $k \in \mathbb{N}_+$ and every $n \in \mathbb{N}$ there exists an isometric C^∞ immersion $f: S^{n+1}_n(c) \to S^{n+2}_{n+1}(c)$, which is not totally geodesic and which maps a generator of the fundamental group $\pi_1(S^{n+1}_n(c)) \cong \mathbb{Z}$ onto k times a generator of the fundamental group $\pi_1(S^{n+2}_{n+1}(c)) \cong \mathbb{Z}$ (see (10)).

We sketch a proof of this Proposition only in case $n = 0$ and $c = 1$, i.e. we present the construction (omitting the more technical computations) of an isometric C^∞ immersion $f: S^1_o(1) \to S^2_1(1)$, which is not totally geodesic and which winds the unit circle S^1 ($= S^1_o(1)$) k times around the rotational (x_1-)axis of $S^2_1(1)$, a hyperboloid of revolution of one sheet with the x_1-axis of $\mathbb{R}^3$ as axis of rotational symmetry (see (8), (7)). The Proposition for arbitrary $n \in \mathbb{N}$

and $c \in \mathbb{R}_+$ follows then exactly as Proposition 5 from the result of section 4 .

The C^∞-map $f: S_0^1(1) \longrightarrow S_1^2(1)$ of the last assertion with $k \in \mathbb{N}_+$ will be obtained - as in section 4 - via a (2π)-periodic space-like C^∞-map $c: \mathbb{R} \longrightarrow S_1^2(1)$ of length 2π and which has now to wind k times (not only once as in section 4) around the x_1-axis during one period $[0,2\pi]$. To construct such a c we observe first: Every (2π)-periodic C^∞-map $c: \mathbb{R} \longrightarrow S_1^2(1)$ with prescribed winding number $k \in \mathbb{N}_+$ around the x_1-axis can be written uniquely as

$$c(t) = \sinh(\zeta(t))(1,0,0) + \cosh(\zeta(t))(0,\cos\varphi(t),\sin\varphi(t)), \tag{38}$$

where

$$\zeta: \mathbb{R} \longrightarrow \mathbb{R} \text{ is a } (2\pi)\text{-periodic } C^\infty\text{-function} \tag{39}$$

and

$$\left.\begin{array}{l} \varphi: \mathbb{R} \longrightarrow \mathbb{R} \text{ is a } C^\infty\text{-function with} \\ \varphi(t+2\pi) - \varphi(t) = k2\pi \quad \text{for all } t \in \mathbb{R} . \end{array}\right\} \tag{40}$$

We want to specialize ζ and φ , satisfying (39), (40), such that c becomes a space-like curve in $S_1^2(1)$ of length 2π. First it follows from (38), (7):

$$\langle c'(t),c'(t)\rangle_1 = \cosh^2(\zeta(t))\cdot(\varphi'(t))^2 - (\zeta'(t))^2 . \tag{41}$$

Choose now for every $\varepsilon \in]0,(\pi/2)[$ a

$$(2\pi)\text{-periodic } C^\infty \text{function} \quad \psi_\varepsilon: \mathbb{R} \longrightarrow [0,+\infty[\tag{42}$$

such that ψ_ε is even with respect to π , i.e.

$$\psi_\varepsilon(\pi+t) = \psi_\varepsilon(\pi-t) \quad \text{for } t \in \mathbb{R} \tag{43}$$

and

$$\psi_\varepsilon(0) = 0 , \qquad \psi_\varepsilon(\pi) \leqq \pi , \tag{44}$$

$$0 < \psi_\varepsilon'(t) \leqq 1 \quad \text{for } t \in]0,\pi[, \tag{45}$$

$$\psi_\varepsilon'(t) = 1 \quad \text{if and only if } t \in [\varepsilon,\pi-\varepsilon] . \tag{46}$$

[Such a function ψ_ε can be explicitely described on $[0,\pi]$ (which is because of (42), (43) sufficient) as follows. Let $\lambda: \mathbb{R} \longrightarrow [0,1[$ denote the C^∞-function

$\lambda(t) := \exp(-t^{-1})$ for $t > 0$ and $\lambda(t) = 0$ for $t \leqq 0$.

Then ψ_ε, defined for $t \in [0,\pi]$ by

$$\psi_\varepsilon(t) := \int_0^t \frac{\lambda(x)\,\lambda(\pi-x)}{\lambda(x)\lambda(\pi-x)+\lambda(\varepsilon-x)+\lambda(x-(\pi-\varepsilon))}dx$$

has the desired properties (44), (45), (46).]

For the given $k \in \mathbb{N}_+$ choose now a fixed $m = m(k) \in \mathbb{N}_+$, such that

$$(2\pi - \frac{4}{m})\sinh(\frac{1+k\pi}{m}) + \frac{4}{m}\cosh(\frac{1+k\pi}{m}) < \frac{2\pi}{k} , \tag{47}$$

and define then for all $\alpha \in \mathbb{R}$ the C^∞-functions

$$\zeta_{k,\alpha}(t) := \frac{1}{m} + \alpha\,\psi_{(1/m)}(mt) \quad \text{for} \quad t \in \mathbb{R} , \tag{48}$$

$$\varphi_k(t) := kt, \quad \text{therefore} \quad \varphi'_k(t) = k \quad \text{for} \quad t \in \mathbb{R} , \tag{49}$$

which satisfy already m.m. (39), resp. (40). Moreover

$$\zeta'_{k,\alpha}(t) = (m\alpha)\,\psi'_{(1/m)}(mt) \quad \text{for} \quad t \in \mathbb{R} \quad (\text{see } (48)) \tag{50}$$

and therefore (see (45), (46), (50)):

$$(\zeta'_{k,\alpha}(t))^2 \leqq (m\alpha)^2 \quad \text{for all} \quad t \in \mathbb{R} , \tag{51}$$

and

$$\left.\begin{array}{l} M_m := \{t \in [0,2\pi] \mid (\zeta'_{k,\alpha}(t))^2 < (m\alpha)^2\} \text{ is a union} \\ \text{of } 2m+1 \text{ disjoint intervals of total length } \frac{4}{m} . \end{array}\right\} \tag{52}$$

If one defines now $c_{k,\alpha}: \mathbb{R} \to S_1^2(1)$ as in (38), where ζ resp. φ are substituted there by $\zeta_{k,\alpha}$ resp. φ_k, then it follows from (41), (42), (51):

$$\left.\begin{array}{l} c_{k,\alpha}: \mathbb{R} \to S_1^2(1) \text{ is a } (2\pi)\text{-periodic } C^\infty\text{-map,} \\ \text{which is space-like for } \alpha \in [0,(k/m)] \end{array}\right\} \tag{53}$$

and evidently

$$\text{Length}(c_{k,0} \mid [0,2\pi]) = k \cdot \cosh(1/m)2\pi > 2\pi , \tag{54}$$

Moreover, using $|\zeta_{k,\alpha}(t)| \leqq (1/m) + \alpha\pi$ for all $t \in \mathbb{R}$ (see (42), (48)) we obtain for $\alpha := (k/m)$ from (41), (49):

$$\text{Length}(c_{k,(k/m)} \mid [0,2\pi]) \leqq$$

$$\leqq \int_0^{2\pi} \sqrt{\cosh^2(\tfrac{1+k\pi}{m})k^2 - (\zeta'_{k,(k/m)}(t))^2}\, dt$$

$$\underset{(51),(52)}{\leqq} \quad k\cdot \int_{[0,2\pi]\setminus M_m} \sqrt{\cosh^2(\tfrac{1+k\pi}{m}) - 1}\, dt + k\cdot \int_{M_m} \cosh(\tfrac{1+k\pi}{m})\, dt$$

$$\underset{(52)}{=} k\left[\sinh(\tfrac{1+k\pi}{m})\cdot(2\pi - \tfrac{4}{m}) + \cosh(\tfrac{1+k\pi}{m})\tfrac{4}{m}\right] \underset{(47)}{<} 2\pi . \tag{55}$$

From (54), (55) follows - as in section 4 - , that there exists an $\alpha \in]0,(k/m)[$ such that

$$\text{Length}(c_{k,\alpha} \mid [0,2\pi]) = 2\pi , \tag{56}$$

and, using $2k < m$ (which follows trivially from (47)), one checks, that

$$c_{k,\alpha}(0),\ c_{k,\alpha}(\tfrac{2\pi}{m}),\ c_{k,\alpha}(\tfrac{4\pi}{m}) \text{ are linearly independent in } \mathbb{R}^3. \tag{57}$$

From (53), (56), (57) follows now by the same procedure as in section 4, the existence of the isometric C^∞-immersion $f: S_0^1(1) \to S_1^2(1)$, which is not totally geodesic and which in addition winds k times around the x_1-axis, q.e.d.

Bibliography.

[1] P. DOMBROWSKI: Jacobi Fields, Geodesic Foliations and Geodesic Differential Forms, Resultate, Math. 1, 156-194 (1978; Zbl. 408.53027).

[2] D. FERUS: Isometric Immersions of Constant Curvature Manifolds, Math. Ann. 217, 155-156, (1975; Zbl. 295.53026).

[3] L. GRAVES: Codimension One Isometric Immersions Between Lorentz Spaces. Ph. D. Thesis, Department of Mathematics, Brown University, (1977).

[4] L. GRAVES and K. NOMIZU: Isometric Immersions of Lorentzian Space Forms, Math. Ann. 233, 125-136, (1978; Zbl. 354.53046).

[5] S. KOBAYASHI and K. NOMIZU: Foundations of Differential Geometry, Vol. II, J. Wiley and Sons, New York, (1969).

[6] J.A. WOLF: Spaces of Constant Curvature, Mc Graw Hill, New York, (1967).

RIEMANNIAN MANIFOLDS WITH HARMONIC CURVATURE

Andrzej Derdziński

SFB 40, Universität Bonn
and Max-Planck-Institut für Mathematik in Bonn

Current address: Mathematical Sciences Research Institute
2223 Fulton Street, Room 603, Berkeley, California 94720, U.S.A.

1. INTRODUCTION

The present paper is a survey of results on manifolds with harmonic curvature, i.e., on those Riemannian manifolds for which the divergence of the curvature tensor vanishes identically. The curvatures of such manifolds occur as a special case of Yang-Mills fields. These manifolds also form a natural generalization of Einstein spaces and of conformally flat manifolds with constant scalar curvature.

After describing the known examples of compact manifolds with harmonic curvature, we give, in Sect. 5, a review of theorems concerning such manifolds. Most of their proofs are either omitted or only briefly sketched. For a complete presentation of the results mentioned in this paper (except for Sect. 3, Sect. 7 and 4.4) the reader is referred to the forthcoming book [5], where one of the chapters deals with generalizations of Einstein spaces.

The preparation of the present article was begun under the program SFB 40 in the Max Planck Institute of Mathematics in Bonn, and completed at the Mathematical Sciences Research Institute in Berkeley. The author is obliged to these institutions for their hospitality and assistance. He also wishes to thank Jerry Kazdan for helpful remarks concerning the topics discussed in 4.4.

2. PRELIMINARIES

2.1. Given a Riemannian vector bundle E over a compact Riemannian manifold (M,g), one studies the *Yang-Mills potentials* in E, i.e., those metric connections ∇ in E which are critical points for the *Yang-Mills functional*

$$YM(\nabla) = \tfrac{1}{2} \int_M |R|^2 ,$$

where R is the curvature of ∇ and the integration is with respect to the Riemannian measure of (M,g) (see [8]). The obvious operator of *exterior differentiation* d as well as its formal adjoint, the *diver-*

gence d^*, can be applied to differential forms on (M,g) valued in Riemannian vector bundles with fixed metric connections (cf. [6]). The Yang-Mills potentials in E now are characterized by $d^*R = 0$. In view of the Bianchi identity $dR = 0$ for the 2-form R, this means that the curvature form of any Yang-Mills potential is *harmonic*.

2.2. A Riemannian manifold (M,g) is said to have *harmonic curvature* if its Levi-Civita connection ∇ in the tangent bundle TM satisfies $d^*R = 0$. If M is compact, this just means that ∇ is a Yang-Mills potential in TM (i.e., a critical point for the Yang-Mills functional in the space of all g-metric connections, where g is *fixed*).

2.3. Let (M,g) be a Riemannian manifold, $\dim M = n \geq 3$. Its curvature tensor R, Ricci tensor Ric, scalar curvature Scal and Weyl conformal tensor W can obviously be viewed as differential forms valued in suitable bundles of exterior forms : $R, W \in \Omega^2(M,\Lambda^2 M)$, $\mathrm{Ric} \in \Omega^1(M,\Lambda^1 M)$, $\mathrm{Scal} = \mathrm{Scal}_g \in \Omega^0(M,\Lambda^0 M)$. The second Bianchi identity $dR = 0$ now easily gives

(1) $$d^*R = -d\,\mathrm{Ric}$$

(in local coordinates : $-\nabla^s R_{sijk} = \nabla_k(\mathrm{Ric})_{ij} - \nabla_j(\mathrm{Ric})_{ik}$),

(2) $$d^*W = -\frac{n-3}{n-2}\, d[\mathrm{Ric} - \frac{1}{2(n-1)}\mathrm{Scal}\cdot g],$$

(3) $$d^*\mathrm{Ric} = -\tfrac{1}{2}\, d\,\mathrm{Scal},$$

(4) $$d^*W = -\,\mathrm{Con}(dW)\;,\quad dW = -(n-3)^{-1} g \circledwedge d^*W \quad (\text{if } n > 3),$$

where, in (1) - (3), we identify $\Omega^i(M,\Lambda^j M)$ with $\Omega^j(M,\Lambda^i M)$, while, in (4), $\circledwedge$ is a natural bilinear pairing and Con is a suitable contraction.

2.4. Riemannian manifolds satisfying $d^*W = 0$ are said to have *harmonic Weyl tensor*, which is justified by the fact that this condition implies $dW = 0$ (cf. (4)). These manifolds are studied here in order to simplify various arguments involving the equation $d^*R = 0$. In particular, *all examples with* $d^*R = 0$ *discussed below arise from natural classes of manifolds having* $d^*W = 0$ *by requiring that their scalar curvature be constant* (cf. 2.6.ii).

2.5. By a *Codazzi tensor* on a Riemannian manifold (M,g) we mean any *symmetric* C^∞ tensor field b of type $(0,2)$ on M which, viewed as a Λ^1-valued 1-form, satisfies the relation $db = 0$; this is clearly nothing else than the *Codazzi equation* $\nabla_i b_{jk} = \nabla_j b_{ik}$.

2.6. Given a Riemannian manifold (M,g), $\dim M \geq 4$, it follows from 2.3 that equation $d^*R = 0$ for (M,g) is equivalent to either of the following two conditions :

i) $d\,\mathrm{Ric} = 0$, i.e., Ric is a Codazzi tensor on (M,g) (hence, by (3), Scal is constant).

ii) (M,g) has harmonic Weyl tensor and constant scalar curvature.

If $\dim M = 3$, $d^*R = 0$ is still equivalent to i) and it characterizes conformally flat 3-manifolds with constant scalar curvature (cf. 2.7).

2.7. In view of (2), a Riemannian manifold (M,g) with $\dim M = n \geq 4$ has harmonic Weyl tensor if and only if $\mathrm{Ric} - (2n-2)^{-1}\mathrm{Scal}\cdot g$ is a Codazzi tensor. If $n = 3$, the latter condition means that (M,g) is conformally flat ([20], p. 306).

2.8. THE SIMPLEST EXAMPLES. By 2.6 and 2.7, the manifolds listed in $1_W, 2_W$ and 3_W (resp., $1_R, 2_R$ and 3_R) of our table of examples actually have harmonic Weyl tensor (resp., harmonic curvature). The problem of finding metrics of type 3_R on compact conformally flat manifolds is obviously related to a special case of Yamabe's conjecture (see [17]).

3. THE MODULI SPACE

3.1. Consider a metric g on a compact manifold M, satisfying the condition $d^*R = 0$. The linearized version of this condition, restricted to a slice through g (see [15]) in a suitable Sobolev space of metrics implies that the corresponding slice vector lies in the kernel of a third order differential operator with injective symbol. Consequently, *the moduli space of all metrics with* $d^*R = 0$ *modulo the group of diffeomorphisms of the underlying compact manifold is locally finite dimensional.*

3.2. The assertion of 3.1 fails to hold for the weaker condition $d^*W = 0$; counterexamples are provided by $3_W, 4_W, 5_W, 6_W$ and 7_W of the table.

4. FURTHER EXAMPLES

4.1. For pointwise conformal metrics g and $\bar{g} = e^{2\sigma}g$ on a manifold M, the tensor d^*W for g and the corresponding quantity $\overline{d^*W}$ for $\bar{g}$ are related by

(5) $$\overline{d^*W} = d^*W - (n-3)\,W(\nabla\sigma,\cdot,\cdot,\cdot)\,, \quad n = \dim M \geq 3.$$

4.2. For Riemannian products of Einstein manifolds like those described in 4_W and 5_W of the table, with any function σ which is constant along the second factor, both terms on the right-hand side of (5) vanish, and so the conformally related metrics of 4_W and 5_W must also have harmonic Weyl tensor. In order that these metrics have constant scalar curvature (and hence harmonic curvature, cf. 2.6.ii), the function F on the first factor manifold must satisfy a second order differential equation, which admits a non-constant positive solution under the conditions stated in 4_R and 5_R (see [9], [12], [5]). If the (N,h) that we use are not of constant curvature, the examples in 4_R and 5_R obtained from this construction are neither locally isometric to those of 1_R, 2_R and 3_R, nor to each other ([9], [12]).

4.3. The Weyl tensor of any *four-dimensional oriented* Riemannian manifold (M,g) can be decomposed into its $SO(4)$-*irreducible components:* $W = W^+ + W^-$, which corresponds to the decomposition $\Lambda^2 M = \Lambda^2_+ M + \Lambda^2_- M$ of $\Lambda^2 M$ into the eigenspace bundles of the Hodge star operator ([2]). If, moreover, (M,g) is a Kähler manifold endowed with the natural orientation, the condition $d^*W^+ = 0$ holds for the conformally related metric $\mathrm{Scal}^{-2}\cdot g$, defined wherever $\mathrm{Scal} = \mathrm{Scal}_g \neq 0$ ([10]). Letting (M,g) now be the Riemannian product of two orientable surfaces, we see that the example described in 6_W of the table has $d^*W = 0$ (since $g = g_1 \times g_2$ is a Kähler metric for two complex structures, corresponding to different orientations of M).

4.4. Let (S^2, g_c) be the sphere of constant curvature $c > 0$ and suppose that we are given a compact Riemannian surface (N,h) with non-constant Gaussian curvature K such that $K + c > 0$ on N and

$$(K + c)^3 - 3(K + c)\cdot\Delta K - 6|dK|^2 - \lambda^3 = 0 \tag{6}$$

for some real $\lambda > 0$, where the quantities $\Delta = - h^{ij}\nabla_i\nabla_j$ and $|\ |^2$ refer to the geometry of (N,h). Then the compact Riemannian 4-manifold $(N \times S^2, (K + c)^{-2}\cdot(h \times g_c))$ has harmonic curvature (since it is of type 6_W and has constant scalar curvature by (6)). However, it is not clear whether this construction really gives new compact 4-manifolds with $d^*R = 0$:

a) For any real c,λ with $c > \lambda > 0$, *the torus* T^2 *admits a metric* h *with non-constant curvature* K *satisfying* $K + c > 0$ *and* (6). Namely, we can define h first on $\mathbb{R}^2$ (with coordinates t,y) by $h = dt^2 + (K(t) + c)^2 dy^2$, taking for $K = K(t)$ any non-constant periodic solution to $3(dK/dt)^2 = -2K^3 - 3cK^2 + c^3 - \lambda^3$ with $K + c > 0$ (which is easily seen to exist), and then project it onto $T^2 = \mathbb{R}^2/\mathbb{Z}^2$ with an appropriate action of $\mathbb{Z}^2$. Unfortunately, the metric with $d^*R = 0$ on $T^2 \times S^2$, obtained as above with this h, turns out to be a Riemannian product (of S^1 and $S^1 \times S^2$ with suitable metrics) and hence it is *nothing really new*. (However, one easily shows that for compact surfaces (N,h) with non-constant K satisfying $K + c > 0$ and (6) which are not locally isometric to a torus as above, the corresponding metrics with $d^*R = 0$ on $N \times S^2$ are never locally isometric to examples $1_R, 2_R$ or 3_R of the table.)

b) Let $\lambda_o > 0$ be a *simple* eigenvalue of the Laplace operator on a compact surface (N, h_o) of constant curvature -1. To any function x on N and to $\lambda \in \mathbb{R}$ we can assign the function $f(x,\lambda)$, de-

TABLE OF EXAMPLES

Known examples of compact Riemannian manifolds with $d^*W = 0$ or $d^*R = 0$ (up to local isometry)	
with $d^*W = 0$	with $d^*R = 0$
1_W. Einstein spaces	1_R. Einstein spaces
2_W. Riemannian products of manifolds with $d^*R = 0$	2_R. Riemannian products of manifolds with $d^*R = 0$
3_W. Conformally flat manifolds	3_R. Conformally flat manifolds with constant scalar curvature
4_W. Warped products $(S^1 \times N, F\cdot(dt^2 \times h))$, F any positive function on S^1, (N,h) any Einstein space	4_R. As in 4_W, with $\mathrm{Scal}_h > 0$ and for a suitable non-constant F (which always exists on S^1 of appropriate length). Also, *twisted warped products* $(\mathbb{R}\times N)/\mathbb{Z}$, where $\mathbb{R}\times N$ has the pull-back metric and the $\mathbb{Z}$-action involves an isometry of h
5_W. Warped products $(M\times N, F\cdot(g\times h))$, (M,g) of constant curvature K, (N,h) Einstein with $\mathrm{Scal}_h = -\dim N\cdot(\dim N - 1)K$, F any positive function on M	5_R. As in 5_W, with $K < 0$ and $\dim N > \dim M - \lambda_1/K$, λ_1 being the lowest positive eigenvalue of Δ in (M,g), for a suitable non-constant F (which must exist)
6_W. $(M_1\times M_2, (K_1+K_2)^{-2}(g_1\times g_2))$, $\dim M_i = 2$, (M_i, g_i) having Gaussian curvature K_i $(i=1,2)$ with $\lvert K_1+K_2\rvert > 0$	6_R ? Not known whether 6_W contains new compact manifolds with $d^*R = 0$ (see 4.4)
7_W. $(M_1\times M_2, \lvert K_1+K_2\rvert^{2/(1-n)}\cdot(g_1\times g_2))$, $\dim M_1 = 2 < \dim M_2 = n$, (M_1,g_1) with non-constant curvature K_1, (M_2,g_2) of constant curvature K_2, $\lvert K_1 + K_2\rvert > 0$	7_R ? Not known whether 7_W gives new compact manifolds with $d^*R = 0$

fined to be the left-hand side of (6) formed with the metric $h = e^{2x}\cdot h_0$ and with $c = \lambda + 1$ (so that f is a fourth order non-linear operator in x with parameter λ). The equation $f(x,\lambda) = 0$ has the curve of *trivial* solutions with $x = 0$ and any λ. At $x = 0$ and $\lambda = \lambda_0$ this equation has a bifurcation point, since the hypotheses of Theorem (4.1.12) of [3] (p. 155) are satisfied. Thus, there is a curve $(x(\varepsilon),\lambda(\varepsilon))$ of non-trivial solutions through $x(0) = 0$, $\lambda(0) = \lambda_0$, and for small $\varepsilon \neq 0$ the metric $h_\varepsilon = e^{2x(\varepsilon)}\cdot h_0$ has non-constant curvature K with $K + c > 0$ and (6) for $c = \lambda(\varepsilon) + 1$. By the above construction, this would lead to new compact Riemannian 4-manifolds with $d^*R = 0$; however, the author does not know whether the Laplace operator of any compact surface with constant negative curvature admits a simple positive eigenvalue.

4.5. For g_1, g_2 as in 7_W, one easily sees ([19]) that the metric $g_1 \times g_2$ satisfies $2|W|^2\cdot\nabla W = d(|W|^2)\otimes W$. Together with (5) and with the fact that, for $g_1 \times g_2$, $|W|$ is proportional to $|K_1 + K_2|$, this implies that the conformally related metric described in 7_W really satisfies $d^*W = 0$.

5. PROPERTIES OF MANIFOLDS WITH HARMONIC CURVATURE

5.1. Most of the known results on manifolds with $d^*R = 0$ are valid in more generality. From $d^*R = 0$ it follows that $d^*W = 0$, which in turn implies the existence of a natural Codazzi tensor b on the manifold (M,g) ($b = \mathrm{Ric} - (2n-2)^{-1}\mathrm{Scal}\cdot g$, $n = \dim M$). On the other hand, $d^*R = 0$ means that $b = \mathrm{Ric}$ is a Codazzi tensor with constant trace (see 2.7, 2.8). The results presented in this section will be stated under appropriate weaker hypotheses, as described above.

5.2. THEOREM (D. DeTurck - H. Goldschmidt, [14]). *Every Riemannian manifold with* $d^*R = 0$ *is analytic in suitable local coordinates.*

5.3. THEOREM ([13]). *Given a Codazzi tensor* b *on* (M,g), $x \in M$ *and eigenspaces* V, V' *of* $b(x)$, *the curvature operator* $R(x) \in \mathrm{End}\,\Lambda^2 T_xM$ *leaves* $V \wedge V' \subset \Lambda^2 T_xM$ *invariant.*

Proof. Assume $\det b \neq 0$ near x (taking $b + tg$ instead of b). Viewing b as a bundle automorphism of TM (near x), we easily conclude from $db = 0$ that $\overline{\nabla} = b^*\nabla$ is the Levi-Civita connection of $\overline{g} = b^*g$ ([16]). Hence $\overline{g}$ has the (0,4) curvature tensor $\overline{R}(X,Y,Z,U) = R(X,Y,bZ,bU)$. Our assertion now follows from the first Bianchi identity for $\overline{R}$.

5.4. BOURGUIGNON'S COMMUTATION THEOREM ([6]). *Any Codazzi tensor* b *on* (M,g) *commutes with* Ric, *while the endomorphism* $g \owedge b$ *of* $\Lambda^2 M$

commutes with R *and* W *(here* ⓧ *is the bilinear pairing of symmetric 2-tensor fields, giving rise to algebraic curvature tensors).*

Proof. At $x \in M$, decompose $\Lambda^2 T_x M$ using a $b(x)$-eigenspace decomposition of $T_x M$ and apply 5.3.

5.5. THEOREM ([13]). *Let* b *be a Codazzi tensor on* (M,g), $\dim M = n$. *If* b *has* n *distinct eigenvalues almost everywhere in* M, *then all Pontryagin forms of* (M,g) *vanish identically.*

5.6. THEOREM (cf. [6], [13]). *If* (M,g) *satisfies* $d^*W = 0$ *and* $x \in M$, *then the inner product* $g(x)$ *and the algebraic curvature tensor* $R(x)$ *in* $T_x M$ *cannot be completely arbitrary; in other words, condition* $d^*W = 0$ *always imposes algebraic restrictions on the curvature. For instance, we have*

i) $\mathrm{Ric}(x)$ *has a multiple eigenvalue,*

or

ii) $T_x M$ *admits an orthonormal basis* $e_1, \ldots, e_n$ *which diagonalizes* $R(x)$ *in the sense that all* $e_i \wedge e_j$ *(i<j) are eigenvectors of* $R(x) \in \mathrm{End}\, \Lambda^2 T_x M$. *This in turn implies that all Pontryagin forms of* (M,g) *vanish at* x.

Proof of 5.5 and 5.6. If $b(x)$ (resp., $\mathrm{Ric}(x)$) has $n = \dim M$ distinct eigenvalues, 5.3 gives rise to 1-dimensional invariant subspaces and hence to eigenvectors of $R(x)$. The assertion concerning the Pontryagin forms now follows immediately.

5.7. COROLLARY. *Let* (M,g) *satisfy* $d^*R = 0$. *If, for some* $x \in M$, *all eigenvalues of* $\mathrm{Ric}(x)$ *are simple, then all real Pontryagin classes of* M *are zero.*

Proof. Immediate from 5.6 and 5.2.

5.8. THEOREM (cf. [6], [13]). *For a Codazzi tensor* b *on* (M,g) *with* $\dim M = 4$, *we have* $P_1 \otimes (b - \frac{1}{4}\mathrm{tr}\, b \cdot g) = 0$, P_1 *being the (first) Pontryagin form of* (M,g). *More precisely, at any point* x *where* b *is not a multiple of* g, *the endomorphisms* $W^+(x)$ *of* $\Lambda^2_+ T_x M$ *and* $W^-(x)$ *of* $\Lambda^2_- T_x M$ *have equal spectra.*

Idea of proof (see [5] for details). We may assume that the number of distinct eigenvalues of b is locally constant at x. The assertion can now be obtained from 5.3 using the algebraic properties of W, except for the case of two double eigenvalues for $b(x)$, where an argument involving differentiation is needed ([13], [5]).

5.9. COROLLARY (J. P. Bourguignon, [6]). *The signature* τ *and the Euler characteristic* χ *of any compact oriented Riemannian four-ma-*

nifold (M,g) *with harmonic curvature satisfy*

$$(2\chi - 3|\tau|)\cdot|\tau| \geq 0 .$$

More precisely, if $\tau \neq 0$, (M,g) *must be Einstein and so* $2\chi \geq 3|\tau|$ *by Thorpe's inequality* ([22]).

Proof. Immediate from 5.8 and 5.2 .

5.10. THEOREM (D. DeTurck - H. Goldschmidt, [14], cf. [5]). *Let* (M,g) *satisfy* $d^*W = 0$. *If, for some point* x , $W(x) = 0$ *and all eigenvalues of* $\mathrm{Ric}(x)$ *are simple, then* $W = 0$ *identically on* M .

Proof. Choose an orthonormal C^∞ frame field diagonalizing Ric (and R, W) near x (cf. 5.6.ii). Equations $d^*W = 0$, $dW = 0$ (see (4)) mean that the components of W in this frame field satisfy a first order system of *linear* differential equations, solved for the directional derivatives. As $W(x) = 0$, $W = 0$ near x . Since W is a solution to the elliptic system $(dd^* + d^*d)W = 0$, $W = 0$ everywhere in view of Aronszajn's unique continuation theorem (cf. [1]).

5.11. THEOREM (M. Berger, cf.[4] and [5]). *Let* b *be a Codazzi tensor with constant trace on a compact Riemannian manifold with sectional curvature* $K > 0$. *Then* b *is a constant multiple of the metric. Thus, a compact manifold* (M,g) *with* $d^*R = 0$ *and* $K > 0$ *must be Einstein.*

Proof. See [4] or [5].

5.12. THEOREM (Y. Matsushima [18], S. Tanno [21]). *Let* b *be a Hermitian Codazzi tensor on a Kähler manifold. Then* b *is parallel. In particular, a Kähler manifold with* $d^*W = 0$ *must have* $\nabla\,\mathrm{Ric} = 0$.

Proof. The expression $(\nabla_X b)(Y,JZ)$, where J is the complex structure tensor, is symmetric in X,Y and skew-symmetric in Y,Z , so that it vanishes.

5.13. THEOREM ([9], cf. [5]). *Let* (M,g) *be compact and satisfy* $d^*R = 0$. *If its Ricci tensor is not parallel and has, at each point, less than 3 distinct eigenvalues, then* (M,g) *admits a finite Riemannian covering by a manifold* $S^1 \times N$ *endowed with a twisted warped product metric as described in* 4_R *of the table. Conversely, all examples of* 4_R *satisfy the above hypotheses.*

Proof. See [9] or [5].

5.14. THEOREM. *Suppose we are given an oriented Riemannian four-manifold* (M,g) *with* $d^*W = 0$, *and a point* x *of* M *at which* $W \neq 0$ *and* $4\,\mathrm{Ric} \neq \mathrm{Scal}\cdot g$. *If the endomorphism* W^+ *of* Λ^2_+M *has less than three distinct eigenvalues at every point, then, in a neighborhood of* *x , g is obtained by a conformal deformation of a product of surface metrics as in* 6_W *of our table.*

Proof. See [5].

6 . SOME OPEN QUESTIONS

6.1. Does there exist a compact simply connected Riemannian manifold with $d^*R = 0$, the Ricci tensor of which is not parallel ? (cf. [7]).

6.2. Are there compact Riemannian 4-manifolds satisfying $d^*R = 0$ and not locally isometric to examples $1_R, 2_R, 3_R$ of our table ([6]; cf. 4.4).

6.3. Must a locally homogeneous Riemannian manifold with $d^*R = 0$ have parallel Ricci tensor ?

6.4. Do the classes 6_W , 7_W of our table contain new compact Riemannian manifolds with harmonic curvature ? (cf. 4.4).

7 . THE CLASSIFICATION PROBLEM IN DIMENSION FOUR

7.1. In this section we present some steps towards a classification of compact Riemannian 4-manifolds with harmonic curvature. By a *classification* we mean a description of all Riemannian universal covering spaces of such manifolds which are different from the "classical" examples (Einstein, conformally flat, products). The results we discuss below consist in excluding certain a priori possible cases. Their proofs, too long to be reproduced here, are available from the unpublished manuscript [11].

7.2. Let us recall the *known examples of compact Riemannian four-manifolds with* $d^*R = 0$. First, we have the types 1_R, 2_R and 3_R of our table (which obviously include many compact manifolds). Explicitly, these are

I. Einstein spaces

II. Conformally flat manifolds with constant scalar curvature

III. Manifolds locally isometric to Riemannian products $\mathbb{R} \times N$, where N is a conformally flat 3-manifold with constant scalar curvature

IV. Manifolds locally isometric to Riemannian products of surfaces with constant curvatures

Examples 4_R and 5_R of the table yield nothing new in dimension four. Finally, it seems convenient to list here also the examples of 6_R :

V. Manifolds locally isometric to $(N \times S^2, (K + c)^{-2}(h \times g_c))$ as described in 4.4.

Although we do not know if new examples with *compact* surfaces N really occur in V., this is clearly true for non-compact N (just take N to be a surface of revolution, which reduces (6) to an ordinary differential equation).

7.3. Let (M,g) now be an oriented Riemannian four-manifold with harmonic curvature. Denote by $r_0 \in \{1,2,3,4\}$ (resp., by $w_0 \in \{1,2,3\}$) the maximal number of distinct eigenvalues of Ric (resp., of W^+ acting on Λ^2_+M), both attained in an open dense subset of M (cf. 5.2). The following cases are possible:

Case A: $r_0 = 1$, i.e., (M,g) is Einstein, as in 7.2.I.

Case B: $r_0 > 1$, $w_0 = 1$. Then (M,g) must be as in 7.2.II. In fact, W^+ is always traceless in Λ^2_+M, which now implies $W^+ = 0$ and, by 5.8, also $W^- = 0$.

Case C: $r_0 > 1$, $w_0 = 2$. Then (M,g) is as in 7.2.IV., or it is of type 7.2.V. (where N may be non-compact and c,λ need not be positive). However, in the latter case $c,\lambda > 0$ and N can be chosen compact, if so is M. See [11] for details.

Case D: $r_0 > 1$, $w_0 = 3$. In an open dense subset M_0 of M we can choose an orthonormal C^∞ frame field $e_1,\dots,e_4$ diagonalizing R in the sense of 5.6.ii. For $l \in \{1,2,3,4\}$ we say that $A_l \neq 0$ if there exist i,j,k with $\{i,j,k,l\} = \{1,2,3,4\}$ and $g(\nabla_{e_i} e_j, e_k) \neq 0$ somewhere in the given connected component of M_0. Define $m_0 \in \{0,1,2,3,4\}$ to be the number of l with $A_l \neq 0$. It turns out that $m_0 \leqq 2$ ([11]) and so three subcases can only occur:

Case D.0 : $m_0 = 0$. See 7.5 below.

Case D.1 : $m_0 = 1$. Then (M,g) is as in 7.2.III. (see [11]).

Case D.2 : $m_0 = 2$. If, moreover, (M,g) is complete, then $|d|Ric|^2|$ is unbounded on M. Thus, *if* M *is compact,* Case D.2 *cannot occur.*

7.4. According to 7.3, our classification problem for compact Riemannian 4-manifolds with $d^*R = 0$ (cf. 7.1) has been reduced to cases C and D.0, since cases A,B and D.1 imply a "classical" situation (I.-IV. of 7.2), and Case D.2 is impossible. In Case C this problem amounts to the question of existence and classification for compact surfaces (N,h) having the properties stated in 4.4 (cf. 4.4.a,b). As for Case D.0

(which may occur for certain *non-compact* manifolds of type 7.2.III.), we do not know whether compact manifolds of this kind exist.

7.5. Some local properties of Riemannian 4-manifolds (M,g) with $d^*R = 0$ in Case D.0. Let $e_1,\dots,e_4$ be the orthonormal C^∞ frame field, defined at the "generic" points of M and diagonalizing R (and Ric). Since $m_0 = 0$, we have $\nabla_{e_i} e_j = F_{ji} e_i$ and $[e_i,e_j] = F_{ji}e_i - F_{ij}e_j$ for $i \neq j$, with certain functions F_{ji} (in particular, the generic subset of M admits local coordinates with mutually orthogonal coordinate lines). Set $b = Ric - \frac{1}{4}\,Scal\cdot g$, $\lambda_i = b(e_i,e_i)$, $\sigma_{ij} = W(e_i,e_j,e_i,e_j)$ $(i \neq j)$, so that $\Sigma_i \lambda_i = 0$ and $\sigma_{ij} = \sigma_{ji} = \sigma_{kl}$ if $\{i,j,k,l\} = \{1,2,3,4\}$. Conditions $db = 0$, $d^*W = 0$ and the fact that $e_1,\dots,e_4$ diagonalizes R (and W) now imply $D_k\lambda_i = (\lambda_k - \lambda_i)F_{ki}$, $D_k\sigma_{ji} = (\sigma_{ki} - \sigma_{ji})F_{kj} + (\sigma_{kj} - \sigma_{ji})F_{ki}$ and $D_k F_{ji} = F_{ki}(F_{jk} - F_{ji})$ for $i \neq j \neq k \neq i$, D_k being the directional derivative along e_k. The integrability conditions for these systems impose certain algebraic relations on the components λ_i, σ_{ij} and F_{ij}.

REFERENCES

1. Aronszajn, N.: A unique continuation theorem for solutions of elliptic partial differential equations or inequalities of second order. J. Math. Pures Appl. 36, 235--249 (1957) (Zbl. 84.304)
2. Atiyah, M.F., Hitchin, N.J., Singer, I.M.: Self-duality in four-dimensional Riemannian geometry. Proc. Roy. Soc. London A362, 425-461 (1978) (Zbl. 389.53011)
3. Berger, M.S.: Nonlinearity and Functional Analysis, Academic Press, 1977 (Zbl.368.4700
4. Berger, M., Ebin, D.: Some decompositions of the space of symmetric tensors on a Riemannian manifold. J. Differential Geometry 3, 379-392 (1969) (Zbl. 194,531)
5. Besse, A.L.: Einstein Manifolds (to appear)
6. Bourguignon, J.P.: Les variétés de dimension 4 à signature non nulle dont la courbure est harmonique sont d'Einstein. Invent. math. 63, 263-286 (1981) (Zbl. 456.53033
7. Bourguignon, J.P.: Metrics with harmonic curvature. Global Riemannian Geometry (edited by T.J. Willmore and N.J. Hitchin), 18-26. Ellis Horwood, 1984
8. Bourguignon, J.P., Lawson, H.B.,Jr.: Yang-Mills theory: Its physical origins and differential geometric aspects. Seminar on Differential Geometry (edited by S.T. Yau), Ann. of Math. Studies No. 102, 395-421 (1982) (Zbl. 482.58007)
9. Derdziński, A.: On compact Riemannian manifolds with harmonic curvature. Math. Ann. 259, 145-152 (1982) (Zbl. 489.53042)
10. Derdziński, A.: Self-dual Kähler manifolds and Einstein manifolds of dimension four. Compos. Math. 49, 405-433 (1983) (Zbl. 527.53030)
11. Derdziński, A.: Preliminary notes on compact four-dimensional Riemannian manifolds with harmonic curvature, 1983 (unpublished)
12. Derdziński, A.: An easy construction of new compact Riemannian manifolds with harmonic curvature (preliminary report). SFB/MPI 83-21, Bonn (1983)
13. Derdziński, A., Shen, C.L.: Codazzi tensor fields, curvature and Pontryagin forms. Proc. London Math. Soc. 47, 15-26 (1983) (Zbl. 519.53015)
14. DeTurck, D.: private communication
15. Ebin, D.G.: The manifold of Riemannian metrics. Proc. of Symposia in Pure Math. 15, 11-40 (1970) (Zbl. 205,537) (Zbl. 135,225)
16. Hicks, N.: Linear perturbations of connexions. Michigan Math. J. 12, 389-397 (1965)
17. Lafontaine, J.: Remarques sur les variétés conformément plates. Math. Ann. 259, 313-319 (1982) (Zbl. 469.53036)

18. Matsushima, Y.: Remarks on Kähler-Einstein manifolds. Nagoya Math. J. 46, 161-173 (1972) (Zbl. 249.53050)
19. Roter, W.: private communication
20. Schouten, J.A.: Ricci Calculus. Springer-Verlag, 1954
21. Tanno, S.: Curvature tensors and covariant derivatives. Ann. Mat. Pura Appl. 96, 233-241 (1973) (Zbl. 277.53013)
22. Thorpe, J.: Some remarks on the Gauss-Bonnet integral. J. of Math. Mech. 18, 779--786 (1969) (Zbl. 183,505)

Structure of Manifolds of Nonpositive Curvature

by Patrick Eberlein*
Chapel Hill, North Carolina

Table of contents

Introduction

This article is an expanded version of an hour lecture given at the Colloquium on Differential Geometry and Global Analysis, which took place at the Technische Universität Berlin in June 1984. This report describes recent work of Werner Ballmann, Misha Brin, Keith Burns, Ralf Spatzier and myself that introduces and applies the idea of the rank of a manifold of nonpositive sectional curvature. The rank of such a manifold M is an integer k with $1 \leq k \leq \dim M$ that measures the flatness of M. M is flat (that is, M has sectional curvature identically zero) if and only if $k = \dim M$, and M has rank 1 if the sectional curvature of M is strictly negative. See section 3 for a precise definition of rank. Striking geometric differences exist between compact manifolds of rank

* Supported in part by NSF Grant MCS-8219609, the Max-Planck-Institut für Mathematik, Bonn, Federal Republic of Germany and the Technische Universität Berlin

1 and compact manifolds of higher rank. For example, the geodesic flow in the unit tangent bundle of a compact manifold M of nonpositive sectional curvature is ergodic if and only if M has rank 1 (Theorem A' of section 3). Compact rank 1 manifolds have the same geometric properties as compact manifolds of strictly negative sectional curvature, but they form a much larger class, and individual rank 1 manifolds may look quite flat. In a sense almost every compact manifold M of nonpositive sectional curvature has rank 1 as one can see from the Rank Rigidity Theorem (Theorem C of section 6) of Ballmann and Burns-Spatzier: Let M be a compact manifold of nonpositive sectional curvature and rank $k \geq 2$. Suppose furthermore that the universal Riemannian covering manifold $\tilde{M}$ is not a Riemannian product manifold. Then $\tilde{M}$ is isometric to a symmetric space of noncompact type.

As a corollary of the rank rigidity theorem one obtains the following structure result: Let $\tilde{M}$ be the universal Riemannian covering of an arbitrary compact Riemannian manifold M of nonpositive sectional curvature. Then $\tilde{M}$ is the Riemannian product of a Euclidean space $\tilde{M}_e$ of dimension $r \geq 0$, a symmetric space $\tilde{M}_s$ of noncompact type, and rank 1 manifolds $\tilde{M}_1, \ldots, \tilde{M}_p$ whose isometry groups are discrete. This result allows one to reduce the proof of many assertions about compact nonpositively curved manifolds M to the separate cases that M has rank 1 or is locally symmetric. In either case a great deal is known about such manifolds, and there exist many techniques with which to test conjectures. On the other hand, the rank rigidity theorem shows that the class of compact rank 1 manifolds of nonpositive sectional curvature is very large, and it will be an interesting challenge to refine further the geometric structure of rank 1 manifolds.

A complete simply connected manifold $\tilde{M}$ of nonpositive sectional curvature is diffeomorphic to a Euclidean space of the same dimension, and it follows that all homotopy groups of a compact quotient manifold M are zero except for the fundamental group. In section 7 of this paper we consider the following problem: how much of the geometry of M is reflected in the algebraic structure of its fundamental group? The existing results show that a great deal of the geometry of M can be seen in $\pi_1(M)$, and many of these results are discussed below. In particular in section 8 we introduce a definition of algebraic rank for an infinite group Γ without torsion elements, and we state the result (Theorem E) that if G is the fundamental group, $\pi_1(M)$, of a compact manifold M with nonpositive sectional curvature, then the algebraic rank of $\pi_1(M)$ equals the geometric rank of M. The definition of algebraic rank is a slight modification of a definition introduced by G. Prasad and M. Raghunathan, who proved the result just stated in the case that M is locally symmetric (Theorem 3.9 of [PR]). Next, one may use the rank rigidity theorem and this characterization of the geometric rank to characterize by purely algebraic data in the fundamental group those compact irreducible manifolds M whose universal Riemannian covering $\tilde{M}$ is a symmetric space of noncompact type and rank $k \geq 2$ (Theorem F).

The contents and organization of this paper are best described by the table of contents, and we add here only a few comments. In section 2 we have included a brief history of the study of geodesic flows on the unit tangent bundle of a compact manifold M of nonpositive sectional curvature. This section is intended as background and motivation for the characterization of the ergodicity of the geodesic flow of such a manifold (Theorem A' of section 3). Much more complete discussions of the research on geodesic flows for such manifolds can be found in the survey articles of G. Hedlund [Hed 7] and Ja. Pesin [Pe 4] and the book of D. Anosov [A 3]. In sections 3 through 6 we briefly summarize the main results and outline proofs of some of these results from the papers [BBE], [BBS], [Ba 2], [Bu S] and [BE].

A geometric understanding of the Riemannian symmetric spaces of noncompact type, and in particular the structure and distribution of their maximal flat, totally geodesic submanifolds (cf. section 4) is an essential ingredient of the papers cited above. Riemannian symmetric spaces have long been studied by algebraic methods that are often somewhat mysterious to differential geometers. For this reason we have included an appendix that presents an outline of the basic structure theory of Riemannian symmetric spaces in a way that is more strongly oriented toward differential geometry than existing treatments. The material presented here is a synthesis of material from various sources, including several discussions with Werner Ballmann and unpublished notes of Ernst Heintze.

Finally I wish to thank the Technische Universität Berlin for its hospitality and financial support during the period of the conference.

Section 1 Notation and Preliminaries

[(General reference [EO])]. In this paper M will always denote a complete, connected Riemannian manifold of finite volume and sectional curvature K satisfying $-a^2 \leq K \leq 0$ for some positive constant a. Let $\tilde{M}$ denote a complete, connected, simply connected Riemannian manifold with $K \leq 0$. Let SM, $S\tilde{M}$ denote the unit tangent bundles of M, $\tilde{M}$, and let $\pi: SM \to M$ or $\pi: S\tilde{M} \to \tilde{M}$ denote the projection map. If $v \in SM$ or $v \in S\tilde{M}$ then γ_v will denote the unique geodesic of M or $\tilde{M}$ with $\gamma'(0)=v$. The geodesic flow on SM, $S\tilde{M}$ will be denoted by $\{g^t\}$. All geodesics of M or $\tilde{M}$ will be assumed to have unit speed. Let d(,) denote the Riemannian metric in M or $\tilde{M}$. This induces a canonical metric d* on $S\tilde{M}$. (see [GKM, pp. 43-46] or [E7, pp. 440-441])

It is known that $\tilde{M}$ is diffeomorphic to the Euclidean space of the same dimension, and moreover for any two distinct points p,q in $\tilde{M}$ there exists a unique geodesic $\gamma_{pq}=\gamma$ of $\tilde{M}$ such that $\gamma(0)=p$ and $\gamma(a)=q$, where $a=d(p,q)$. Two unit speed geodesics γ,σ of M are said to be <u>asymptotic</u> if the function $t \to d(\gamma t, \sigma t)$ is bounded above for $t \geq 0$. The relation of being asymptotic is an equivalence relation on the unit speed geodesics of $\tilde{M}$, and $\tilde{M}(\infty)$ denotes the set of these equivalence classes or "points at infinity". Given a point $p \in \tilde{M}$ and a point $x \in \tilde{M}(\infty)$ there exists a unique geodesic $\gamma_{px}=\gamma$ such that $\gamma(0)=p$ and γ belongs to x. One may therefore identify $\tilde{M}(\infty)$ with the unit sphere $S_p\tilde{M} \subseteq T_p\tilde{M}$ for any point p in $\tilde{M}$. The resulting topology on $\tilde{M}(\infty)$ is independent of the point p and in fact is induced from a natural topology on $\bar{\tilde{M}}=\tilde{M} \cup \tilde{M}(\infty)$, which contains $\tilde{M}$ as a dense open subset in its original topology. The space $\bar{\tilde{M}}$ is homeomorphic to a closed n-disk, where n=dim $\tilde{M}$. If M is the model of the hyperbolic plane consisting of the open unit disk $D \subseteq \mathbb{R}^2$ together with a suitable metric, then $\bar{\tilde{M}}$ becomes the closed unit disk $\bar{D}$ and $\tilde{M}(\infty)$ becomes the unit circle S^1.

Unit vectors $v,w \in S\tilde{M}$ are said to be asymptotic if the corresponding geodesics γ_v,γ_w are asymptotic. For each point $q \in \tilde{M}$ and each vector $v \in S\tilde{M}$ then by the discussion above there is a unique unit vector in $T_q\tilde{M}$ that is asymptotic to v. We let v(q) denote this vector.

Busemann functions and horospheres

Given a unit vector $v \in S\tilde{M}$ one defines a Busemann function $f_v: \tilde{M} \to \mathbb{R}$ by

$f_v(q)=\lim_{t\to+\infty} d(q,\gamma_v t)-t$. Busemann functions are C^2 and also convex; that is, $(f_v \circ \sigma)''(t) \geq 0$ for any geodesic $\sigma:\mathbb{R}\to\tilde{M}$. If $v' \in S\tilde{M}$ is asymptotic to v, then $f_{v'}-f_v$ is constant in $\tilde{M}$ so that Busemann functions are determined up to an additive constant by the asymptote class $x \in \tilde{M}(\infty)$ of γ_v. It follows that the level sets of $f_v, f_{v'}$ are the same, and these are called the horospheres determined by x. Given $v \in S\tilde{M}$ we let $H(v)$ denote the horosphere determined by $v=\{q \in \tilde{M}: f_v(q)=0\}$. Clearly if $q \in H(v)$ then $H(v)=H(v(q))$.

Section 2 A brief history of geodesic flows on manifolds of nonpositive curvature

Geodesic flows have traditionally played an important role in the study of the global behavior of the trajectories of a dynamical system, beginning perhaps with the studies of Poincaré [Po] on convex surfaces and of Hadamard [Ha] on open surfaces of negative curvature. During the 1920's it became clear that compact surfaces with constant negative Gaussian curvature offered a rich field of investigation for the study of the dynamical properties of the geodesic flow. The object of most of the works in this period was to establish topological dynamical properties of the geodesic flow such as the density of periodic trajectories (= closed geodesics) and the existence of a dense trajectory (topological transitivity). Occasionally metric dynamical properties such as ergodicity or mixing were also considered. The properties just mentioned were established in special cases by many people including Artin, Myrberg, Nielsen, G.D. Birkhoff, Koebe and Löbell (see the bibliography of [Hed 7]). The latter two authors were the first to succeed in showing that for a general Fuchsian group Γ of the first kind the geodesic flow in the unit circle bundle of the surface H/Γ, where H denotes the hyperbolic plane with Gaussian curvature $K\equiv-1$, admits dense orbits as well as a dense set of vectors whose orbits are periodic. Fuchsian groups Γ of the first kind include but are not limited to discrete groups Γ of isometries of the hyperbolic plane H such that the quotient surface H/Γ has finite hyperbolic area. Important contributions were also made then and later by M. Morse [Mor 1,2,3].

The study of metric dynamical properties such as ergodicity and metrical mixing of the geodesic flow for surfaces of constant negative curvature was seriously begun and carried out in the 1930's primarily by G. Hedlund [Hed 1,2,3,4,5,6] and E. Hopf [Ho 1,2,3,4]. Hedlund was the first to recognize the importance of the close relationship between the horocycle and geodesic flows and to use this relationship to derive properties of the geodesic flow. This viewpoint led in 1939 to the first proof that the geodesic flow on the unit tangent bundle of a surface of finite area and constant negative curvature was metrically mixing [Hed 6]. This result was extended to arbitrary dimensions for manifolds of constant negative sectional curvature by Hopf in [Ho 3]. Earlier in 1936 Hopf [Ho 1] had proved using harmonic functions that the geodesic flow for such surfaces is ergodic.

During the 1930's the study of the geodesic flow on surfaces of variable negative curvature and finite area was also begun. Following historical precedent the first results were topological in nature and involved proving the existence of dense orbits [Mor 3] and the stronger property of permanent regional transitivity [Gra]. Again horocycles played a key role in this study. Toward the end of the 1930's the emphasis shifted toward metric problems of the geodesic flow and in [Hed 5] Hedlund showed that almost all vectors in the unit tangent bundle of a surface of finite area and variable negative curvature had dense orbits (topometric transitivity). Finally in 1939 Hopf [Ho 3] showed that the geodesic flow is ergodic for a surface of finite area and variable negative curvature whose curvature and first derivative of curvature are uniformly bounded in absolute value. Hopf's proof of ergodicity adds to the earlier methods a systematic use of the properties of asymptotic geodesics in a simply connected space.

It should be remarked that the results of the papers [Mor 3] and [Ho 4] were actually valid under certain growth conditions on the Jacobi vector fields that permitted the existence of small regions of positive curvature. In fact the conditions considered by Hopf implied that the surface had no focal points and its geodesic flow satisfied what are now called the Anosov conditions. For an explicit construction of such examples see [Gu].

After two decades of steady progress the study of geodesic flows on manifolds of negative sectional curvature came practically to a halt after the mixing and ergodicity results of Hedlund and Hopf in the late 1930's. The difficulty occurred when one tried to generalize the methods that worked on surfaces to manifolds of arbitrary dimension and negative sectional curvature. The difficulty can be seen in Hopf's proof of ergodicity for surfaces of variable negative curvature [Ho 3, section 14,

299-303] and in fact is pointed out by Hopf on page 299 of that article. Hopf constructed certain C^1 coordinate systems $x=(x_1,x_2,x_3)$ in the unit tangent bundles of negatively curved surfaces in which translation in the first coordinate corresponds to the action of the geodesic flow. Being differentiable these coordinate systems have the property that motion in a fixed coordinate direction is absolutely continuous; more precisely, a set $A \subseteq SM$ is a null set if and only if $x(A) \subseteq \mathbb{R}^3$ is a null set with respect to Lebesque measure, and in particular if A is a null set then so is $A_t = x^{-1}(x(A)+te_i)$ for $t \in \mathbb{R}$ and $1 \leq i \leq 3$. For manifolds of arbitrary dimension $n \geq 3$ and variable negative sectional curvature one can construct analogous coordinate systems in the unit tangent bundle, but these coordinate systems are no longer C^1 in general. Since absolute continuity of these coordinate systems is essential to Hopf's proof it was unclear for over 20 years if one could extend his method in a natural way to higher dimensions.

What defeated the geometric methods of the 1930's was overcome by a new analytical approach to dynamical systems initiated around 1960 by Anosov [A 1,2,3] and Sinai [Sin 1,2], [AS] who considered special dynamical systems (U-systems) that abstracted the key properties of the geodesic flow on the unit tangent bundle of a compact manifold of negative sectional curvature. A flow $\{\phi^t\}$ on a compact Riemannian manifold N is said to be a U-flow (now called more commonly an Anosov flow) if there is a uniformly hyperbolic splitting of the tangent bundle TN; that is, there exist positive constants a and c and a direct sum splitting of each tangent space

$$T_xN = E^s(x) \oplus E^u(x) \oplus V(x)$$

such that:

1) $V(x)$ is the 1-dimensional subspace of T_xN determined by the flow $\{\phi^t\}$

2) For every $v \in E^s(x)$ and $t \geq 0$

$$\|d\phi^t(v)\| \leq ae^{-ct}\|v\|$$

3) For every $v \in E^u(x)$ and $t \geq 0$

$$\|d\phi^t(v)\| \geq (1/a)e^{ct}\|v\|$$

The condition of being a U-flow does not depend on the Riemannian metric of N.

One of the important features of manifolds satisfying these axioms is that the bundles E^s,E^u,V are integrable and their integral manifolds determine coordinate systems in N that turn out to be absolutely continuous, allowing one to prove the strongest possible metric dynamical properties for $\{\phi^t\}$ including ergodicity, mixing and the Bernoulli property. See [A 3]. Specializing to the case that N=SM, the unit tangent bundle of a compact manifold M of negative sectional curvature, and $\{\phi^t\}$= the geodesic flow on SM one overcomes the problem encountered by Hopf.

Despite the beauty of this application of U-flows to geodesic flows one does not obtain complete satisfaction if one encounters compact manifolds with sectional curvature that is nonpositive but not strictly negative. Such examples are easy to construct, and the geodesic flows are no longer examples of U-flows. For example, consider a compact surface of constant negative curvature and change the metric tensor in a neighborhood U of a simple closed geodesic γ so that the new Gaussian curvature is zero on γ, strictly negative on $U-\{\gamma\}$ and constant negative outside U. The Anosov conditions do not hold along the vectors tangent to γ although is is fairly clear that the new metric must have dynamical properties of its geodesic flow that are just as strong as those of the geodesic flow of the old metric. The unit tangent bundle of the new metric possesses a nonuniform hyperbolic splitting of its tangent bundle T(SM) that is defined on the complement of a null set in SM (those vectors tangent to γ).

The next step in the ergodic theory of geodesic flows on manifolds of nonpositive sectional curvature was carried out by Pesin in the 1970's [Pe 1,2,3], [BP]. Pesin considered compact Riemannian manifolds N with a nonuniform hyperbolic splitting of T_xN that was defined for almost all points x in N. By refining the methods of Anosov, Pesin was able to obtain essentially the same conclusions under these weaker hyptheses, and the theory of these systems with a nonuniform hyperbolic splitting can then be applied to a far greater class of nonpositively curved manifolds than could the theory of the original U-systems. For example, Pesin shows in [Pe 3] that the geodesic flow on the unit tangent bundle of any compact surface with arbitrary nonpositive Gaussian curvature and negative Euler characteristic is an example of a manifold with such a nonuniform hyperbolic splitting. His theory then shows that the geodesic flow of such a surface is ergodic, mixing and Bernoulli. His survey article [Pe 4] provides an excellent overview of his general theory as well as a summary of the progress on geodesic flows (also horocycle and frame flows) through the late 1970's.

Nonuniform hyperbolic splittings of the unit tangent bundle relative to the geodesic flow occur, roughly speaking, when the sectional curvature of the manifold is mostly strictly negative and only rarely zero. The problem is to find relatively simple criteria for the existence of this nonuniform hyperbolic splitting in dimensions $n \geq 3$. More generally, can one give some kind of precise measurement of the amount of zero sectional curvature in an arbitrary manifold of nonpositive sectional curvature? The work of Mautner in [Ma] shows that one cannot expect an ergodic geodesic flow if the manifold has too much zero sectional curvature. Mautner considered the geodesic flow on a compact locally symmetric space M of nonpositive sectional curvature. Each such space has a rank $k \geq 1$, where k is the largest integer such that M admits a k-flat, i.e. an isometric, totally geodesic immersion of a Euclidean space $\mathbb{R}^k$ with its standard flat metric. If a locally symmetric space M has rank $k \geq 2$ then every vector in the unit tangent bundle of M is tangent to at least one k-flat, and a dense set of vectors are tangent to an isometric, totally geodesic immersion of a flat k-torus T^k. Mautner showed that the geodesic flow on the unit tangent bundle of a compact locally symmetric space M is ergodic if M has rank 1 (strictly negative sectional curvature) and is never ergodic if M has rank $k \geq 2$. In fact he described the ergodic components of the flow in SM.

An important step in the final characterization of ergodicity for geodesic flows in finite volume manifolds of bounded nonpositive sectional curvature was made in the 1978 dissertation of W. Ballmann [Ba 1]. Ballmann considered manifolds of nonpositive sectional curvature that admitted a geodesic bounding no immersed flat half plane (a geodesic is said to bound an immersed flat half plane if there exists an isometric, totally geodesic immersion $F: \mathbb{R} \times [0,\infty) \to M$ with $F(z,0)=\gamma(t)$ for all $t \in \mathbb{R}$). This no flat half plane condition is a much weaker restriction than the "Visibility" axiom introduced in [E3],[EO]. Ballmann was able to show that all of the important topological properties of the geodesic flow that are true for manifolds satisfying the Visibility axiom are also true for manifolds satisfying the no flat half plane condition if every vector in the unit tangent bundle is nonwandering relative to the geodesic flow. These properties include density of the vectors with periodic orbits as well as the existence of a dense orbit or, more generally, topological mixing (= permanent regional transitivity).

The next step was made in 1981 by Ballmann and M. Brin [BB] who studied the geodesic flow on compact manifolds of nonpositive sectional curvature that admit a geodesic γ whose vector space of parallel Jacobi vector fields is 1-dimensional; that is, spanned by the tangent vector field of γ. Under this condition they showed that the geodesic flow in the unit tangent bundle possesses the nonuniform hyperbolic splitting considered by Pesin, and it then follows from the work of Pesin that the geodesic flow is ergodic (and in fact mixing and Bernoulli). This Jacobi vector field condition is weaker than the hypothesis of strictly negative sectional curvature but apparently stronger than the no flat half plane condition described above. Later developments in fact showed that the Jacobi vector field condition is equivalent to the no flat half plane condition for manifolds of arbitrary dimension, finite volume and bounded nonpositive sectional curvature. In 1982 K. Burns [Bu] in his dissertation showed that compact manifolds of nonpositive sectional curvature that satisfy the Visibility axiom also satisfy the Jacobi vector field condition above, and he also deduced independently the ergodicity of the geodesic flow from the work of Pesin. Ballmann and Brin also succeeded in 1982 in proving the ergodicity of the geodesic flow for compact manifolds satisfying the no flat half plane condition.

Even after the substantial progress just described the characterization of finite volume manifolds with bounded nonpositive curvature and ergodic geodesic flow was still incomplete. Could there be such manifolds that did not satisfy the Jacobi vector field condition? The answer to this question came as a byproduct of a program involving W. Ballmann, M. Brin, K. Burns, R. Spatzier and myself whose results were the subject of my talk in Berlin.

Section 3 Rank and flatness

The Jacobi vector field condition considered by Ballmann and Brin in [BB] suggested a natural way to proceed. Why not try to measure the flatness of a compact manifold M of nonpositive sectional curvature by looking at the vector spaces of parallel Jacobi vector fields along the geodesics of M? Specifically for each vector v in SM, the unit tangent bundle of M, one defines r(v) (=the rank of v) to be the dimension of the space of parallel Jacobi vector fields along γ_v, the unique geodesic with initial velocity v. One then defines r(M) (= the rank of M) to be the smallest of the integers r(v), $v \in SM$. This definition of r(M) works equally well for any complete manifold of nonpositive sectional curvature, not necessarily compact. The manifolds considered by Ballmann and Brin are now precisely the manifolds of rank 1 in this terminology.

We define a vector $v \in SM$ to be <u>regular</u> if $r(v)=r(M)$, and we let $R \subseteq SM$ denote the set of regular vectors. Since $r(w) \leq r(v)$ for all w in a neighborhood of a given vector $v \in SM$ it follows that R is an open subset of SM. In fact R is dense in SM if M has finite volume and nonpositive curvature [BBE, Theorem 2.6], but this fact at the moment has no simple proof.

Note that $r(v) \geq 1$ for all $v \in SM$ and hence $r(M) \geq 1$ since the velocity vector field of the geodesic γ_v is always a parallel Jacobi vector field on γ_v for every $v \in SM$. If Y is a parallel Jacobi vector field defined on $\mathbb{R}$ and perpendicular to a geodesic γ, then it follows from the Jacobi equation $Y''(t)+R(Y,\gamma')\gamma'(t)=0$ that the sectional curvature $K(Y,\gamma')(t)$ determined by Y and γ' is identically zero on $\mathbb{R}$. Here $Y''(t)$ denotes the second covariant derivative of Y along γ, and R denotes the curvature tensor of M. One may therefore think of a parallel Jacobi vector field on γ that is perpendicular to γ as a kind of infinitesimal flat strip. If M has rank $r(M)=k \geq 2$, then each geodesic of M admits a family of parallel vector fields along γ of dimension at least $k-1 \geq 1$ that determine zero sectional curvatures along γ. One may hope (and with justification) that if $v \in SM$ is a regular vector, then one may integrate the infinitesimal flat strips on geodesics γ_w close to γ_v into pieces of flat manifolds of dimension $r(v)=r(M)$.

It is easy to verify from the definition that $1 \leq r(M) \leq \dim (M)$ and $r(M)=\dim (M)$ if and only if M is flat. Moreover the rank of a Riemannian product $M_1 \times M_2$ is the sum of the ranks of M_1 and M_2, and ranks are unchanged by passing to Riemannian coverings.

Using the idea of rank we may now characterize ergodicity of the geodesic flow. For a somewhat different formulation of the result see Theorem A of section 5.

<u>Theorem A'</u> [BBE, BBS] Let M be complete with finite volume and bounded nonpositive sectional curvature. Then the following properties are equivalent:

1) M has rank $r(M) = 1$
2) The geodesic flow in SM is Bernoulli.
3) The geodesic flow in SM is ergodic.
4) The geodesic flow has a dense orbit in SM.

It is shown in [E4] that if M_1, M_2 are compact manifolds of nonpositive sectional curvature and isomorphic fundamental groups, then $r(M_1)=1$ if and only if $r(M_2)=1$. Hence in some mysterious way the ergodicity or nonergodicity of the geodesic flow in SM can be determined from the fundamental group of M if M is compact. We give later an explicit criterion from [BE] for determining the rank of M in terms of algebraic data of the fundamental group of M.

It is clear from the remarks above that the rank r(M) does indeed give some measure of the flatness of M. Moreover in view of the result just stated it follows that manifolds with finite volume and rank 1 are more general than but very similar in spirit to compact manifolds of strictly negative sectional curvature. (Clearly if all sectional curvatures at a point p in M are negative, then $r(v)=1$ for all unit vectors v in T_pM.) In fact, the condition of having rank $k \geq 2$ is such a stringent global condition on M that one would expect that a "random" or "generic" compact manifold of nonpositive sectional curvature would have rank 1. This expectation is supported by the known examples of rank 1 manifolds and is explicitly justified later.

Examples with rank 1

If M is a compact surface of nonpositive Gaussian curvature and negative Euler characteristic, then M must have rank 1 by a remark above since Gaussian curvature must be negative somewhere on M according to the Gauss-Bonnet theorem. Such surfaces can look very flat as one can see by considering a connected sum of two flat tori. The curvature can be made identically zero except for a small region around both ends of the tube joining the two flat tori. A construction of Heintze yields compact manifolds M of dimension $n \geq 3$ whose sectional curvature is constant and negative except in the neighborhood of finitely many (n-1)-dimensional flat tori imbedded isometrically and totally geodesically in M. Unlike the 2-dimensional examples one cannot perturb the metric above in dimensions $n \geq 3$ to obtain another metric with strictly negative sectional curvature; the fundamental group of M contains subgroups isometric to $\mathbb{Z}^{n-1}$ corresponding to the imbedded (n-1)-tori, and a theorem of A. Preissmann [Pr] says that all abelian subgroups of the fundamental group of M must be infinite cyclic if the sectional curvature of M is strictly negative. Even more complicated examples of compact rank 1 manifolds have been constructed by M. Gromov. These examples also contain imbedded flat tori and have the additional property that every vector in SM is tangent to a 2-plane with zero sectional curvature. See the introduction to [BBE] for further details about these examples.

Section 4 Symmetric spaces and manifolds of rank at least two

Compact manifolds M of nonpositive sectional curvature with rank at least two are more difficult to construct except in trivial cases. If the universal Riemannian cover $\tilde{M}$ of M is a Riemannian product $\tilde{M}_1 \times \tilde{M}_2$, then $r(M)=r(\tilde{M})=r(\tilde{M}_1)+r(\tilde{M}_2)\geq 2$. If M is flat then $r(M)=\dim(M)=n\geq 2$. The final obvious large class of compact higher rank examples consists of those (locally symmetric) M for which $\tilde{M}$ is a globally symmetric space of noncompact type and rank at least two. These trivial cases as we will see later nevertheless constitute a complete list of compact manifolds of nonpositive curvature and rank at least two, giving precise form to the feeling mentioned earlier that "most" compact manifolds of nonpositive sectional curvature should have rank 1.

Symmetric Spaces

To begin the discussion of manifolds of nonpositive curvature and rank at least two it will be both useful and appropriate to describe briefly the Riemannian symmetric spaces of nonpositive curvature. These spaces are the nontrivial models for the manifolds of higher rank, and many of the definitions and methods used later will seem more natural when one sees how they arise in symmetric spaces. For the convenience of the reader we include an appendix that describes in some detail the basic elementary algebraic and geometric structure of a symmetric space of noncompact type.

A complete simply connected manifold $\tilde{M}$ of nonpositive sectional curvature is called <u>(globally) symmetric</u> if for each point $p \in \tilde{M}$ the geodesic symmetry $s_p:\tilde{M}\to\tilde{M}$ that fixes p and takes $\exp_p(v)$ into $\exp_p(-v)$ for all $v \in T_p\tilde{M}$ is an isometry of $\tilde{M}$. A manifold $\tilde{M}$ is called (globally) symmetric of noncompact type if $\tilde{M}$ is not the Riemannian product of a Euclidean space with another manifold. A globally symmetric space $\tilde{M}$ is homogeneous for in order to get from a point $p \in \tilde{M}$ to another point $q \in \tilde{M}$ by an isometry of $\tilde{M}$ one need only apply the geodesic symmetry s_r, where r is the midpoint of the unique geodesic from p to q. It follows that $G=I_0(M)$ is transitive on $\tilde{M}$, and $\tilde{M}$ may be identified with the coset space G/K equipped with a G-invariant Riemannian metric, where K is the maximal compact subgroup of G consisting of all isometries fixing a certain point $p \in \tilde{M}$. If $\tilde{M}$ is symmetric of noncompact type, then G is a semisimple Lie group with no compact factors and with trivial center. Conversely if G is a semisimple Lie group with finite center and with no compact factors and if K is a maximal compact subgroup of G, then the coset space G/K is simply connected and carries a G-variant complete metric of nonpositive sectional curvature that makes G/K into a symmetric space of noncompact type. A case of particular interest is when $G=SL(n,\mathbb{R})$ and $K=SO(n,\mathbb{R})$ for $n\geq 2$. Much of the structure of G/K can be described by elementary and familiar linear algebra. Moreover every irreducible symmetric space of noncompact type can be imbedded isometrically and totally geodesically as a submanifold of $SL(n,\mathbb{R})/\ SO(n,\mathbb{R})$, where $n=\dim I_0(M)$. (See the appendix)

Flats

An r-flat of a globally symmetric space $\tilde{M}$ of nonpositive sectional curvature is defined to be a complete imbedded totally geodesic submanifold F of $\tilde{M}$ that is isometric to a flat r-dimensional Euclidean space. The maximum dimension of a flat in $\tilde{M}$ is then traditionally called the rank of $\tilde{M}$, and this definition of rank agrees with the Jacobi vector field definition of rank given earlier. Every vector $v \in S\tilde{M}$ is tangent to at least one k-flat, where k=rank (M), and $v \in S\tilde{M}$ is tangent to exactly one k-flat if and only if $v \in \mathcal{R}$, the open dense subset of $S\tilde{M}$ consisting of regular vectors w, i.e. $r(w)=k$ (see the appendix). For a symmetric space $\tilde{M}$, $\mathcal{R}$ has full measure in $S\tilde{M}$ and if $v \in \mathcal{R}$ then $v' \in \mathcal{R}$ for any asymptote v'. We remark that for symmetric spaces k=1 if and only if the sectional curvature of $\tilde{M}$ is strictly negative. In this case $\mathcal{R}=S\tilde{M}$ and a 1-flat is a complete geodesic.

Strong Stable Manifolds

Given a vector $v \in S\tilde{M}$ we define $H^s(v) = \{q \in \tilde{M} : d^*(g^t v, g^t v(q)) \to 0$ as $t \to +\infty\}$. Here $\{g^t\}$ denotes the geodesic flow in $S\tilde{M}$ and d^* denotes the Riemannian metric in $S\tilde{M}$ induced from that of $\tilde{M}$. The set $H^s(v)$ is a complete submanifold of $H(v)$, the horosphere determined by v, and is called the strong stable manifold determined by v. In fact there exist positive constants $c = c(v,q)$ and $\lambda = \lambda(v)$ such that $d^*(g^t v, g^t v(q)) \leq ce^{-\lambda t}$ for all $t \geq 0$ and all $q \in H^s(v)$.

$H^s(v)$ has maximal dimension n-k, where k=rank $(\tilde{M})$, precisely when $r(v)=k$ i.e. $v \in R$. If $v \in R$, then $H^s(v)$ meets the k-flat $F(v(q))$ orthogonally at q for each point $q \in H^s(v)$, where for $w \in R \subseteq S\tilde{M}$ we let $F(w)$ denote the unique k-flat in $\tilde{M}$ that contains the geodesic γ_w. For $v \in R$ the strong stable manifolds $H^s(v)$ are locally constant in $R \subseteq TF(v)$; in fact if $v \in T_p\tilde{M}$ and $U \subseteq R \cap T_pF$ is a sufficiently small neighborhood of v, then $H^s(v) = \{\cap H(w^*) : w^* \in U\} = H^s(w)$, for all $w \in U$.

Weyl chambers

Every vector $v \in R$ determines a <u>Weyl chamber</u> $C(v) \subseteq R \cap T_pF(v)$, where p is the footpoint of v, and the relation $v \sim w$ if $w \in C(v)$ defines an equivalence relation on R. Let $S_pF(v)$ denote the set of unit vectors at p tangent to $F(v)$, and let $S^* = S_pF(v) - R$, the singular vectors at p tangent to $F(v)$. It is known (see the appendix) that $\mathbb{R}S^* = \{tv : t \in \mathbb{R},\ v \in S^*\}$ is a finite union of hyperplanes through the origin in $T_pF(v)$. Hence S^* is a finite union of great hyperspheres in $S_pF(v)$ and $S_pF(v) - S^* = S_pF(v) \cap R$ has finitely many connected components in $S_pF(v)$ called the <u>Weyl chambers</u> determined by v. The connected component containing v is denoted by $C(v)$. It follows from the definition that $C(w) = C(v)$ if $w \in C(v)$.

In the case k=1 $F(v)$ is just the geodesic γ_v and $S_pF(v) = \{v, -v\}$. There are two Weyl chambers $C(v) = \{v\}$ and $C(-v) = \{-v\}$.

We describe some important properties of Weyl chambers

1) (Convexity) The Weyl chamber $C(v)$ is an open spherically convex subset of an open hemisphere in $S_pF(v)$.
2) (Rigidity) Given regular vectors $v \in T_p\tilde{M}$ and $v^* \in T_{p^*}\tilde{M}$ there is a linear isometry ϕ of $T_p\tilde{M}$ onto $T_{p^*}\tilde{M}$ such that $\phi C(v) = C(v^*)$.

3) (Asymptotic Property) Given a regular vector $v \in S\tilde{M}$ and a point $q \in \tilde{M}$ the map $w \to w(q)$ carries $C(v)$ onto $C(v(q))$ and is the restriction of a linear isometry from $T_p\tilde{M}$ onto $T_q\tilde{M}$, where $v \in T_p\tilde{M}$.
4) (Rigidity of strong stable manifolds) Let $v, w \in R$ be vectors with $w \in C(v)$. Then $H^s(w) = H^s(v)$.

The convexity property follows immediately from the fact that each connected component of $T_pF(v) - \mathbb{R}S^*$ is an open convex subset of a half space in $T_pF(v)$. The rigidity property follows from the fact that for any two regular unit vectors v, v^* in R there exists an isometry ψ of $\tilde{M}$ such that $(d\psi)(v) \in C(v^*)$. (See appendix). The asymptotic property follows from the fact that the subgroup of isometries of $\tilde{M}$ that fixes some point $x \in \tilde{M}(\infty)$ is transitive on $\tilde{M}$. Hence given $v \in S\tilde{M}$ and $q \in \tilde{M}$ there exists an isometry ψ of M such that $(d\psi)(v) = v(q)$. (It is not immediately clear why $(d\psi)(w) = w(q)$ for all $w \in C(v)$. See the appendix for an explanation). The rigidity of strong stable manifolds follows from (33) in the appendix. This property sharpens the statement made earlier that strong stable manifolds $H^s(v)$ are locally constant with respect to regular vectors in $S_pF(v)$.

The asymptotic property 3) of Weyl chambers says that the k-flats $F(v)$ and $F(v(q))$ are asymptotic to each other along the convex open subregions determined by $C(v)$ and $C(v(q))$. More precisely $F(v(q)) = F(w(q))$ if $w \in C(v)$ and hence the geodesic γ_w, which lies in $F(v)$, is asymptotic to the geodesic $\gamma_{w(q)}$, which lies in $F(v(q))$. The asymptotic behavior of $F(v)$ and $F(v(q))$ along convex open subregions determined by Weyl chambers other than $C(v)$, $C(v(q))$ may be completely unrelated. For example, in the case $k=1$ the 1-flats are geodesics and asymptotic geodesics can have prescribed (nonasymptotic) behavior in the backward direction. For a more detailed discussion of flats and Weyl chambers see [Mos, sections 4,7].

Arbitrary manifolds of rank at least two

Now let $\tilde{M}$ denote an arbitrary simply connected complete manifold of nonpositive sectional curvature and rank $k \geq 2$ that admits a smooth quotient manifold $\tilde{M}$ with finite volume. It is natural to try to build up in $\tilde{M}$ as much as possible of the geometric structure of symmetric spaces and to impose further restrictions on $\tilde{M}$ where needed to obtain characterizations of the symmetric spaces. For details of the results sketched below see [BBE] and[BBS].

Construction of k-flats

Keeping in mind the properties of symmetric spaces one would like to show in the general case that every regular vector v in $S\tilde{M}$ is tangent to exactly one k-flat F in $\tilde{M}$. If such a k-flat F exists then it must be contained in the union of all geodesics σ in $\tilde{M}$ that are parallel to γ_v; two unit speed geodesics σ_1, σ_2 in $\tilde{M}$ are said to be parallel if one (hence both) of the functions $t \to d(\sigma_1 t, \sigma_2)$ or $t \to d(\sigma_2 t, \sigma_1)$ is bounded on $\mathbb{R}$. Since these two functions are convex on $\mathbb{R}$ they are constant and it follows that σ_1 and σ_2 bound a flat strip in $\tilde{M}$; that is there exists a totally geodesic, isometric imbedding $F: \mathbb{R} \times [0,a] \to \tilde{M}$ with $F(t,0) = \sigma_1(t)$ and $F(t,a) = \sigma_2(t)$, where $a \equiv d(\sigma_1 t, \sigma_2) \equiv d(\sigma_2 t, \sigma_1)$.

Given a vector $v \in SM$ (respectively a geodesic γ in $\tilde{M}$) we define $F(v)$ (respectively $F(\gamma)$) to be the union of all geodesics σ parallel to γ_v (respectively parallel to γ). Since $F(\gamma)$ $(F(v))$ is a union of flat strips by the discussion above one may hope that if $\gamma'(0)$ (respectively v) lies in R then $F(\gamma)$ (respectively $F(v)$) is the unique k-flat containing γ (respectively γ_v). The sets $F(v)$, $F(\gamma)$ contain γ_v, γ and are closed and convex in $\tilde{M}$. They are not in general smooth submanifolds of $\tilde{M}$ although by the results of [CG] they do have the structure of a totally geodesic submanifold of M with possibly nonempty boundary. However, if $\tilde{M}$ admits a quotient manifold of finite volume or more generally if $I(M)$ satisfies the "duality condition" [CE], [E3,5] then the sets $F(\gamma)$, $F(v)$ are smooth flat k-manifolds if γ, v are regular.

The restriction that γ, v be regular for $F(\gamma)$, $F(v)$ to be flat is necessary. In a symmetric space $\tilde{M}$ the sets $F(\gamma)$, $F(v)$ are the Riemannian product of a Euclidean space with dimension $k_0 < k$ and a symmetric space $\tilde{M}^*$ of noncompact type if γ, v are singular (i.e. not regular). The proof that the convex sets $F(\gamma)$, $F(v)$ are k-flats if $\gamma'(0)$, v are regular vectors involves both a mixture of standard integrablity arguments and somewhat ad hoc geometric arguments that depend strongly on the hypothesis that M admits a quotient manifold M of finite volume.

A vector $w \in S\tilde{M}$ is said to be parallel to a vector $v \in S\tilde{M}$ if the corresponding geodesics γ_v, γ_w are parallel in $\tilde{M}$. Let $P(v)$ denote the set of vectors $w \in S\tilde{M}$ that are parallel to v. It is not difficult to show that the projection $\pi: S\tilde{M} \to \tilde{M}$ maps $P(v)$ isometrically onto $F(v)$.

The first step in building k-flats in $\tilde{M}$ is to construct an integrable C^∞ foliation F defined on $R\subseteq S\tilde{M}$ whose maximal integral manifold through $v\in R$ is an open subset of $P(v)$. This shows that $F(v)$ is smooth near any point q such that $v(q)$ lies in R. Unfortunately, it might a priori be true that $v(q)$ does not lie in R for some point $q\in F(v)$ and to prove the smoothness of $F(v)$ at q one then requires (at the moment) a separate geometric argument.

Given a vector $v\in SN$, the unit tangent bundle of a complete Riemannian manifold N, there is a natural isomorphism between $T_v(SN)$ and $J^*(\gamma_v)=$ {Jacobi vector fields Y on γ_v such that $\langle Y'(t),\gamma_v'(t)\rangle\equiv 0$}. If $\xi\in T_v(SN)$ is given let $v(t)$, $|t|<\varepsilon$, be a curve in SN with initial velocity ξ and let Y_ξ be the Jacobi vector field on γ_v that arises as the variation vector field $(\alpha)_*\frac{\partial}{\partial s}(t,0)$ of the geodesic variation $\alpha(t,s)=\gamma_{v(s)}(t)$ defined on $\mathbb{R}\times(-\varepsilon,\varepsilon)$. The correspondence $\xi\to Y_\xi$ is a well defined isomorphism of $T_v(SN)$ onto $J^*(\gamma_v)$.

Returning now to M we let $v\in R$ be chosen arbitrarily and define $J^p(\gamma_v)$ to be the set of Jacobi vector fields on γ_v that are parallel on $\mathbb{R}$; that is, $Y'(t)\equiv 0$ on $\mathbb{R}$. Clearly $J^p(\gamma_v)$ is a k-dimensional subspace of $J^*(\gamma_v)$, and we define $F(v)\subseteq T_v(S\tilde{M})$ to be the k-dimensional subspace that corresponds to $J^p(\gamma_v)$ under the isomorphism just described. It is not difficult to show that the k-dimensional distribution F is C^∞ and integrable and that its maximal integral manifold through a vector $v\in R$ is an open subset of $P(v)$.

Now let $q\in F(v)$ by any point. The set $F(v)$ is a smooth k-manifold near p, where $v\in T_p\tilde{M}$, by the remarks above and hence for some $\varepsilon>0$ $F(v)$ contains the smooth k-disk $B_\varepsilon(v)=\{q^*\in F(v):d(q,q^*)<\varepsilon\}$ of radius ε. The fact that $\tilde{M}$ admits a quotient manifold M of finite volume implies that there exist sequences $\{\phi_n\}\subseteq I(\tilde{M})$, $\{t_n\}\subseteq\mathbb{R}$ and $\{v_n\}\subseteq S\tilde{M}$ such that $t_n\to+\infty$, $v_n\to v$ and $v_n^*=(d\phi_n\circ g^{t_n})(v_n)\to v(q)$ as $n\to+\infty$. For large n the vectors v_n lie in R and the sets $F(v_n)$ contain the smooth imbedded k-disks $B_{\varepsilon/2}(v_n)$. It follows that the sets $F(v_n^*)$ contain the smooth imbedded k-disks $B_{\varepsilon/2}(v_n^*)$ and these k-disks converge to a smooth imbedded k-disk $B_{\varepsilon/2}(v(q))\subseteq F(v(q))=F(v)$, which proves that $F(v)$ is smooth near q. We have shown that $F(v)$ is a complete k-dimensional totally geodesic submanifold of $\tilde{M}$ if $v\in R$, and an additional short argument shows that $F(w)=F(v)$ if $w\in T_pF(v)$ is sufficiently close to v. Hence nearby vectors w determine globally parallel vector fields on $F(v)$ and it follows that $F(v)$ is flat.

Only at this stage do we show that R is a dense subset of $S\tilde{M}$. Finally using this density we conclude that every geodesic γ of $\tilde{M}$ is tangent to at least one k-flat in $\tilde{M}$ and exactly one k-flat $F=F(\gamma)$ if the tangent vectors of γ are regular.

Construction of Strong Stable Manifolds

Up to this point we have not used the assumption that $\tilde{M}$ has a quotient manifold M of finite volume in an essential way. To construct k-flats as above one needs only the hypothesis that for every vector $v \in SM$ there exist sequences $\{v_n\} \subseteq S\tilde{M}$, $\{t_n\} \subseteq \mathbb{R}$ and $\{\phi_n\} \subseteq I(\tilde{M})$ such that $v_n \to v$, $t_n \to +\infty$ and $(d\phi_n \circ g^{t_n})(v_n) \to v$. This condition (equivalent to the "duality" condition of [EO], [CE]) that every vector in $S\tilde{M}$ be nonwandering relative to the geodesic flow modulo the action of $I(\tilde{M})$ is implied by but much weaker than the existence of a finite volume quotient manifold M.

At the present time further progress, and in particular the construction of strong stable manifolds, requires the existence of a dense set of "uniformly recurrent" vectors in $S\tilde{M}$ as well as a lower bound for the sectional curvatures of $\tilde{M}$. It is not clear if these conditions are necessary. It may even be the case that the duality condition described above implies the existence of a dense set of uniformly recurrent vectors in $S\tilde{M}$.

A vector $v \in S\tilde{M}$ is said to be <u>uniformly recurrent</u> if for any neighborhood $\tilde{U}$ of v in $S\tilde{M}$ and any constant $T>0$ there exist sequences $\{\phi_n\} \subseteq I(\tilde{M})$, $\{t_n\} \subseteq \mathbb{R}$ and a constant $\sigma>0$ so that $(d\phi_n \circ g^{t_n})(v) \in \tilde{U}$, $t_n \to +\infty$, $t_{n+1}-t_n > T$ and $t_n < n/\sigma$ for every n. The simplest example of a uniformly recurrent vector is a (periodic) vector $v \in S\tilde{M}$ such that $(\phi \circ \gamma_v)(t) = \gamma_v(t+\omega)$ for all $t \in \mathbb{R}$, some $\phi \in I(\tilde{M})$ and some $\omega \neq 0$. If $v \in S\tilde{M}$ is uniformly recurrent and if the sequence $\{\phi_n\}$ above can be chosen to lie inside a discrete group $\Gamma \subseteq I(\tilde{M})$ such that $M=\tilde{M}/\Gamma$ is a smooth manifold, then the $\{g^t\}$-orbit of the projected vector $\rho_*(v) \in SM$ meets every neighborhood of $\rho_*(v)$ in SM with positive frequency in forward time; that is

$$\lim_{T\to+\infty} \inf \frac{1}{T} \int_0^T \chi_U(g^t \rho_* v)\,dt \geq \alpha > 0$$

for every neighborhood $U=\rho(\tilde{U})$ of $\rho_*(v)$ in SM. Here $\tilde{U}$ and σ refer to the notation in the definition of uniformly recurrent vector. Conversely if $\rho_*(v)$ satisfies the positive frequency condition above for some vector $v \in SM$, then v is uniformly recurrent.

If $\tilde{M}$ admits a quotient manifold M of finite volume, then the Birkhoff Ergodic Theorem (see for example [Ho 2, pp. 49-56]) implies that almost every vector in SM satisfies the positive frequency condition above, and hence the set of uniformly recurrent vectors in $S\tilde{M}$ has full measure in $S\tilde{M}$ and in particular is dense in $S\tilde{M}$.

In [BBE] and [BBS] it is convenient to carry out many of the constructions needed for strong stable manifolds and Weyl chambers first in the case that v is uniformly recurrent and then in the general case by approximating arbitrary vectors in R by uniformly recurrent vectors in R.

We now outline the construction of strong stable manifolds. Given $v \in S\tilde{M}$ we recall that $J^p(\gamma_v)$ denotes the vector space of parallel Jacobi vector fields on γ_v. Let $SJ^*(\gamma_v)$ denote the vector space of Jacobi vector fields on γ_v that arise as variation vector fields $(\alpha)_*\frac{\partial}{\partial s}(t,0)$ of horospherical geodesic variations

$$\alpha(t,s) = \gamma_{v(s)}(t)$$

defined on $\mathbb{R} \times (-\varepsilon,\varepsilon)$, where $v(s)=v(\alpha(s))$ and $\alpha(s)$ is a curve in the horosphere $H(v)$ starting at the footpoint of v. If $SJ(\gamma_v)$ denotes the subspace of $J(\gamma_v)$ spanned by $SJ^*(\gamma_v)$ and the velocity vector field of γ_v, then one may show easily that $J^p(\gamma_v) \subseteq SJ^*(\gamma_v) \subseteq SJ(\gamma_v) = \{Y \in J(\gamma_v) : \|Y(t)\| \leq c$ for all $t \geq 0$ and some $c>0\}$. The n-dimensional vector space $SJ(\gamma_v)$ is called the space of <u>stable Jacobi vector fields</u> on γ_v.

Given $v \in S\tilde{M}$ let $J^s(\gamma_v) = \{Y \in SJ(\gamma_v) : \langle Y(0), Z(0)\rangle = 0$ for all $Z \in J^p(\gamma_v)\}$. The vector space $J^s(\gamma_v)$ is referred to as the space of <u>strong stable Jacobi vector fields</u> on γ_v. The space $SJ(\gamma_v)$ is the direct sum of $J^p(\gamma_v)$ and $J^s(\gamma_v)$ and $\langle Y,Z\rangle(t) \equiv 0$ for all $t \in \mathbb{R}$, all $Y \in J^s(\gamma_v)$ and all $Z \in J^p(\gamma_v)$.

If v lies in R we may define a distribution D^s_v on the horosphere $H(v)$ by $D^s_v(q) = \{Y(0) : Y \in J^s(\gamma_{v(q)})\}$. The distribution D^s_v is C^1 and integrable of dimension n-k on a neighborhood $U \subseteq H(v)$ of the footpoint of v. We let $H^s(v)$ denote the unique maximal connected integral manifold of D^s_v that contains the footpoint of v. If $v \in R$ is uniformly recurrent then $U=H(v)$ and for each $q \in H(v)$ the corresponding maximal connected integral manifold through q, $H^s(v(q))$, is a complete closed submanifold of H(v). In this case we may characterize $H^s(v(q)) = \{q^* \in \tilde{M}: d^*(g^t v(q^*), g^t v(q)) \to 0 \text{ as } t \to +\infty\}$, and in fact there exist positive constants $C^* = C^*(v,q^*)$ and $\lambda = \lambda(v,q)$ such that $d^*(g^t v(q^*), g^t v(q)) \leq C^* e^{-\lambda t}$ for all $t \geq 0$ and for all $q^* \in H^s(v(q))$.

Construction of Weyl Chambers

The definition given earlier of Weyl chambers for a symmetric space $\tilde{M}$ seems unsuitable in the general case of an arbitrary manifold $\tilde{M}$ of nonpositive curvature with a finite volume quotient M. Given a point p in a k-flat F of $\tilde{M}$ there is no obvious reason why the singular (=not regular) unit vectors in T_pF should be contained in a finite union of hyperplanes. If we define a Weyl chamber to be a connected component of $R \cap T_pF$, then it is not clear how to show that such Weyl chambers have any of the four nice properties described earlier for symmetric spaces. Instead we choose an alternate definition of Weyl chamber from which it is easier to establish these properties or slightly weakened versions of them.

To motivate the new definition of Weyl chamber we need to discuss further properties of uniformly recurrent vectors in R. Given an arbitrary vector $v \in S\tilde{M}$ we define the maximal exponent $\lambda(v)$ to be $\sup\{\lambda \geq 0: \limsup_{t \to +\infty} \frac{\ln \|Y(t)\|}{t} \leq -\lambda \text{ for all } Y \in J^s(\gamma_v)\}$. The function $\lambda: S\tilde{M} \to \mathbb{R}$ is nonnegative, measurable and invariant under the geodesic flow $\{g^t\}$. If $v \in R$ is uniformly recurrent then $\lambda(v) > 0$ by Lemma 3.4 of [BBE]. Moreover we have the Angle Lemma (Lemma 4.4 of [BBE]). Let $\tilde{M}$ have a lower bound $-a^2 < 0$ for the sectional curvature, and let $\tilde{M}$ admit a quotient manifold M of finite volume. Let $v \in R$ be uniformly recurrent, and let p denote the footpoint of v. Choose $0 < \alpha \leq \pi/4$ such that $a \tan \alpha < \lambda(v)$ and choose $\beta > 0$ such that $\beta < \lambda(v)\cos\alpha - a\sin\alpha$. Then the following hold for any $w \in T_pF(v)$ with $\measuredangle(v,w) \leq \alpha$

1) For every $q \in H^s(v)$ the vectors $v(q)$, $w(q)$ lie in R and $F(v(q)) = F(w(q))$. Moreover $\measuredangle(v(q), w(q)) = \measuredangle(v,w)$.
2) For every $q \in H^s(v)$ we have $d^*(g^t w, g^t w(q)) \leq C e^{-\beta t} d^*(w, w(q))$, where $C = C(q) > 0$ is bounded on compact subsets of $H^s(v)$ and d^* denotes the canonical metric in $S\tilde{M}$.

Under the conditions on v and w stated above it follows from the Angle

Lemma that $F(v(q))=F(w(q))$ is a k-flat for every point $q \in \tilde{M}$. In fact, given a point $q \in \tilde{M}$ we may choose a point $q^* \in H^s(v)$ such that q lies in the k-flat $F^*=F(v(q^*))=F(w(q^*))$. Since $v(q)$, $w(q)$ are obtained from $v(q^*)$, $w(q^*)$ by parallel translation in F^* the assertion above follows. It also follows that $q \to \measuredangle(v(q), w(q))$ is a constant function in $\tilde{M}$. Note that $v(q)$ need not be regular a priori even though it is parallel to a regular vector $v(q^*)$ for some $q^* \in H^s(v)$ (In fact all vectors $v(q)$ <u>are</u> regular by an argument using the uniform recurrence of v). Nevertheless the set $F(v(q))$, the union of all geodesics of $\tilde{M}$ parallel to $\gamma_{v(q)}$, is still a k-flat $F^*=F(v(q^*))$.

We now define the Weyl chamber $C(v)$ for an arbitrary vector $v \in R$. For technical reasons that appear already in the Angle Lemma we need to assume that $\tilde{M}$ has a lower bound for the sectional curvatures as well as a quotient manifold M of finite volume. In [BBS] one also introduces a notion of <u>p-regularity</u> for vectors in $S\tilde{M}$ to deal with the technical difficulty mentioned above that a vector parallel to a regular vector is not a priori regular itself.

A vector $w \in S\tilde{M}$ is defined to be <u>p-regular</u> if it is parallel to a regular vector v; that is, the geodesics γ_w, γ_v are parallel in $\tilde{M}$ for some $v \in R$. Equivalently w is p-regular if $w=v(q)$ for some $v \in R$ and some point $q \in F(v)$. Note that $F(\gamma_w)=F(\gamma_v)=F(v)$ if w is parallel to v, and hence one may speak of the k-flat $F(\gamma_w)$ determined by a p-regular vector w.

For a p-regular vector $w \in S\tilde{M}$ one defines $A(w)=\{q \in \tilde{M} : w(q) \text{ is p-regular}\}$. Finally if $v \in T_q\tilde{M}$ is a p-regular unit vector for some point $q \in \tilde{M}$ we define

$$C(v) = \{w \in T_q\tilde{M} : w \text{ is p-regular, } A(w) = A(v) \text{ and } F(w(q^*)) = F(v(q^*)) \text{ for all points } q^* \in A(v) = A(w)\}.$$

$C(v)$ is called the <u>Weyl chamber</u> determined by v, and Weyl chambers determine an equivalence relation on p-regular unit vectors at a fixed point q. Clearly the definition of $C(v)$ is motivated by the behavior of uniformly recurrent regular vectors, and for such vectors it follows from the Angle Lemma and subsequent remarks that $C(v)$ is a neighborhood of v in $S_pF(v)$, the unit vectors of $T_pF(v)$.

Under the hypothesis that $\tilde{M}$ admits a compact quotient manifold M it is shown in [BBS, Section 2 and Proposition 4.9] that Weyl chambers have in general the same nice properties that they have in symmetric spaces. From these properties it is later shown in [BBS, Corollary 4.11] that every p-regular vector in $S\tilde{M}$ is actually regular.

To be more precise we have the following properties:

1) (Convexity) For every $v \in R$ the set $C(v)$ is a spherically convex open subset of a hemisphere in $S_pF(v)$, the (k-1) sphere of unit vectors in $T_pF(v)$, where p is the footpoint of v.
2) (Local Rigidity) For every $v \in R$ there exists a neighborhood $U \subseteq R$ of v such that $C(w)$ is isometric in a canonical way to $C(v)$ for all $w \in U$. (In fact one can show that all Weyl chambers $C(v)$ are isometric as in [Bu S] or one can deduce this fact from the Rigidity Theorem of [Ba 2] and [Bu S]).
3) (Asymptotic Property) Let $v \in R$ and let $q \in \tilde{M}$ be a point such that $v(q)$ is p-regular. Then $C(v)$ is isometric to $C(v(q))$ under the map $w \to w(q)$, $w \in C(v)$. In particular if q lies in the k-flat $F(v)$ then this isometry is ordinary parallel translation in $F(v)$.
4) (Rigidity of strong Stable Manifolds) Let $v \in R$. Then $H^s(w)=H^s(v)$ for all $w \in C(v)$.

If $\tilde{M}$ admits a noncompact quotient manifold M with finite volume and sectional curvature bounded below, then property 3) remains true as stated and properties 1), 2) and 4) are true on a dense open subset R^{**} of R.

Section 5 First application of rank

In this section we discuss two applications of the results of the previous section. We first state the results and then give a sketch of their proof. Again we refer the reader to [BBE] and [BBS] for details.

Theorem A Let M be a compact Riemannian manifold of nonpositive sectional curvature. Then the geodesic flow $\{g^t\}$ in SM is ergodic if and only if the rank of M is 1. Moreover if M has rank $k \geq 2$ then there exists a dense open $\{g^t\}$-invariant subset $R^* \subseteq SM$ and $\{g^t\}$-invariant C^1 functions $f_i : R^* \to \mathbb{R}$, $1 \leq i \leq k-1$, whose gradients are linearly independent on R^*.

Remark: The $\{g^t\}$ invariant functions $f_1,\ldots,f_{k-1}$ are called <u>first integrals</u> of the flow $\{g^t\}$ and the existence of even one nonconstant C^1 first integral implies that the geodesic flow $\{g^t\}$ is not ergodic; in fact $\{g^t\}$ does not even have a dense orbit in SM since continuous first integrals are constant on the closure of any orbit.

<u>Theorem B</u> Let M be a compact Riemannian manifold of nonpositive sectional curvature with rank $k\geq 1$ and universal cover $\tilde{M}$. Let $R=\{v\in S\tilde{M}:r(v)=r(\tilde{M})\}$. Then

1. R is a dense, open $\{g^t\}$-invariant subset of $S\tilde{M}$ and $F(\gamma_v)$ is a k-flat for every $v\in R$.
2. The vectors $v\in SM$ for which γ_v is a smooth closed geodesic are dense in SM.
3. The vectors $v\in SM$ tangent to an immersed, flat, totally geodesic k-torus in M form a dense subset of SM.

Remark: If $k=1$ then assertions 2 and 3 are identical. If $k\geq 2$ then assertion 3 implies assertion 2 but we state both since assertion 2 is of basic interest for any flow. We recall that $F(\gamma_v)$ in assertion 1 denotes the union of all geodesics in $\tilde{M}$ that are parallel to γ_v.

We now sketch proofs of Theorems A and B, beginning with Theorem A. If M has rank 1, then the geodesic flow $\{g^t\}$ in SM is ergodic by [BB] as we already observed in section 2. Conversely suppose that the geodesic flow in SM is ergodic, and let k denote the rank of M. If $k\geq 2$ then the existence of the independent first integrals $f_1,\ldots,f_{k-1}$ as described in the statement of Theorem A contradicts the ergodicity of the geodesic flow. Hence $k=1$. However, the construction of the first integrals $f_1,\ldots,f_{k-1}$ is somewhat complicated and uses all of the properties of Weyl chambers as developed in [BBS]. One can actually show as in [BBE, Theorem 4.5] that $k=1$ if $\{g^t\}$ is ergodic, without considering first integrals or even Weyl chambers. One proceeds as follows. Since $\{g^t\}$ is ergodic we know that $R\subseteq SM$ is an open subset of full measure and the maximal exponent $\lambda:SM\to[0,\infty)$ is constant almost everywhere in SM. Moreover we can choose a subset $L\subseteq R$ of full measure in SM with the following properties:

1) $L=-L$
2) $\lambda|L$ is a constant function λ'
3) Every vector $v\in L$ is uniformly recurrent.

It is not difficult to show that there is a vector $v \in L$ such that $L \cap S_{\pi(v)}F(v)$ has full measure in $S_{\pi(v)}F(v)$, the unit sphere at $\pi(v)$ in $T_{\pi(v)}F(v)$. It follows from the Angle Lemma that there exists a positive number $\alpha = \alpha(\lambda')$ such that

$$H^s(w_1) = H^s(w_2)$$

if w_1, w_2 lie in $L \cap S_{\pi(v)}F(v)$ and $\measuredangle(w_1, w_2) < \alpha$ (This is a special case of the stability of strong stable manifolds and does not require the elementary properties of Weyl chambers). Moving within $L \cap S_{\pi(v)}F(v)$ from v to $-v$ in steps of angular length at most α we conclude that

$$H^s(v) = H^s(-v) \quad ,$$

but this contradicts the fact that $H^s(v) \cap H^s(-v) = \{\pi(v)\}$ since both $H^s(v), H^s(-v)$ are submanifolds of $H(v)$, $H(-v)$ orthogonal to $F(v)$ and $H(v) \cap H(-v) \subseteq F(v)$.

We now sketch the construction of the first integrals $f_1, \ldots, f_{k-1}$ in the case that M has rank $k \geq 2$. To avoid technical details and complications we assume that M has no Euclidean de Rham factor and we omit any discussion of the construction of the dense open $\{g^t\}$-invariant subset $R^* \subseteq R \subseteq SM$ on which the functions $\{f_i\}$ are defined. Assume therefore that we have somehow constructed a "nice" dense open subset R^* of the regular vectors R in $S\tilde{M}$ that is invariant under the geodesic flow $\{g^t\}$ and the action of isometries of $\tilde{M}$. If v is a vector in R^* then the fact that the Weyl chamber $C(v)$ is a spherically convex open subset of a hemisphere in $S_{\pi(v)}F(v)$ means that $C(v)$ has a unique "center of gravity" $m(v)$ that depends continuously on v. The local rigidity property of Weyl chambers implies that for each $v \in R^*$ we can find an open neighborhood $\tilde{U} \subseteq R$ of v, an integer N and a positive number δ such that for each vector w in $\tilde{U}$ there are exactly N Weyl chambers $C_1(w), \ldots, C_N(w)$ in $S_{\pi(w)}F(w)$ whose volumes are at least δ. If $m_i(w)$ denotes the center of gravity of $C_i(w)$ then we define functions $f_i : \tilde{U} \to \mathbb{R}$, $1 \leq i \leq N$, by $\tilde{f}_i(v) = \langle v, m_i(v) \rangle$. Replacing $\tilde{U}$ by its image under the geodesic flow $\{g^t\}$ and isometries of M we obtain extended functions $\tilde{f}_1, \ldots, \tilde{f}_N : \tilde{U} \to \mathbb{R}$ that are invariant under $\{g^t\}$ and isometries of $\tilde{M}$. Hence the functions $\{\tilde{f}_i\}$ induce $\{g^t\}$-invariant C^1 functions $\{f_i\} : U \to \mathbb{R}$, where U is the $\{g^t\}$ invariant open set $\rho_*(\tilde{U}) \subseteq SM$, and $\rho : \tilde{M} \to M$ is the covering projection. We now take $k-1$ of the functions $\{f_i\}$ with linearly independent gradients and do this consistently over a family $\{U_\alpha\}$ of $\{g^t\}$-invariant overlapping open subsets of R to obtain k-1 first integrals $f_1,\ldots,f_{k-1}:R^*\to\mathbb{R}$ for the geodesic flow $\{g^t\}$.

We begin the sketch of the proof of Theorem B. The proof of assertion 1 has already been sketched in the previous section in the discussion of the construction of k-flats. Assertion 2 follows from an application of the Closing Lemma (see Lemma 4.5 of [BBS]), which in this case says roughly the following: Let $v \in R\subseteq SM$ be a vector such that $g^t v$ is close to v for some large positive number t. Then there exists a vector v' in R close to v and a number t' close to t such that $g^{t'}(v')=v'$. The lemma applies in particular to the uniformly recurrent (or even just recurrent) vectors v in R, which are dense in SM, and it follows that the periodic vectors v' in R are dense in SM.

To sketch the proof of assertion 3 we introduce the notion of a Γ-compact k-flat in $\tilde{M}$, where $\Gamma\subseteq I(\tilde{M})$ is the deck transformation group of $M=\tilde{M}/\Gamma$. Given a k-flat F in $\tilde{M}$ we let Γ_F denote $\{\phi \in \Gamma:\phi(F)=F\}$, the stabilizer of F in Γ. A k-flat F is said to be Γ-compact if Γ_F is a uniform lattice in F, that is, F/Γ_F is a smooth compact flat manifold of dimension of k. If $\tilde{M}$ is a symmetric space of noncompact type, then it is known that the Γ-compact k-flats are dense in all k-flats; that is, given a k-flat F in $\tilde{M}$ there exists a sequence of Γ-compact flats F_n that converges to F uniformly on compact subsets (see [Mos, Lemma 8.3]). In [BBS] one uses ergodic theory and geometry to prove that this result is still true if $\tilde{M}$ is the universal cover of an arbitrary compact manifold M of nonpositive sectional curvature and rank $k\geq 2$. More precisely it is shown by Lemmas 4.5 and 4.7 of [BBS] that the k-flat F(v) is Γ-compact if $v\in R\subseteq S\tilde{M}$ is the lift of a periodic vector in $R\subseteq SM$, and by the discussion of assertion 2 the periodic vectors of R are dense in SM. If F(v) is a Γ-compact k-flat then $\rho(F(v))\subseteq M$ is an immersed compact flat k-manifold, where $\rho:\tilde{M}\to M$ is the projection. By a theorem of Bieberbach any compact flat k-manifold has a finite Riemannian covering by a flat k-torus and hence $\rho(F(v))$ is also an isometrically immersed flat, totally geodesic k-torus. This completes the sketch of assertion 3 and hence of Theorem B.

Section 6 The Rank Rigidity Theorem

The properties of k-flats and Weyl chambers built up in [BBE] and [BBS] proved to be sufficient to yield a striking characterization of irreducible symmetric spaces of rank at least two among the class of all complete manifolds of finite volume and bounded nonpositive sectional curvature. Independent proofs have been obtained by W. Ballmann [Ba 2] and K. Burns - R. Spazier [Bu S] although the methods used are quite different and the proof of Burns-Spatzier is at present valid only for compact manifolds.

Theorem C (Rank Rigidity Theorem) Let M be a complete manifold of finite volume and sectional curvature $-a^2 \leq K \leq 0$ for some constant $a>0$. Let $\tilde{M}$ denote the universal Riemannian cover of M, and suppose that $\tilde{M}$ is irreducible and of rank $k \geq 2$. Then $\tilde{M}$ is isometric to a symmetric space of noncompact type.

Here $\tilde{M}$ irreducible means that $\tilde{M}$ is not a Riemannian product $\tilde{M}_1 \times \tilde{M}_2$ of two manifolds of positive dimension. In particular $\tilde{M}$ has no Euclidean de Rham factor.

Using Proposition 4.1 of [E 5]one obtains from Theorem C the following corollary:

Theorem D Let M be a complete manifold of finite volume and sectional curvature $-a^2 \leq K \leq 0$. If $\tilde{M}$ denotes the universal Riemannian cover of M, then $\tilde{M}$ is a Riemannian product $\tilde{M}_e \times \tilde{M}_s \times \tilde{M}_1 \times \ldots \times \tilde{M}_r$, where $\tilde{M}_e$ is a flat Euclidean space, $\tilde{M}_s$ is a symmetric space of noncompact type and $\tilde{M}_i$ has rank 1 for every $1 \leq i \leq r$. (Of course, any r+1 of the r+2 factors above may be missing).

In practice this corollary allows one to prove many results about compact nonpositively curved manifolds M by reducing separately to the cases that M is locally symmetric or M has rank 1. In each of these two special cases many techniques are available and a great deal is known about the geometric structure of M and $\tilde{M}$.

A key role in Ballmann's proof of the Rank Rigidity Theorem is played by the following holonomy theorem due to M. Berger [Be] which follows also from later work of J. Simons [Sim]:

Theorem (Berger) Let N be an irreducible Riemannian manifold, and let Φ_x denote the holonomy group at a point x in M. Then either Φ_x acts transitively on the unit sphere in T_xM or M is a locally symmetric space of rank $k \geq 2$.

We now give a brief sketch of Ballmann's proof of the Rank Rigidity Theorem. Let $\tilde{M}$ admit a quotient M of finite volume and bounded sectional curvature $-a^2 \leq K \leq 0$. On the set $R \subseteq S\tilde{M}$ of regular vectors we define a function $f: R \to \mathbb{R}$ by f(v)=cosine of the angular distance in $T_{\pi(v)}\tilde{M}$ between v and the boundary of its Weyl chamber $C(v)$. Ballmann observed that if M is a symmetric space of noncompact type and rank $k \geq 2$, then f is C^1 on R and invariant under parallel translation along arbitrary curves in M; that is, if $\gamma: [a,b] \to \tilde{M}$ is any smooth curve with $\gamma(a) = \pi(v)$, then $f(v(b)) = f(v)$, where v(b) denotes the parallel translate of v along γ to $\gamma(b)$. Remarkably, by using the results of [BBE] and [BBS] and in particular the structure theory of Weyl chambers for vectors in $S\tilde{M}$, Ballmann was able to show that the function f defined above is C^1 on a suitable domain R_0 if $\tilde{M}$ satisfies the hypotheses of the theorem, and moreover f is invariant under parallel translation of vectors in R_0 along arbitrary curves of $\tilde{M}$. If M is compact then the domain R_0 of f is the set of all vectors asymptotic to some vector in R, while if M is noncompact then R_0 is somewhat smaller than the set just described. In either case R_0 is a $\{g^t\}$-invariant dense open subset of $S\tilde{M}$, and if v is any vector in R_0 then the parallel translate of v along any curve in $\tilde{M}$ remains in R_0. Finally since f is invariant under parallel translation any orbit of a vector v in $R_0 \cap T_pM, p \in M$, under the holonomy group Φ_p is contained in a level set of f. Since f is nonconstant on $T_pM \cap R_0$ the holonomy theorem of Berger implies that $\tilde{M}$ is locally symmetric. Hence $\tilde{M}$ is symmetric since it is simply connected.

Section 7 A general rigidity problem

If M is a compact Riemannian manifold of nonpositive sectional curvature, then the homotopy group $\pi_k(M)$ is zero for each $k \geq 2$ since $\tilde{M}$ is diffeomorphic to Euclidean space. Hence two such manifolds M_1, M_2 are homotopically equivalent if and only if they have isomorphic fundamental groups. In this context we define a geometric property to be rigid if whenever it holds for a compact manifold M of nonpositive sectional curvature it also holds for every other compact manifold M' of nonpositive sectional curvature that is homotopically equivalent to M.

A problem of interest is to find rigid geometric properties in the category of compact manifolds of nonpositive sectional curvature. The problem itself is suggested by the striking Rigidity Theorem of Mostow, which states that if two compact locally symmetric manifolds of nonpositive sectional curvature have no hyperbolic planes as (local) de Rham factors and have isomorphic fundamental groups then they are isometric up to constant multiples of the metric.

The most ambitious attempt to carry out this rigidity program is to describe rigid geometric properties of M in terms of algebraic properties of the fundamental group of M. To find potential rigid geometric properties it is reasonable to consider any result that relates the geometry of a compact nonpositively curved manifold M to the algebra of $\pi_1(M)$. One of the first results of this nature was a theorem of A. Preissmann [Pr] that says if M is a compact manifold with strictly negative sectional curvature, then every abelian subgroup of $\pi_1(M)$ is infinite cyclic. It is interesting to ask whether a compact manifold M admits a Riemannian metric of strictly negative sectional curvature if it admits a metric of nonpositive sectional curvature and if every abelian subgroup of $\pi_1(M)$ is infinite cyclic. An affirmative answer to this unresolved question would show that the property of admitting a metric of strictly negative sectional curvature is a rigid geometric property in the category of compact nonpositively curved manifolds.

It follows from the Preissmann theorem that if M is compact with nonpositive sectional curvature and if $\pi_1(M)$ admits a subgroup isomorphic to $\mathbb{Z}^k$ for some integer $k\geq 2$ (for example if M is a Riemannian product manifold $M_1\times M_2$) then M cannot admit a Riemannian metric with strictly negative sectional curvature. It was shown by Wolf [W 2] and Lawson-Yau [LY] that if M is compact with nonpositive sectional curvature and if $\pi_1(M)$ admits a subgroup isomorphic to $\mathbb{Z}^k$ for some integer $k\geq 2$, then this subgroup is the fundamental group of a totally geodesic flat k-torus T^k isometrically immersed in M. Hence the property of admitting a totally geodesic, isometric immersion of a flat k-torus is a rigid geometric property of M. If M is flat then it has a finite Riemannian covering by an n-torus and hence $\pi_1(M)$ admits a finite index subgroup isomorphic to $\mathbb{Z}^n$. It follows from the result described above that being flat ($K\equiv 0$) is a rigid geometric property among compact nonpositively curved manifolds that can be described by algebraic conditions in π_1.

In the early 1970's Lawson-Yau [LY] and Gromoll-Wolf [GW] proved that if M is compact with nonpositive sectional curvature and if $\pi_1(M)$ is a direct product $A_1\times A_2$ and has trivial center, (or more generally if M has no Euclidean factor and $\pi_1(M)=A_1\times A_2$), then $\tilde{M}$ is a Riemannian product $M_1\times M_2$, where $\pi_1(M)=A_1$ for i=1,2. This splitting result is perhaps the most striking early example of the relationship between the geometry of M and the algebra of $\pi_1(M)$. Combined with the main theorem of [E2] it follows that the property of admitting no Euclidean de Rham factor and being isometric to a Riemannian product is also a rigid geometric property that can be described by algeraic conditions in the fundamental group.

This splitting result, which plays a key role in the proofs of Theorems E and F below, has recently been extended by V. Schroeder in [Sc 1] to all manifolds (compact or noncompact) with nonpositive curvature and finite volume.

Other rigid geometric properties of the class of compact nonpositively curved manifolds M than can be described by algeraic properties of $\pi_1(M)$ include the following:

1. The dimension of the Euclidean de Rham factor of M, which equals the rank of the unique maximal abelian normal subgroup of $\pi_1(M)$ [E 2].
2. The rank of M, which equals the algebraic rank of $\pi_1(M)$ (see Theorem E of section 8)
3. The property of being an irreducible Riemannian quotient of a symmetric space of noncompact type and rank at least two (see Theorem F of section 8)
4. Various geometric properties of the free homotopy classes of closed curves of M (see [E 4]).

A result clearly fitting into this general framework is the following rigidity example of [Sc 2].

<u>Theorem</u> Given a positive integer n and positive constants b>a consider the class of complete Riemannian manifolds of dimension n with sectional curvatures satisfying $-b^2 \leq K \leq -a^2 < 0$. Then for a manifold M in this class the property of being noncompact and having finite Riemannian volume can be described by algebraic conditions on the fundamental group of M.

There exist other known rigid geometric properties of the class of compact nonpositively curved manifolds that at the moment cannot be described by algeraic conditions on $\pi_1(M)$. Two of these rigid geometric properties are the following:

5. The property of being a "Visibility" manifold [E 4] (see also [E0] for a definition)
6. The number and dimensions of the non-Euclidean de Rham factors of $\tilde{M}$ [E 2,6].

It is still possible that the properties 5 and 6 can be described by algebraic data in $\pi_1(M)$. For example, one could demonstrate this for property 5 by proving the conjecture that a compact manifold M of nonpositive sectional curvature is a Visibility manifold if and only if $\pi_1(M)$ satisfies the Preissmann property: every abelian subgroup of $\pi_1(M)$ is infinite cyclic. An interesting offshoot of problem 6 would be to identify the de Rham factors of a compact locally symmetric manifold M of nonpositive sectional curvature by properties of $\pi_1(M)$. For example, can one tell just by looking at $\pi_1(M)$ that M is an irreducible quotient of the Riemannian product of $k \geq 2$ hyperbolic planes?

The present nonalgebraic method used for proving the existence of rigid geometric properties involves the use of "quasi-isometries", which we now describe. If M,M* are homotopically equivalent manifolds of nonpositive sectional curvature with universal Riemannian covers $\rho:\tilde{M}\to M$ and $\rho^*:\tilde{M}^*\to M^*$ then any homotopy equivalence $f:M\to M^*$ lifts to a "quasi-isometry" $\tilde{f}:\tilde{M}\to\tilde{M}^*$; that is, $\rho^*\circ\tilde{f}=f\circ\rho$ and there exist positive numbers k, r such that

$$d(\tilde{f}(p),\ \tilde{f}(q)) \leq k d(p,q) \qquad \text{for all } p,q \in \tilde{M}$$

and

$$d(\tilde{f}(p),\ \tilde{f}(q)) \geq (1/k)\ d(p,q)$$

for all $p,q \in \tilde{M}$ with $d(p,q)\geq r$. The quasi-isometry $\tilde{f}$ allows one to make correspondences between various kinds of geometric objects in $\tilde{M}$ and those in $\tilde{M}^*$. This technique was first used by M. Morse [Mor 2] and has since then been used in modified form by many authors including Mostow in his proof of the rigidity theorem [Mos, section 13]. Morse showed that if g_1,g_2 are two Riemannian metrics on $\mathbb{R}^2$ with nonpositive curvature such that $\tilde{M}^*=(\mathbb{R}^2,g_1)$ has $K\equiv-1$ and $1/c\leq g_1/g_2\leq c$ for some $c>0$, then there exists a positive constant A such that for every geodesic γ in $\tilde{M}=(\mathbb{R}^2,g_2)$ there exists a unique geodesic γ^* in $\tilde{M}^*$ whose Hausdorff distance in $\tilde{M}^*$ from γ is at most A. If the curvature in $\tilde{M}$ is also strictly negative, then the correspondence $\gamma\to\gamma^*$ between the geodesics of $\tilde{M},\tilde{M}^*$ is a bijection. This bijective correspondence of geodesics also holds in arbitrary dimensions on $\mathbb{R}^n$ for metrics g_1,g_2 of strictly negative (possibly nonconstant) sectional curvature. In this case one can even define a homeomorphism of the unit tangent bundles $S\tilde{M}$, $S\tilde{M}^*$

that carries geodesic flow orbits into geodesic flow orbits and commutes with isometries [Gr 2]. The geodesic correspondence just described for two negatively curved metrics on $\mathbb{R}^n$ carries over if one has any quasi-isometry $\tilde{f}$ between the universal Riemannian covers $\tilde{M},\tilde{M}^*$ of M, M*. Quasi-isometries are used to prove the rigid properties 5 and 6 stated above.

The method of quasi-isometries is used in part to prove a rigidity theorem of Gromov [Gr 1], [GS] which generalizes but uses the rigidity theorem of Mostow. See also [E 6] for a special case of this result.

Theorem (Gromov) Let M, M* be compact manifolds of nonpositive sectional curvature with isomorphic fundamental groups. Suppose that M* has no finite Riemannian cover that splits as a Riemannian product and that its universal Riemannian cover is a symmetric space of noncompact type and rank $k \geq 2$. Then M* and M are isometric up to constant multiples of the metric of M*.

In particular it follows from this result that on a locally symmetric manifold (M*,g*) without Euclidean de Rham factor and with rank $k \geq 2$ there is an essentially unique metric of nonpositive sectional curvature: the given locally symmetric metric and constant multiples of it when restricted to each local de Rham factor. This is one way to establish rigidity property 3 as stated above. However, this proof fails to describe property 3 in terms of explicit algebraic conditions on the fundamental group. A description of property 3 along algebraic lines is given in the next section, and this description yields a proof of the Gromov rigidity theorem as a corollary.

Section 8 Algebraic rank of the fundamental group

Let G be an abstract group such that every nonzero element has infinite order. For example we may let G be the fundamental group of a complete manifold of nonpositive sectional curvature. For such groups G we can define an abstract "algebraic rank" and it is the main result of [BE] that if M is compact with nonpositive sectional curvature, then the geometric rank of M, described earlier in terms of Jacobi vector fields on M, equals the algeraic rank of $\pi_1(M)$. This result has several strong consequences as we shall see below.

The algebraic rank of a group without torsion elements may be regarded as the "generic rank" of the centralizer in G of a random element ϕ in G. More precisely for each integer $k\geq 1$ let $A_k(G)=\{\phi \in G: Z(\phi)$ contains $\mathbb{Z}^r$ as a finite index subgroup for some $1\leq r\leq k\}$. Here $Z(\phi)=\{\psi\in G:\phi\psi=\psi\phi\}$, the centralizer of ϕ in G. Now let $\alpha(G)=\min\{k\geq 1: G=\bigcup_{i=1}^{m} \phi_i\cdot A_k(G)\}$, where $\{\phi_1,\phi_2,\ldots,\phi_m\}$ is a finite subset of G. Finally define

$$r(G)=\max\{\alpha(G^*): G \text{ is a finite index subgroup of } G\}.$$

The integer r(G) is the <u>algebraic rank</u> of G and has the following properties:

1) $r(G)=r(G^*)$ if G^* is any finite index subgroup of G
2) $r(G_1\times\ldots\times G_k) \leq \sum_{i=1}^{k} r(G_i)$ with equality if $r(G_i)\geq 2$ for at most one integer i.
3) $r(\mathbb{Z}^k\times G) = k+r(G)$ for any integer $k\geq 1$.

For an arbitrary group G without torsion elements r(G) and $\alpha(G)$ may be infinite, but r(G) is always finite if $G=\pi_1(M)$, where M has finite volume and bounded nonpositive sectional curvature. Prasad and Raghunathan in Theorem 3.9 of [PR] showed that $\alpha(\pi_1(M))=r(M)$, the geometric rank of M, if M is a compact quotient of a symmetric space $\tilde{M}$ of noncompact type. Our definition of the algebraic rank is clearly based upon this result. Unfortunately one cannot use $\alpha(G)$ as a definition of algebraic rank if one wishes the algeraic rank to be invariant under passage to a finite index subgroup (or, on a practical level, if one wishes to prove Theorem E below). For example, if G is the fundamental group of a Klein bottle and if G^* is the index 2 subgroup isomorphic to $\mathbb{Z}\times\mathbb{Z}$, then $\alpha(G^*)$ $=r(G^*)=r(G)=2$, but $\alpha(G)=1$. In fact if one regards G as a discrete group of isometries of $\mathbb{R}^2$, then $A_1(G)$ is the set of orientation reversing isometries of G and $Z(\phi)$ is infinite cyclic for each $\phi\in A_1(G)$. It follows that $\alpha(G)=1$ since $G=A_1(G)\cup\phi\cdot A_1(G)$, where $\phi\in A_1(G)$ is arbitrary.

Combining Theorems C and D above we now obtain

<u>Theorem E</u> $r(\pi_1(M))$, the algebraic rank of $\pi_1(M)$, equals r(M), the geometric rank of M for any compact manifold M of nonpositive sectional curvature.

Remarks: 1. If M has no Euclidean de Rham factor, then the result above remains true for manifolds M of finite volume and bounded nonpositive sectional curvature.

2. There are other possible definitions of the algebraic rank that also yield the theorem above, but at the moment these look somewhat clumsier than the definition chosen. Nevertheless it is not clear that our definition is even the best possible one of its kind. For example, it would be more satisfying to define $A_k(G)=\{\phi \in G:Z(\phi)$ contains $\mathbb{Z}^k$ as a finite index subgroup$\}$. However, it is not clear even in locally symmetric spaces M if the corresponding definition of the algebraic rank of $\pi_1(M)$ would equal the geometric rank of M.

As an immediate corollary of Theorem E and Theorem A we obtain the following

<u>Corollary E</u> In the category of compact manifolds of nonpositive sectional curvature the following are rigid geometric properties determined by algebraic data in the fundamental group of M:

1. The geometric rank of M
2. The property of having ergodic geodesic flow in the unit tangent bundle SM.

Using the Rank Rigidity Theorem of Ballmann and Burns-Spatzier and Theorem E we obtain the following:

<u>Theorem F</u> Let M be a complete manifold with finite volume and sectional curvature $-a^2 \leq K \leq 0$ for some positive constant a. Then the following properties are equivalent:

1. The universal cover $\tilde{M}$ is a symmetric space of noncompact type with rank $k \geq 2$. Moreover, M is irreducible; that is, no finite covering of M is a Riemannian product manifold.
2. The fundamental group $\pi_1(M)$ satisfies the following conditions:

   a) $\pi_1(M)$ contains no proper abelian normal subgroup
   b) $\pi_1(M)$ contains no direct product subgroup $A \times B$ of finite index.
   c) The algebraic rank of $\pi_1(M)$ is $k \geq 2$.

Condition 2a) is equivalent to the condition that $\tilde{M}$ have no Euclidean de Rham factor by the main theorem of [E 2]. If M is compact, then condition 2b) implies condition 2a) by the remark just made and Corollary 2 of [E 2]. Using the splitting theorem of Gromoll-Wolf and Lawson-Yau one concludes that conditions 2a) and 2b) hold simultaneously if and only if $\tilde{M}$ has no Euclidean de Rham factor and M is irreducible. If in addition $\tilde{M}$ is a symmetric space of noncompact type with rank $k \geq 2$, then condition 2c) is satisfied by Theorem 3.9 of [PR] or Theorem E. Conversely, if conditions a, b, c of 2) are satisfied then 1) holds by the previous discussion, Theorems C and E, and Proposition 4.5 of [E 5].

We conclude this paper by deriving the Gromov rigidity theorem as a corollary of the result above. One may also derive this as a corollary of Theorem C.

Corollary F (Gromov rigidity theorem)
(see the end of section 7 for a formulation).

Proof The hypotheses on M* imply that $\pi_1(M^*)$ satisfies conditions 2a), 2b) and 2c). Since $\pi_1(M)$ also satisfies these conditions M is locally symmetric by Theorem F. The result now follows from the Mostow rigidity theorem.

Appendix

Table of Contents

Part I General Information

Part II Symmetric spaces of noncompact type

The material in this appendix is known and can be found somewhere in the literature though perhaps not in this form, with the possible exception of the discussion on Jacobi vector fields. On the other hand, the basic facts about Riemannian symmetric spaces most directly relevant to differential geometers are widely scattered throughout the standard sources and are sometimes presented in a way that is unnecessarily difficult. This appendix is an attempt to give a concise exposition of some useful elementary facts. Proofs of assertions here are often omitted. General references for this section are the following: [Hel: pp. 121-125; 156-159; 173-174; 205; 214-219]; [Ka: 55-79], [O: pp. 219-232; 300-331], [Mi: pp. 109-117] and [Mos: pp. 10-30].

Part I General information

Riemannian symmetric spaces

In the first half of this appendix we describe some basic properties of an arbitrary Riemannian symmetric space, and in the second half we specialize to the case of a symmetric space of noncompact type, which is complete and simply conncted with nonpositive sectional curvature.

A (connected) Riemannian manifold M is said to be symmetric if for each point p in M there is an isometry s_p of M that fixes p and whose differential map at p is $-I$ on T_pM. Symmetric manifolds are complete since all geodesics can be extended over $\mathbb{R}$ by using the maps s_p.

Since s_p^2 and I have the same value and derivative at p we have $s_p^2=I$ on M and $(s_p o\gamma)(t)=\gamma(-t)$ for all geodesics γ with $\gamma(0)=p$ and for all numbers t. The isometry s_p is called the geodesic symmetry at p.

Symmetric spaces as coset manifolds G/H

If $G=I_0(M)$, the connected component of the isometry group I(M) that contains the identity, then G is a Lie group in the compact-open topology and moreover G acts transitively on M. Given distinct points p and q in M let $\gamma:[0,1]\to M$ be a geodesic with $\gamma(0)=p$ and $\gamma(1)=q$. For each $t\in\mathbb{R}$ let s_t denote the geodesic symmetry $s_{\gamma(t)}$ and let $p_t=s_{t/2}os_0$. The curve $\{p_t\}$ is a 1-parameter group of isometries of M such that $p_0=I$ and $p_1(p)=q$. This shows that G(p)=M. The isometries $\{p_t\}$ are called transvections and have the property that $dp_t:T_{\gamma(s)}M\to T_{\gamma(s+t)}M$ is exactly parallel translation along γ from $\gamma(s)$ to $\gamma(s+t)$.

Now fix a point $p\in M$ and let $H=G_p=\{g\in G:gp=p\}$. Then H is a compact subgroup of G, and M is diffeomorphic to the coset manifold G/H under the map $\Phi:G/H\to M$ given by

$$\Phi(gH) = g(p)$$

If <,> is the metric on M and if Φ^*<,> denotes the pullback metric on G/H that makes Φ an isometry the Φ^*<,> is left invariant; that is, the transformation $\zeta(g):g^*H\to gg^*H$ is an isometry of $(G/H,\Phi^*<,>)$.

The point $p\in M$ also defines an involutive automorphism $\sigma_p:G\to G$ given by $\sigma_p(g)=s_p o g o s_p^{-1}=s_p o g o s_p$. If F is the closed subgroup of G consisting of the elements fixed by σ_p, then it is elementary to show that $F_0\subseteq H\subseteq F$, where F_0 denotes the identity component of F.

Examples

Symmetric spaces can be obtained from coset spaces G/H that satisfy the conditions above. More precisely, let G be a connected Lie group that admits an involutive automorphism σ(i.e. $\sigma^2=I$). Let H be a compact subgroup of G such that $F_0\subseteq H\subseteq F$, where F denotes the subgroup of G consisting of elements fixed by σ. Then the coset space G/H admits left invariant Riemannian metrics and G/H is a Riemannian symmetric space with respect to any such metric. Moreover, the geodesic symmetry S at p=eH is given by $S\circ\pi=\pi\circ\sigma$, where $\pi:G\to G/H$ is the projection.

A proof of this result can be found in [Hel: pp. 174-175]. Actually H can be any closed subgroup of G such that $Ad_G(H)$ is a compact subgroup of $GL(\mathfrak{g})$, where $\mathfrak{g}$ denotes the Lie algebra of G and $Ad_G:G\to GL(\mathfrak{g})$ is the adjoint representation [Hel: pp. 116-120].

We now present some classical examples of coset spaces that satisfy the hypotheses of the result just stated.

Example 1 Let $G=SL(n,\mathbb{R})=\{n\times n$ matrices A over $\mathbb{R}$ with det A=1}. Let $H=SO(n,\mathbb{R})=\{A\in G:A\cdot A^t=I\}$. Let $\sigma:G\to G$ be given by $\sigma(g)=(g^t)^{-1}$. Then $F=Fix(\sigma)=SO(n,\mathbb{R})$ and $H=F_0=F$. The coset space G/H can be identified with the set M_n of all n×n positive definite symmetric matrices over $\mathbb{R}$ with determinant 1. M_n with a canonical metric is the "largest" of the symmetric spaces M with nonpositive sectional curvature and no Euclidean de Rham factor; if M is any such space then there exists a natural map $F:M\to M_n$, $n=\dim I_0(M)$, such that F is a diffeomorphism onto a totally geodesic submanifold of M_n and the metric on M pulled back by F agrees with the metric on M up to a constant multiple λ_i on each irreducible de Rham factor M_i. For further details see the discussion later in this section.

Example 2 Let G be a closed self adjoint subgroup of $SL(n,\mathbb{R})$ (i.e. $G=G^t$) and let $H=G\cap SO(n,\mathbb{R})$. Let $\sigma:G\to G$ be also given by $\sigma(g)=(g^t)^{-1}$. Then G/H becomes a Riemannian symmetric space and a totally geodesic submanifold of $SL(n,\mathbb{R})/SO(n,\mathbb{R})$ if the sectional curvature of G/H is nonpositive.

Example 2a $G=SO(p,q)$, $p+q=n$, where $SO(p,q)=\{A\in SL(n,\mathbb{R}):A$ leaves invariant the symmetric bilinear form $\phi:\mathbb{R}\times\mathbb{R}\to\mathbb{R}$ given by

$\phi(x,y)=-\sum_{i=1}^{p}x_iy_i+\sum_{j=p+1}^{n}x_jy_j$. Alternately, $SO(p,q)=\{A\in SL(n,\mathbb{R}):A^tSAS=I\}$,

where $$S=\left(\begin{array}{c|c}-I_p & 0\\ \hline 0 & I_q\end{array}\right)$$

Example 2b $G=Sp(n,\mathbb{R})=\{A\in SL(2n,\mathbb{R}):A$ leaves invariant the symmetric bilinear form $\phi:\mathbb{R}^{2n}\times\mathbb{R}^{2n}\to\mathbb{R}$ given by

$\phi(x,y)=\sum_{i=1}^{n} x_i y_{n+i}-\sum_{j=1}^{n} y_j x_{n+j}\}$. Alternately $Sp(n,\mathbb{R})=\{A\in SL(2n,\mathbb{R}):A^t SAS=I\}$, where $S=\left(\begin{array}{c|c} 0 & -I_n \\ \hline I_n & 0 \end{array}\right)$.

Example 3 (Spheres) $G=SO(n+1)$, $H=\{1\}\times SO(n)=\{g\in SO(n+1):ge_1=e_1\}$, where $e_1=(1,0,\ldots,0)$ is the first standard basis vector in $\mathbb{R}^{n+1}$. Let $\sigma:G\to G$ be given by $\sigma(g)=SgS^{-1}=SgS$ where S is the diagonal matrix $\mathrm{diag}\,(-1,\underbrace{1,\ldots 1}_{n})$. The coset space G/H with a left invariant metric is a standard n-sphere with a metric of constant sectional curvature.

Example 4 (Oriented Grassmann manifolds) Let p,q be positive integers with $n=p+q$. Let $G=SO(n)$, $H=SO(p)\times SO(q)$ where $SO(p)\times SO(q)=\{A\in GL(n,\mathbb{R}):$

$$A=\left(\begin{array}{c|c} A_1 & 0 \\ \hline 0 & A_2 \end{array}\right) \text{ with } A_1\in SO(p),\ A_2\in SO(q)\}\ .$$

Let S be the diagonal matrix $\mathrm{diag}\ (\underbrace{-1,-1,\ldots,-1}_{p},\underbrace{1,\ldots,1}_{q})$, and let $\sigma:G\to G$ be given by $\sigma(g)=SgS^{-1}=SgS$. The coset space G/H may be identified with the collection of oriented p-dimensional subspaces of $\mathbb{R}^n=\mathbb{R}^{p+q}$.

The orthogonal involutive Lie algebra associated to a symmetric space

Let $p\in M$ be a point and $\sigma_p:G\to G$, $G=I_o(M)$, be the involutive automorphism $\sigma_p(g)=s_p\circ g\circ s_p$, where s_p is the geodesic symmetry at p. Identifying the Lie algebra $\mathfrak{g}$ with T_eG we obtain an involutive Lie algebra automorphism $\Theta_p=d\sigma_p:\mathfrak{g}\to\mathfrak{g}$ characterized by the equation

$$\sigma_p(e^{tX})=e^{t(\Theta_p(X))} \quad \text{for all } X\in\mathfrak{g} \text{ and } t\in\mathbb{R}$$

where e^{tX} denotes the 1-parameter subgroup of G determined by X. Since $\Theta_p^2=I$ we may write

(1) $\mathfrak{g}=\mathfrak{h}+\mathfrak{m}$ (direct sum)

where $\mathfrak{h}=\{X\in\mathfrak{g}:\Theta_p(X)=X\}$ and $\mathfrak{m}=\{X\in\mathfrak{g}:\Theta_p(X)=-X\}$. The fact that Θ_p preserves brackets implies that

(2) $[\mathfrak{h},\mathfrak{h}]\subseteq\mathfrak{h}$, $[\mathfrak{h},\mathfrak{m}]\subseteq\mathfrak{m}$ and $[\mathfrak{m},\mathfrak{m}]\subseteq\mathfrak{h}$

The first relation shows that h is a subalgebra and in fact is the Lie algebra corresponding to $H=\{g \in G: gp=p\}$.

The point p induces a map $p: G \to M$ given by $p(g)=g(p)$, and identifying g with T_eG we obtain a homomorphism $dp: g \to T_pM$ whose kernel is precisely h. In particular $dp: m \to T_pM$ is an isomorphism, and we may define an inner product Q on m by

$$(3) \quad Q(X,Y) = \langle dp(X), dp(Y) \rangle_p$$

where $\langle , \rangle_p$ is the inner product in T_pM. From the definition of H it follows that

$$Q(Ad(\phi)X, Ad(\phi)Y) = Q(X,Y)$$

for all $X, Y \in m$ and $\phi \in H$. ($Ad(\phi): g \to g$ equals dc_ϕ, where $c_\phi: G \to G$ is given by $c_\phi(g)=\phi g \phi^{-1}$. This is the adjoint representation of G). Differentiating the constant function $t \to Q(Ad(e^{tZ})X, Ad(e^{tZ})Y)$, $Z \in h$, $X \in m$, $Y \in m$ we obtain the relation

$$(4) \quad Q(ad(Z)X,Y) + Q(X,ad(Z)Y) = 0$$

for all $Z \in h$ and all $X, Y \in m$($ad(h)$-invariance). (Here $adZ: g \to g$ is given by $adZ(X)=[Z,X]$. It is known that $Ad(e^{tX})=e^{tadX}$). Now extend Q arbitrarily to an $ad(h)$-invariant inner product on g so that $Q(h,m)=0$.

Definition A finite dimensional Lie algebra g over $\mathbb{R}$ together with an involutive Lie algebra automorphism Θ and an $ad(h)$-invariant inner product Q on g satisfying $Q(h,m)=0$ (cf. 1), 2) and 4)) is called an orthogonal involutive Lie algebra. These were studied and classified by E. Cartan. For a recent treatment see [Hel: pp. 193-205] or [W 1: pp. 232-247]. We outline some of the structure theory of such algebras below.

Geodesics and curvature tensor

Let $p \in M$ be a point, and let $\sigma_p: G \to G$ and $\{\Theta_p, h, m\}$ be the data of the orthogonal involutive Lie algebra determined by p. We exploit the isomorphism $dp: m \to T_pM$ to describe some of the geometry of M. Since M is homogeneous much of its geometry can be described by what happens at p or algebraically by what happens in m.

Geodesics Every geodesic γ in M with $\gamma(0)=p$ determines a 1-parameter family of transvections $\{p_t\}$, where $p_t=s_{t/2}\circ s_0$ and s_t denotes the geodesic symmetry $s_{\gamma(t)}$. Moreover for all $s,t \in \mathbb{R}$ we have $p_t\gamma(s)=\gamma(t+s)$ and $dp_t:T_{\gamma(s)}M \to T_{\gamma(s+t)}M$ is just parallel translation along γ. It is not difficult to show that $p_t=e^{tX}$ for some $X \in m$ and hence we obtain

(5) The geodesics γ of M with $\gamma(0)=p$ are given by $\gamma(t)=e^{tX}(p)$, $X \in m$.

If M arises as a coset space G/H with a left invariant metric, where $\sigma:G \to G$ is an involutive automorphism with $F_0 \subseteq H \subseteq F = \mathrm{Fix}(\sigma)$, then

(5') The geodesics γ of G/H with $\gamma(0)=eH$ are given by $\gamma(t)=e^{tX}H$ where $X \in m = \{Z \in g : \Theta(Z)=-Z\}$ and $\Theta = d\sigma : g \to g$.

Curvature tensor of M

Identifying T_pM with m by means of the isomorphism $dp: m \to T_pM$ the curvature tensor of M at p is given by

(6) $R(X,Y)Z = -\mathrm{ad}[X,Y](Z) = -[[X,Y],Z]$

for all X, Y, $Z \in m$. For a proof see [W 1: pp. 245-246].

From this expression for the curvature tensor one can derive formulas for sectional curvature. See [W 1: p. 246] or [Hel: pp. 205-206]. In particular one obtains

(7) A locally irreducible symmetric space M has constant Ricci curvature (i.e. M is an Einstein manifold). In fact, the Ricci tensor equals - 1/2 the restriction of the Killing form to m.

Killing form and classification of symmetric spaces

Given a finite dimensional Lie algebra g (here over $\mathbb{R}$) one has the Killing form

$$B : g \times g \to \mathbb{R}$$

given by $B(X,Y) = \text{Trace}\,(\text{ad}X \circ \text{ad}Y)$.

It is easy to show

(8) $B(\Theta X, \Theta Y) = B(X,Y)$
and $B(\text{ad}Z(X),Y) = -B(X,\text{ad}Z(Y))$

for all X, Y, $Z \in g$ and any Lie algebra automorphism Θ of g.

Given a point p in a symmetric space M let $G = I_0(M)$ and let $g = h + m$ be the decomposition of g into eigenspaces of the involution $\Theta_p : g \to g$. Let B_m denote the restriction of the Killing form B to $m \times m$

(9) M is said to be of <u>compact</u> (<u>noncompact</u>, <u>Euclidean</u>) type if B_m is <u>negative definite</u> (<u>positive definite</u>, <u>identically zero</u>).

Since $H = \{g \in G : gp = p\}$ is compact it follows that B restricted to $h \times h$ is always negative definite. Moreover from the bracket relations (2) between h and m it follows that $B(h,m) = 0$. This definition is independent of the point p chosen.

From (6) and (7) above it follows that M has sectional curvature $K \geq 0$ and positive Ricci curvature if M is of the compact type; M has sectional curvature $K \leq 0$ and negative Ricci curvature if M is of the noncompact type; M has zero sectional curvature (and hence zero Ricci curvature) if M is of the Euclidean type. If M is of the Euclidean type then M is isometric to Euclidean space. If M is of the noncompact type, then M is simply connected and diffeomorphic to Euclidean space. If M is of the compact type then M is compact and the fundamental group of M is finite and abelian. If M is an arbitrary simply connected symmetric space, then M is a Riemannian product $M_0 \times M_1 \times M_2$, where M_0 is a Euclidean space, M_1 is a symmetric space of noncompact type and M_2 is a symmetric space of compact type (see [W 1: p. 243]).

Semisimple Lie algebras and the decomposition of orthogonal involutive Lie algebras

A Lie algebra g is said to be semisimple if the Killing form B is non-degenerate on $g \times g$. A Lie algebra g is simple if it is semisimple and contains no proper ideals. For semisimple Lie algebras one has the following structure result (see [Hel: pp. 121-122])

(10) Proposition Let g be a finite dimensional semisimple Lie algebra (over $\mathbb{R}$). Then g has a direct sum decomposition

$$g = g_1 \oplus g_2 \oplus \cdots \oplus g_m$$

such that

a) $\{g_1,\ldots,g_m\}$ are ideals of g that are also simple Lie algebras

b) Every ideal of g is a direct sum of some subcollection of the $\{g_i\}$

c) $B(g_i,g_j)=\{0\}$ if $i\neq j$

d) $[g_i,g_j] = \{0\}$ if $i\neq j$

Moreover

e) $z(g) = \{X \in g : [X,Y] = 0 \text{ for all } Y \in g\} = \{0\}$.

Now consider an arbitrary symmetric space M and let p be a point of M. Let $\{g,\Theta_p,h,m,Q\}$ be the orthogonal involutive Lie algebra data determined by p as described above in (1), (2), (3) and (4). We consider the ad(h)-invariant inner product Q restricted to m (see (4)).

Let $m_0=\{X \in m : B(X,Y)=0 \text{ for all } Y \in m\}$, where B is the Killing form of g. By the skew symmetry of adX on g relative to B (property (8)) it follows that m_0 is invariant under ad(h). Since ad(h) is a family of skew symmetric transformations of m relative to Q (property (4)) it follows that there exists an ad(h)-invariant subspace m^* of m such that

$$m_0 + m^* = m \quad \text{(Q-orthogonal direct sum)}$$

Proceeding inductively we write

$$m^* = m_1^* + \dots + m_k^* \quad \text{(Q-orthogonal direct sum)}$$

where each m_i^* is ad(h)-invariant and has no proper ad(h)-invariant subspace. This can be done by property (4) above. This decomposition of m^* is in fact unique up to order and the decomposition

$$m = m_0 + m_1^* + \dots + m_k^*$$

corresponds precisely to the decomposition of T_pM under the holonomy group of M at p.

Now define $g_0=h_0+m_0$, where $h_0=\{X \in h: \text{ad}X(m) \subseteq m_0\}$. For $1 \leq i \leq k$ we define $g_i=h_i+m_i^*$, where $h_i=[m_i^*,m_i^*]$. The Lie algebra g_0 is Euclidean; that is, $[m_0,m_0]=\{0\}$ and in fact $[m_0,m]=\{0\}$. Moreover we have the following

(11) Proposition The Lie algebra g is a direct sum $g=g_0+g_1+\dots+g_k$ and moreover

a) each g_i, $0 \leq i \leq k$, is a Θ_p-invariant ideal of g. If $i \neq j$ then $B(g_i,g_j)=Q(g_i,g_j)=0$ and $[g_i,g_j]=\{0\}$.

b) g_0 is the unique maximal Θ_p-invariant Euclidean ideal of g.

c) For $1 \leq i \leq j$ either the ideal g_i is a simple Lie algebra of noncompact type or $g_i=a+a$, where a is a simple Lie algebra of compact type. In each case $B=\lambda_i Q$ restricted to $m_i^* \times m_i^*$, where g_i is a nonzero constant. The ideal g_i is of compact (noncompact) type if $\lambda_i<0$ $(\lambda_i>0)$.

d) If $g_0=\{0\}$, then g is semisimple. The direct sum decomposition $g=g_1+\dots+g_k$ is precisely the decomposition of (10) above if each g_i is of noncompact type.

de Rham decomposition of symmetric spaces

If M is any Riemannian manifold, then the holonomy group Φ_p of M at p decomposes T_pM into an orthogonal direct sum $T_pM=V_0+V_1+\dots+V_k$, where $V_0=\{v \in T_pM: \phi(v)=v \text{ for all } \phi \in \Phi_p\}$. The spaces $V_1,\dots,V_k$ are uniquely determined up to order. If M is simply connected then we obtain integrable distributions $V_0,V_1,\dots,V_k$ on M and M splits as a Riemannian product $M=M_0 \times M_1 \times \dots \times M_k$ where M_0 is a Euclidean space, the Euclidean de Rham factor of M. If M is a symmetric space, then each of the

M_i, $1 \le i \le k$ is also a symmetric space. If M is a simply conncted symmetric space with no Euclidean factor M_0, then the holonomy group Φ_p at p consists precisely of the differential maps of elements of H, where $H=\{g \in G: gp=p\}$. Finally, if $M=M_0 \times M_1 \times \ldots \times M_k$ is an arbitrary simply connected symmetric space, then setting $G=I_0(M)$, $G_0=I_0(M_0)$, $G_i=I_0(M_i)$ for $1 \le i \le k$ we have $G=G_0 \times G_1 \times \ldots \times G_k$ (direct product) and $\mathfrak{g}=\mathfrak{g}_0+\mathfrak{g}_1+\ldots+\mathfrak{g}_k$ (direct sum), where $\mathfrak{g}$ denotes the Lie algebra of $\mathfrak{g}$ and $\mathfrak{g}_i$ the Lie algebra of G_i for $0 \le i \le k$. This direct sum decomposition of $\mathfrak{g}$ is precisely the decomposition obtained above in (11).

Totally geodesic submanifolds of a symmetric space

Let M be a Riemannian symmetric space with $G=I_0(M)$, and let $\mathfrak{g}=\mathfrak{h}+\mathfrak{m}$ be the decomposition of $\mathfrak{g}$ induced by a point $p \in M$ and the corresponding involution $\Theta_p: \mathfrak{g} \to \mathfrak{g}$. A subspace $\mathfrak{m}^*$ of $\mathfrak{m}$ is said to be a Lie triple system if $[X,[Y,Z]] \in \mathfrak{m}^*$ for all elements X, Y, $Z \in \mathfrak{m}^*$. Useful examples of Lie triple systems include the following

1) $\mathfrak{m}^*$ is an abelian subspace of $\mathfrak{m}$; that is, $[X,Y]=0$ for all $X,Y \in \mathfrak{m}^*$
2) $\mathfrak{m}^*=\{X \in \mathfrak{m}: [X,Y]=0$ for some fixed $Y \in \mathfrak{m}\}$

(12) Proposition Let M^* be any complete, totally geodesic submanifold of M that contains $p \in M$. Let $\mathfrak{m}^* \subseteq \mathfrak{m}$ be the subspace corresponding to T_pM^* under the isomorphism $dp: \mathfrak{m} \to T_pM$. Then $\mathfrak{m}^*$ is a Lie triple system and $M^*=\exp(\mathfrak{m}^*)(p)=\{e^X(p): X \in \mathfrak{m}^*\}$. Conversely if $\mathfrak{m}^* \subseteq \mathfrak{m}$ is any Lie triple system, then $M^*=\exp(\mathfrak{m}^*)(p)$ is a complete, totally geodesic submanifold of M.

In particular of $\mathfrak{m}^* \subseteq \mathfrak{m}$ is an abelian subspace, then $M^*=\exp(\mathfrak{m}^*)(p)$ is a flat ($K \equiv 0$) complete, totally geodesic submanifold by the formula (6) for the curvature tensor of M (and hence by restriction to $\mathfrak{m}^*$, of M^*).

Proof (See also [Hel: pp. 189-191]). Let M^* be totally geodesic, and let $X,Y,Z \in \mathfrak{m}^*$ be arbitrary elements. Since M^* is totally geodesic at p its curvature tensor is the restriction to T_pM^* of the curvature tensor of M at p. In particular $R(X,Y)Z=-[[X,Y]Z] \in \mathfrak{m}^*$ by (6) and $\mathfrak{m}^*$ is a Lie triple system. Conversely let $\mathfrak{m}^*$ be a Lie triple system and let

$\mathfrak{g}^*=\mathfrak{h}^*+\mathfrak{m}^*$, where $\mathfrak{h}^*=[\mathfrak{m}^*,\mathfrak{m}^*]\subseteq\mathfrak{h}$ The fact that $\mathfrak{m}^*$ is a Lie triple system means that $\mathfrak{g}^*$ is a subalgebra of $\mathfrak{g}$, $[\mathfrak{h}^*,\mathfrak{h}^*]\subseteq\mathfrak{h}^*$, $[\mathfrak{h}^*,\mathfrak{m}^*]\subseteq\mathfrak{m}^*$ and $[\mathfrak{m}^*,\mathfrak{m}^*]\subseteq\mathfrak{h}^*$. Let G^* denote the connected Lie subgroup of $G=I_0(M)$ that corresponds to $\mathfrak{g}^*$, and let $H^*=\{g\in G^*:gp=p\}$. Then $M^*=G^*(p)$ is diffeomorphic to the coset space G^*/H^* and hence is a submanifold of M. If $X\in\mathfrak{m}^*$, then $t\to e^{tX}(p)$ is a geodesic of M contained in M^* and hence M^* is totally geodesic at p and equal to $\exp(\mathfrak{m})(p)$. Finally M^* is totally geodesic at all of its points since G^* operates transitively on M^* and carries geodesics of M to geodesics of M.

A useful criterion for finding complete totally geodesic submanifolds of M is given by the following

(13) Proposition Let M be a symmetric space and let $G=I_0(M)$. Let $p\in M$ be a point, and let $\sigma_p:G\to G$ be the involutive automorphism determined by p. Let G^* be a connected Lie subgroup of G invariant under σ_p. Then $M^*=G^*(p)$ is a complete totally geodesic submanifold of M.

Proof If $H^*=\{g\in G^*:gp=p\}$, then H^* is a closed subgroup of G^*, and M^* is diffeomorphic to the coset space G^*/H^* and hence is a submanifold of M. Now let $\mathfrak{g}^*$ denote the Lie algebra of G^*. The fact that G^* is invariant under σ_p means that $\mathfrak{g}^*$ is invariant under $\Theta_p=d\sigma_p:\mathfrak{g}\to\mathfrak{g}$. Hence

$$\mathfrak{g}^* = \mathfrak{h}^* + \mathfrak{m}^*$$

where $\mathfrak{g}^*\cap\mathfrak{h}=\mathfrak{h}^*$ and $\mathfrak{g}^*\cap\mathfrak{m}=\mathfrak{m}^*$. The subspace $\mathfrak{m}^*$ corresponds to T_pM^* under the isomorphism $dp:\mathfrak{m}\to T_pM$. We can now verify directly that $\mathfrak{m}^*$ is a Lie triple system since it follows from the bracket relations (2) for $\mathfrak{h},\mathfrak{m}$ that $[\mathfrak{h}^*,\mathfrak{h}^*]\subseteq\mathfrak{h}^*$, $[\mathfrak{h}^*,\mathfrak{m}^*]\subseteq\mathfrak{m}^*$ and $[\mathfrak{m}^*,\mathfrak{m}^*]\subseteq\mathfrak{h}^*$. We can also use the argument above to show directly that $M^*=G^*(p)$ is a complete, totally geodesic submanifold of M.

(14) If $\mathfrak{g}^*$ is a subalgebra of $\mathfrak{g}$ invariant under Θ_p, then the connected Lie subgroup G^* of G with Lie algebra $\mathfrak{g}^*$ is invariant under σ_p. Hence $M^*=G^*(p)$ is a complete, totally geodesic submanifold of M.

Part II Symmetric spaces of noncompact type

For the remainder of this appendix we consider only symmetric spaces M of the noncompact type, which as we remarked earlier are simply connected with nonpositive sectional curvature and no Euclidean de Rham factor.

Let $p \in M$ be a point and let σ_p, Θ_p be the corresponding involutions of $G, \mathfrak{g}$. In this context it is customary to write

$$\mathfrak{g} = \mathfrak{k} + \mathfrak{P}$$

that is, we use $\mathfrak{k}, \mathfrak{P}$ instead of $\mathfrak{h}, \mathfrak{m}$ for the eigenspaces of Θ_p. This is called a Cartan decomposition of $\mathfrak{g}$. We also use $K = \{g \in G : gp = p\}$.

Relationship to semisimple Lie groups

If M is a symmetric space of noncompact type, then $G = I_0(M)$ is a semi-simple Lie group. If $\mathfrak{g} = \mathfrak{g}_1 + \ldots + \mathfrak{g}_k$ is the direct sum decomposition (10) into simple ideals of $\mathfrak{g}$, then each $\mathfrak{g}_i$ is noncompact; that is, B is not negative definite on each $\mathfrak{g}_i$.

Conversely if $\mathfrak{g}$ is a finite dimensional semisimple Lie algebra over $\mathbb{R}$ and if each of its simple ideals $\mathfrak{g}_i$ in the decomposition (10) is non-compact, then there exists a symmetric space M of noncompact type such that $\mathfrak{g}$ is the Lie algebra of $G = I_0(M)$.

The inner product ϕ_p on $\mathfrak{g}$

Given a symmetric space M of noncompact type and a point p of M we consider the corresponding Cartan decomposition $\mathfrak{g} = \mathfrak{k} + \mathfrak{P}$ together with the bracket relations (2): $[\mathfrak{k}, \mathfrak{k}] \subseteq \mathfrak{k}$, $[\mathfrak{k}, \mathfrak{P}] \subseteq \mathfrak{P}$ and $[\mathfrak{P}, \mathfrak{P}] \subseteq \mathfrak{k}$. Since M is of non-compact type we know by definition that the Killing form B is negative definite on $\mathfrak{k}$, positive definite on $\mathfrak{P}$ and that $B(\mathfrak{k}, \mathfrak{P}) = 0$.

We define $\phi_p : \mathfrak{g} \times \mathfrak{g} \to \mathbb{R}$ by

$$(16) \quad \phi_p(X,Y) = -B(\Theta_p X, Y).$$

Equivalently $\phi_p = B$ on $\mathfrak{P}$, $\phi_p = -B$ on $\mathfrak{k}$ and $\phi_p(\mathfrak{P}, \mathfrak{k}) = 0$. From the skew symmetry of adX on $\mathfrak{g}$ relative to B (property (8)) and the bracket relations (2) we obtain

(17) a) adX is symmetric on $\mathfrak{g}$ relative to ϕ_p for every $X \in \mathfrak{P}$

b) adX is skew symmetric on $\mathfrak{g}$ relative to ϕ_p for every $X \in \mathfrak{k}$

c) The inner product ϕ_p is invariant under Ad(K).

The example $SL(n,\mathbb{R})/SO(n,\mathbb{R})$

For each integer $n \geq 2$ let $P(n,\mathbb{R})$ denote the set of all positive definite, symmetric n×n matrices with determinant 1. The group $SL(n,\mathbb{R})$ acts transitively on $P(n,\mathbb{R})$ as follows: given $g \in SL(n,\mathbb{R})$ and $A \in P(n,\mathbb{R})$ let $g(A) = gAg^t$. The subgroup of $SL(n,\mathbb{R})$ that fixes the identity $I_n \in P(n,\mathbb{R})$ is precisely $SO(n,\mathbb{R})$, and hence we may identify $P(n,\mathbb{R})$ with the coset space $SL(n,\mathbb{R})/SO(n,\mathbb{R})$.

If M_n is the space of n×n symmetric matrices with trace zero, then the matrix exponential $\exp: M_n \to P(n,\mathbb{R})$ is a bijection where $\exp(A) = e^A = I + \sum_{k=1}^{\infty} \frac{A^k}{k!}$. Hence $P(n,\mathbb{R})$ is naturally diffeomorphic to a Euclidean space of dimension $n(n+1)/2$. We use exp to identify M_n with $T_{I_n} P(n,\mathbb{R})$; $X \in M_n$ is associated with the initial velocity of $t \to \exp(tX)$. Any $SO(n,\mathbb{R})$-invariant metric on M_n determines an $SL(n,\mathbb{R})$-invariant metric on $P(n,\mathbb{R})$ and $P(n,\mathbb{R})$ becomes a Riemannian symmetric space of noncompact type in this metric. Here $SO(n,\mathbb{R})$ acts on M_n by conjugation. The geodesics of $P(n,\mathbb{R})$ that start at I_n in any such metric are precisely the curves $t \to \exp(tX)$ for $X \in M_n$.

For a further discussion of the geometry of $P(n,\mathbb{R})$ see section 3 of [Mos]. We mention here only one $SO(n,\mathbb{R})$-invariant inner product on M_n, and in fact this inner product and scalar multiples of it are the only $SO(n,\mathbb{R})$-invariant inner products on M_n. Let

(18) $Q(X,Y) = \text{Trace}\ (X \cdot Y)$

define an inner product on M_n. This is just the restriction of the Euclidean inner product $(X,Y) \to \text{Trace}\ (X \cdot Y^t)$ on $\mathbb{R}^{n^2}$. Using the exponential map we obtain an inner product on $T_{I_n}(P(n,\mathbb{R}))$.

The imbedding theorem

The next result shows that every symmetric space M of noncompact type can be realized isometrically as a complete, totally geodesic submanifold of $P(n,\mathbb{R})$, $n=\dim I_0(M)$, provided that one multiplies the metric on each irreducible de Rham factor M_i of M by a suitable positive constant λ_i.

(19) Proposition Let (M,g_0) be a symmetric space of noncompact type, and let $p \in M$ be a point. Let ϕ_p be the induced inner product on $\mathfrak{g}$, the Lie algebra of $G=I_0(M)$. Then

1) Ad(G) is a self adjoint subgroup of $SL(\mathfrak{g})$ relative to the adjoint operation on $\mathfrak{g}$ determined by ϕ_p
2) If $n=\dim G$, let $P(n,\mathfrak{g},\phi_p)$ denote the corresponding space of $n\times n$ positive definite symmetric matrices of determinant 1, equating $(\mathfrak{g},\phi_p)$ with $(\mathbb{R}^n, \langle , \rangle)$. Let $P(n,\mathfrak{g},\phi_p)$ be given the symmetric space structure arising as above from the inner product $Q(X,Y)=\text{Trace}(X\cdot Y)$ on M_n. Then

a) The map $F_p : M \to P(n,\mathfrak{g},\phi_p)$ given by $F_p(g(p))=\text{Ad}(g)\text{Ad}(g)^t$, $g\in G$, is a diffeomorphism onto a complete, totally geodesic submanifold of $P(n,\mathfrak{g},\phi_p)$.
b) If g^* is the metric on M pulled back by F_p from the metric on $P(n,\mathfrak{g},\phi_p)$, then on each irreducible de Rham factor M_i of M the metric g^* is a constant multiple $\lambda_i g_0$ of the original metric g_0 for some positive constant λ_i.

Remark: If $q=g_0^*(p)$, $g_0^*\in G$, then a computation shows that given $x\in M$ we have $F_q(g_0^* x)=\text{Ad}(g_0^*)\cdot F_p(x)\cdot \text{Ad}(g_0^{*-1})$. Hence the image $F_p(M)$ is uniquely determined up to conjugacy by an element of Ad(G).

Proof This result does not seem to be stated in the literature in precisely this form (although compare [Mos: section 2.6, pp. 14-16]). We therefore provide a brief sketch. To see that Ad(G) is a subgroup of $SL(\mathfrak{g})$ we use the nondegeneracy of B on $\mathfrak{g}\times\mathfrak{g}$ to choose a basis $E_1,\dots,E_n$ of $\mathfrak{g}$ such that $B(E_i,E_j)=\varepsilon_i\delta_{ij}$, where $\varepsilon_i=\pm 1$ for each $1\le i\le n$. Hence for $X\in\mathfrak{g}$ we have trace (adX) =

$$= \sum_{i=1}^{n} \varepsilon_i B(\text{ad}X(E_i),E_i)=0$$ by the skew symmetry of adX relative to B.

Therefore $\det(\mathrm{Ad}(e^{tX}))=\det(e^{t\,\mathrm{ad}X})=e^{t(\mathrm{Trace\ ad}X)}=1$ for all $X\in g$, $t\in\mathbb{R}$ and it follows that $\mathrm{Ad}(G)\subseteq SL(g)$. Note also that G has trivial center: for any central element ψ in $G=I_0(M)$ the function $d_\psi:p\to d(p,\psi p)$ is constant on M since G is transitive on M. Since M has no Euclidean de Rham factor it follows by [W 2] that ψ must be the identity. Hence

$$\mathrm{Ad} : G\to SL(g)$$

is an injection.

To show that Ad(G) is self adjoint (i.e. $\mathrm{Ad}(G)=\mathrm{Ad}(G)^t$) relative to the inner product ϕ_p on g, observe that $\mathrm{ad}(g)=\mathrm{ad}(k)+\mathrm{ad}(P)$ is the Lie algebra of Ad(G). By the discussion above $\mathrm{ad}(k)$ consists of skew symmetric transformations and $\mathrm{ad}(P)$ of symmetric transformations on g relative to ϕ_p. Hence $\mathrm{ad}(g)$ is invariant under the involution $\Theta^*:Z\to -Z^t$ of g^* into g^* where g^* is the Lie algebra of $G^*=SL(g)$. Hence Ad(G), which is generated by $\exp(\mathrm{ad}(g))$, is invariant under the adjoint operation.

Finally we consider $F_p:M\to P(n,g,\phi_p)$ given by $F_p(g(p))=\mathrm{Ad}(g)\mathrm{Ad}(g)^t$. Note that Ad(K) leaves ϕ_p invariant since it leaves the Killing form B invariant and leaves the subspaces k and P of g invariant. Hence $\mathrm{Ad}(K)\subseteq SO(g,\phi_p)$ and the map F_p is well defined. Regarding $SL(g)=G^*$ as $I_0(P(n,g,\phi_p))$ the involution σ of G^* determined by the point $I_n\in P(n,g,\phi_p)$ is given by $\sigma(g^*)=[(g^*)^t]^{-1}$ and has $SO(g,\phi_p)$ as its fixed point set. The group Ad(G) is self adjoint in $SL(g)$ and hence invariant under σ. It follows from (13) that the orbit $\mathrm{Ad}(G)(I_n)=F_p(M)$ is a complete totally geodesic submanifold of $P(n,g,\phi_p)$.

We compute the differential map of $F=F_p$. Given $q\in M$ and vectors $v,w\in T_qM$ we choose $g\in G$ and $X,Y\in P$ so that $g(p)=q$, v is the initial velocity of $s\to ge^{sX}(p)$ and w is the initial velocity of $s\to ge^{sY}(p)$. (Here $g=k+P$ is the Cartan decomposition determined by $p\in M$). Hence $F_*(v)$ is the initial velocity of $s\to \mathrm{Ad}(g)\mathrm{Ad}(e^{sX})\mathrm{Ad}(e^{sX})^t\mathrm{Ad}(g)^t$ with a similar expression for $F_*(w)$. The map $Z\to \mathrm{Ad}(g)Z\,\mathrm{Ad}(g)^t$ is an isometry of $P(n,g,\phi_p)$ and $\mathrm{Ad}(e^{sX})=e^{s\,\mathrm{ad}X}$ is symmetric since $X\in P$ and hence adX is symmetric relative to ϕ_p. It follows that

$$\langle F_*(v),\ F_*(w)\rangle \;=\; 4\ \mathrm{Trace}\ (\mathrm{ad}X\circ\mathrm{ad}Y) \;=\; 4\ B(X,Y)$$

by definition of the inner product $\langle\ ,\ \rangle$ on $P(n,g,\phi_p)$ obtained from (18). Hence F is nonsingular since B is positive definite on P.

One can show directly without difficulty that F is injective, and hence F is a diffeomorphism onto its image. Finally let $g^*=F^*\langle,\rangle$ denote the metric pulled back by F and induced by the Ad(K)-invariant inner product 4B on $\mathcal{P}$. The metric g^* is invariant under the action of $G=I_0(M,g_0)$ as one can see from the fact that $F_p(gx)=\mathrm{Ad}(g)F_p(x)\mathrm{Ad}(g)^t$ and hence (M,g^*) is also a Riemannian symmetric space of noncompact type.

Let Q_0 be the ad(k)-invariant metric on $\mathcal{P}$ induced by the original metric g_0 of M. Let $\mathcal{P}=\mathcal{P}_1+\ldots+\mathcal{P}_k$ be the decomposition of $\mathcal{P}$ into irreducible ad(k)-invariant subspaces described above in (11) and the discussion preceding it. By (11)a and (11)c it follows that $B(\mathcal{P}_i,\mathcal{P}_j)=Q_0(\mathcal{P}_i,\mathcal{P}_j)=0$ if $i\neq j$ and $B=\lambda_i Q_0$ on $\mathcal{P}_i$, where λ_i is a positive constant (M has noncompact type). Since $g^*=F^*\langle,\rangle$ is induced by 4B it follows that $g^*=4\lambda_i g_0$ on the subspace of T_pM corresponding to $\mathcal{P}_i$. This completes the proof since the subspaces $\mathcal{P}_i$ correspond to the irreducible factors of M in the de Rham decomposition.

<u>Rootspace decompositions and rank of g</u>

Let M be a symmetric space of noncompact type, and let $g=k+\mathcal{P}$ be a Cartan decomposition determined by a point p in M. If A_1,A_2 are any two maximal abelian subspaces of $\mathcal{P}$, then we shall see below that $\mathrm{Ad}(k)(A_1)=A_2$ for some $k\in K$. If $q=g(p)$ is another point of M, then $G_q=gG_pg^{-1}$ and $g=\mathrm{Ad}(g)k+\mathrm{Ad}(g)\mathcal{P}$ is the Cartan decomposition of g corresponding to q. Hence the dimension of the largest abelian subspace of $\mathcal{P}$ is an integer k_0 that depends neither on the abelian subspace nor on $\mathcal{P}$. This integer k_0 is called the <u>rank of g</u>.

Alternatively we can describe the rank of g as the dimension of the largest flat Euclidean space isometrically imbedded in M as a complete, totally geodesic submanifold. By the transitivity of G on M there is such a largest flat subspace that contains p. Totally geodesic submanifolds M* containing p are described by Lie triple systems m^* contained in $\mathcal{P}$, and by the curvature tensor formula (6) a Lie triple system m^* determines a flat submanifold M* if and only if m^* is abelian.

Now consider a fixed maximal abelian subspace A of $\mathcal{P}$. Since $[A,B]=0$ for all $A,B\in A$ the Jacobi identity and the fact that adX is symmetric on g relative to the canonical inner product ϕ_p imply that the linear operators $\{ad(A): A \in A\}$ form a commutative family of symmetric linear operators relative to ϕ_p. We may therefore decompose g into a direct sum of common eigenspaces $\{g_\alpha : \alpha \in \Lambda\}$ that are orthogonal relative to ϕ_p. Specifically we write

$$(20)\ g = g_0 \oplus \sum_{\alpha \in \Lambda} g_\alpha \qquad (\phi_p\text{-orthogonal direct sum})$$

where by definition $ad(A) \equiv 0$ on g_0 for every $A \in A$ and $ad(A) = \alpha(A)I$ on g_α for every $A \in A$. Here $\alpha(A)$ is defined to be the eigenvalue of $ad(A)$ on g_α. Note that $A \subseteq g_0$ since A is abelian. Moreover $\alpha : A \to \mathbb{R}$ is linear since $ad(A+B) = ad(A) + ad(B)$ and $ad(\lambda A) = \lambda ad(A)$ for all $A, B \in A$ and all $\lambda \in \mathbb{R}$.

(21) Remark: If $X \in A$ then since $Ad(e^{tX}) = e^{t\,ad(X)} = I + \sum_{n=1}^{\infty} \frac{t^n}{n!}[ad(X)]^n$ it follows that $Ad(e^{tX}) = e^{t\alpha(X)} I$ on g_α.

The decomposition (20) is called the <u>rootspace decomposition of g determined by A</u>. The linear functionals $\alpha : A \to \mathbb{R}$ are called the <u>roots</u> determined by A, and Λ denotes the set of these roots. Note that if α is a root then $-\alpha$ is also a root. In fact if $\alpha \in \Lambda$, $Z \in g_\alpha$ and $A \in A$ are given, it is routine to show that $ad(A)(\Theta_p Z) = -\alpha(A)(\Theta_p Z)$, where $\Theta_p : g \to g$ is the Cartan involution determined by p. It follows that $\Theta_p : g_\alpha \to g_{-\alpha}$ is an isomorphism for any nonzero root α, and similarly $\Theta_p : g_0 \to g_0$ is an isomorphism whose +1 eigenspace is $k \cap g_0$ and whose -1 eigenspace is $g_0 \cap P = A$.

The root space decomposition of g is a very useful tool for studying among other things the geometry of symmetric spaces M of noncompact type. We shall see some applications of it in the remainder of this section.

<u>Regular elements and Weyl chambers</u>

Let $g = k + P$ be the Cartan decomposition determined by a point p of M. We say that a nonzero element X of P is a <u>regular</u> element if $Z(X) \cap P$ is an abelian subspace of P, where $Z(X) = \{Y \in g : [X,Y] = 0\}$. Clearly if $Z(X) \cap P$ is abelian, then it is a maximal abelian subspace of P. Hence $X \neq 0$ is a regular element of P if and only if X is contained in a unique maximal abelian subspace of P. If we identify T_pM with P by means of the isomorphism $dp : P \to T_pM$, then we shall see later in the discussion of Jacobi vector fields that any element $X \neq 0$ in P is a regular element of

P if and only if the vector $v=dp(X) \in T_pM$ is a regular vector as defined in section 3.

(22) <u>Proposition</u> Let $A \subseteq P$ be a maximal abelian subspace. Then a nonzero element $X \in A$ is a regular element of P if and only if $\alpha(X) \neq 0$ for every nonzero root $\alpha \in \Lambda$ that occurs in the root space decomposition of g determined by A.

<u>Proof</u> If $X \neq 0$ is a regular element of A, then $Z(X) \cap P = A$ by the discussion above. Given a root $\alpha \in \Lambda$ let $Z=K^*+P^*$ be a nonzero element of g_α, where $K^* \in k$ and $P^* \in P$. By definition $\alpha(A)(K^*+P^*)=ad(A)K^*+ad(A)P^*$ for every $A \in A$, which implies that

$$(*) \qquad \begin{aligned} ad(A)K^* &= \alpha(A)P^* \\ ad(A)P^* &= \alpha(A)K^* \end{aligned}$$

since $[k,P] \subseteq P$ and $[P,P] \subseteq k$. If $\alpha(X)=0$ then by setting $A=X$ in the equations (*) we see that $P^* \in Z(X) \cap P = A$. Hence $\alpha(A)K^*=0$ for all $A \in A$ by the second equation in (*). This implies that $K^*=0$ since $\alpha \not\equiv 0$ on A and it now follows from the first equation in (*) that $P^*=0$, contradicting the hypothesis that $Z=K^*+P^*$ is nonzero. Hence $\alpha(X) \neq 0$ for every $\alpha \in \Lambda$.

Conversely suppose that $\alpha(X) \neq 0$ for all roots $\alpha \in \Lambda$ and some $X \neq 0$ in A. Let $Y \in Z(X) \cap P$ be given and write $Y=Y_0+\sum_{\alpha \in \Lambda} Y_\alpha$, where $Y_0 \in g_0$ and $Y \in g_\alpha$. Then $0=adX(Y)=\sum_{\alpha \in \Lambda} \alpha(X)Y_\alpha$, which shows that $Y_\alpha=0$ for every $\alpha \in \Lambda$. Hence $Y \in g_0 \cap P = A$, which shows that $Z(X) \cap P = A$ and X is regular.

<u>Weyl chambers</u>

Let A be a maximal abelian subspace of P, and let $R(A)$ denote the set of regular elements of A. The result above shows that $R(A)$ is the complement in A of the union of the finite collection of hyperplanes $\alpha=0, \alpha \in \Lambda$. We define an equivalence relation $\sim$ in $R(A)$ by setting

$$(23) \quad A_1 \sim A_2 \quad \text{if } \alpha(A_1)\cdot\alpha(A_2)>0 \quad \text{for every } \alpha \in \Lambda$$

The equivalence classes in $R(A)$ are called the <u>Weyl chambers</u> determined by A, and $C(A)$ will denote the equivalence class or Weyl chamber determined by $A \in R(A)$. It is easy to see that a Weyl chamber $C(A)$ is a connected component of $R(A)$ and is an open subset of a half space in A.

Let $A \in R(A)$ be a unit vector relative to the inner product Q in P given by $Q(X,Y)=\langle dp(X), dp(Y)\rangle$. If $C_1(A)$ denotes the unit vectors of A that lie in $C(A)$, then $C_1(A)$ is an open spherically convex subset of a hemisphere in the unit sphere in A. If $F=\exp(A)(p)$, then F is a flat, totally geodesic submanifold of M by formula (6) for the curvature tensor and the fact that A is a Lie triple system in P. If $v=dp(A) \in T_pM$, then $C(v) \overset{(def)}{=} dpC_1(A)$ is the Weyl chamber determined by v as defined in section 4. From its definition it is clear that $C(v)$ is an open spherically convex subset of a hemisphere in the unit sphere of $T_pF=dp(A)$.

Conjugacy of abelian subspaces and Weyl chambers

The main result that we wish to indicate is the following (compare [Hel: Lemma 6.3, pp. 211-212]):

(24) Proposition Let $A, A^* \subseteq P$ be two maximal abelian subspaces, and let X, X^* be regular elements of A, A^*. Let $K=G_p=\{g \in G : gp=p\}$, and let $f: K \to \mathbb{R}$ be the differentiable function given by

$$f(\phi) = B(\mathrm{Ad}(\phi)X^*, X) .$$

1) If $\phi_0 \in K$ is a critical point of f, then $\mathrm{Ad}(\phi_0)X^* \in A$ and $\mathrm{Ad}(\phi_0)(A^*)=A$.
2) If $\phi_0 \in K$ is a maximum point for f, then $\mathrm{Ad}(\phi_0)X^* \in C(X)$ and $\mathrm{Ad}(\phi_0)C(X^*)=C(X)$.

This result says that any two abelian subspaces or Weyl chambers in P are conjugate by an element of $\mathrm{Ad}(K)$ since K is compact. Moreover since any two Cartan decompositions of g differ by an element of $\mathrm{Ad}(G)$ it follows from 2) that any two Weyl chambers are isometric. This translates directly into the Rigidity Property of Weyl chambers $C(v)$, where v is a unit vector in M, as defined in section 4.

Proof Let $\phi_0 \in K$ be a critical point for $f: K \to \mathbb{R}$. For any $Z \in k$ define the function $g_Z(t)=f(\phi_0 e^{tZ})=B(\mathrm{Ad}(\phi_0 e^{tZ})X^*, X)$. Using the fact that ϕ_0 is a critical point of f and properties of the Killing form B expressed in (8) a routine computation shows that

$$0 = g_Z'(0) = B(\mathrm{Ad}(\phi_0)Z,\ [\mathrm{Ad}(\phi_0)X^*,X])$$

Since Ad(K) leaves P invariant and $[P,P]\subseteq k$ it follows that $[\mathrm{Ad}(\phi_0)X^*,X]\in k$. Since $\mathrm{Ad}(\phi_0):k\to k$ is an isomorphism and $Z\in k$ in the equation above is arbitrary it follows that $[\mathrm{Ad}(\phi_0)X^*,X]=0$ from the fact that B is negative definite on k. Hence $\mathrm{Ad}(\phi_0)X^*\in Z(X)\cap P=A$ since X is a regular element of A. The maximal abelian subspaces $\mathrm{Ad}(\phi_0)(A^*)$ and A both contain $\mathrm{Ad}(\phi_0)X^*$ and hence they must be equal since X^* (and hence $\mathrm{Ad}(\phi_0)X^*$) is a regular element of P by hypothesis.

Now suppose that f has a maximum value at $\phi_0\in K$. We show that $\mathrm{Ad}(\phi_0)X^*$ lies in the Weyl chamber of A determined by X. This implies that $\mathrm{Ad}(\phi_0)C(X^*)=C(\mathrm{Ad}(\phi_0)X^*)=C(X)$. Given $Z\in k$ we define again $g_Z(t)=f(\phi_0e^{tZ})=B(\mathrm{Ad}(\phi_0e^{tZ})X^*,X)$. The fact that f has a maximum at ϕ_0 means that $g_Z'(0)=0$ and $g_Z''(0)\leq 0$ for all $Z\in k$.

We compute $g_Z''(0)$. Now $\mathrm{Ad}(e^{tZ})X^*=e^{t\,\mathrm{ad}Z}(X^*)=X^*+t\,\mathrm{ad}Z(X^*)+\frac{t^2}{2}(\mathrm{ad}Z)^2(X^*)+$ higher order terms in t. It follows that $g_Z(t)=B(\mathrm{Ad}(\phi_0)X^*,X)+$ $t\,B(\mathrm{Ad}(\phi_0)\circ\mathrm{ad}Z(X^*),X)+\frac{t^2}{2}B(\mathrm{Ad}(\phi_0)\circ(\mathrm{ad}Z)^2(X^*),X)+$higher order terms in t. Hence

$$g_Z''(0) = B((\mathrm{ad}Z')^2(X'),X) = B(\mathrm{ad}X\circ\mathrm{ad}X'(Z'),Z')$$

where $Z'=\mathrm{Ad}(\phi_0)Z\in k$ and $X'=\mathrm{Ad}(\phi_0)X^*\in A$ by the first part of the proposition. We use also the properties of (8). Since $Z\in k$ was arbitrary the condition $g_Z''(0)\leq 0$ becomes

(25) $0\geq B(\mathrm{ad}X\circ\mathrm{ad}X'(Z),Z)$ for every $Z\in k$.

Since X,X' belong to A by the first part of the proposition we have $[X,X']=0$ and it follows from the Jacobi identity that the transformations adX,adX' commute. Since both adX and adX' are symmetric relative to B it follows that $T=\mathrm{ad}X\circ\mathrm{ad}X'$ is symmetric on g relative to B. Moreover T leaves invariant both k and P since $[k,P]\subseteq P$ and $[P,P]\subseteq k$. Since B is negative definite on k the eigenvalues of T on k are all real and condition (25) implies that all eigenvalues of T on k are nonnegative.

To show that X,X' determine the same Weyl chamber in A it suffices by the discussion above to show that for every root $\alpha\in\Lambda$ the number

$\alpha(X)\cdot\alpha(X')$ is an eigenvalue of T on k. The numbers $\alpha(X)\cdot\alpha(X')$ are always nonzero by (22) since both X and X' are regular elements of P by hypothesis. Given a root $\alpha \in \Lambda$ and a nonzero element $Z_\alpha \in g_\alpha$ we let $X_\alpha = Z_\alpha + \Theta_p(Z_\alpha)$, where $\Theta_p: g \to g$ is the Cartan involution induced by p. Since Θ_p fixes X_α it follows that $X_\alpha \in k$. Finally since $\Theta_p(Z_\alpha) \in g_{-\alpha}$ it follows that $T(X_\alpha) = \mathrm{ad}X \circ \mathrm{ad}X'(X_\alpha) = \alpha(X)\cdot\alpha(X')X_\alpha$. This concludes the proof of (24).

Distribution of k-flats in M

Given a symmetric space M of noncompact type let k be the rank of g, where g is the Lie algebra of $G=I_0(M)$; that is, k is the dimension of a maximal abelian subspace $A\subseteq P$, where $g=k+P$ is a Cartan decomposition induced by some point p of M. We observed earlier that k is also the largest dimension of a complete, totally geodesic, flat submanifold of M. We call any k-dimensional complete, totally geodesic, flat submanifold of M a k-flat of M.

(26) Proposition Let M be any symmetric space of noncompact type, and let $G=I_0(M)$. Let v be any unit vector tangent to M at a point p. Then there exists a k-flat F through p that contains the geodesic of M determined by v. If F_1, F_2 are two k-flats that contain p, then there exists an element $\phi \in K=G_p=\{g \in G: gp=p\}$ such that $\phi(F_1)=F_2$. Finally if F_1, F_2, are any two k-flats in M and p_1, p_2 are arbitrary points of F_1, F_2 then there exists an element $g \in G$ such that $g(p_1)=p_2$ and $g(F_1)=F_2$.

Remark: The proof will actually show that given two regular unit vectors v_1, v_2 tangent to M at points p_1, p_2 there exists an element $g \in G$ such that $g(p_1)=p_2$ and $(g)_*C(v_1)=C(v_2)$, where $C(v_1), C(v_2)$ denote the Weyl chambers consisting of unit vectors of M that are defined in section 4.

Proof of the proposition Let v be a unit vector at p in M, and let $g=k+P$ be the Cartan decomposition determined by p. Choose $X \in P$ so that $dp(X)=v$ and hence $\gamma(t)=e^{tX}(p)$ is the geodesic of M with initial velocity v. Let $A\subseteq P$ be a maximal abelian subspace that contains X. Let $F=\exp(A)(p)=\{e^Z(p): Z \in A\}$. Then F is a k-flat by the curvature tensor formula (6) since A is a Lie triple system in P. Clearly F contains the geodesic $\gamma(t)=e^{tX}(p)$.

Suppose now that F_1,F_2 are two k-flats that contain p. Choose maximal abelian subspaces A_1,A_2 in P such that $F_1=\exp(A_1)(p)$ and $F_2=\exp(A_2)(p)$. By (24) there exists an element $\phi \in K$ such that $\mathrm{Ad}(\phi)(A_1)=A_2$. This means precisely that $\phi(F_1)=F_2$. The final assertion of (26) now follows immediately from the fact that G operates transitively on M.

Structure of the space $P(n,\mathbb{R})=SL(n,\mathbb{R})/SO(n,\mathbb{R})$

For this canonical example of a symmetric space of noncompact type we describe explicitly the algebraic objects discussed above.

In this case $G=I_0(P(n,\mathbb{R}))=SL(n,\mathbb{R})$ with the action $g(A)=gAg^t$ for $g \in G$, $A \in P(n,\mathbb{R})$, the set of n×n positive definite symmetric matrices of determinant 1. Hence $\mathfrak{g}$ is the set of n×n matrices with real entries and trace zero. A computation shows that the Killing form is given by

$$B(X,Y) = 2n\,\mathrm{Trace}(X\cdot Y)$$

for all $X,Y \in \mathfrak{g}$.

Cartan decompositions

Let $\sigma: SL(n,\mathbb{R}) \to SL(n,\mathbb{R})$ be the involution given by $\sigma(g)=(g^t)^{-1}$. Then $SO(n,\mathbb{R})$ is the fixed point set of σ and $\Theta=d\sigma:\mathfrak{g}\to\mathfrak{g}$ is the map $\Theta(X)=-X^t$. It follows that $\mathfrak{k}=\{X \in \mathfrak{g}:\Theta X=X\}$ consists of the skew symmetric n×n matrices and $P=\{X \in \mathfrak{g}:\Theta X=-X\}$ consists of the symmetric n×n matrices of trace zero. The corresponding Cartan decomposition

$$\mathfrak{g} = \mathfrak{k}+P$$

is precisely the decomposition of a matrix into a sum of symmetric and skew symmetric matrices. All other Cartan decompositions of $\mathfrak{g}$ have the form $\mathfrak{k}^*=g\mathfrak{k}g^{-1}$ and $P^*=gPg^{-1}$ for some $g \in G$.

Canonical inner product corresponding to $\mathfrak{g}=\mathfrak{k}+P$

The canonical inner product ϕ was defined in (16) to be B on P and -B n k. In this case $\phi(X,Y)=2n$ Trace $(X\cdot Y^t)$, which is 2n times the usual uclidean inner product in $\mathbb{R}^{n^2}$.

axima1 abelian subspaces of P

f A^* is an abelian subspace of P, then $0=[A_1,A_2]=A_1A_2-A_2A_1$ for all ele-ents $A_1,A_2 \in A^*$. Since A^* is a commuting family of symmetric n×n matri-es we can find an orthonormal basis $\{f_1,\dots,f_n\}$ of $\mathbb{R}^n$ consisting of ommon eigenvectors of elements of A^*. If $\{e_1,\dots,e_n\}$ denotes the tandard basis in $\mathbb{R}^n$, then choose $\phi \in SO(n,\mathbb{R})$ so that $\phi(f_i)=\pm e_i$ for ll $1\le i\le n$. Clearly $\phi A^*\phi^{-1}$ is an abelian subspace of P that has $e_1,\dots,e_n$ s a basis of common eigenvectors. In other words $\phi A^*\phi^{-1}\subseteq A$, the set of iagonal matrices of trace zero. Clearly A is an abelian subspace of P nd this argument shows that A is a maximal abelian subspace of P. An bvious extension of this argument shows that if A^* itself is a maxi-al abelian subspace of P, then $\phi A^*\phi^{-1}=A$. Hence any two maximal abelian ubspaces of P are conjugate under an element (not unique!) of $SO(n,\mathbb{R})$.

ootspace decompositions and rank

he rank of g is the dimension of A, the diagonal matrices, which equals -1 since every matrix in A must have trace zero.

or $1\le i,j\le n$ let E_{ij} denote the n×n matrix with a 1 in position (i,j) nd zeros elsewhere. Let $H_i=E_{ii}-E_{i+1,i+1}$, $1\le i\le n-1$. Then $\{E_{ij}$, $i\ne j$ and $_1,\dots,H_{n-1}\}$ is a basis for g. If $A\in A$ is the diagonal matrix $\lambda_1,\lambda_2,\ \dots\ ,\lambda_n) = \sum_{i=1}^{n}\lambda_i E_{ii}$, then it is easy to see that

$$\mathrm{ad}(A)(E_{ij}) = (\lambda_i-\lambda_j)E_{ij} \quad \text{if } i\ne j$$

$$\mathrm{ad}(A)(H_i) = 0 \text{ for all } i\ .$$

ence we obtain n(n-1) nonzero roots $i\ne j$ $\{\alpha_{ij}\}$ given by $\alpha_{ij}(A)=A_{ii}-A_{jj}$ nd n(n-1) 1-dimensional root spaces g_{ij} spanned by E_{ij} for $i\ne j$. The pace A is spanned by $\{H_1,\dots,H_{n-1}\}$ and $\mathrm{ad}(A)=\alpha_{ij}(A)I$ on g_{ij} for $A\in A$ nd $i\ne j$. Also $A=g_0$.

Regular elements of P

The regular elements of P are precisely those symmetric matrices in P whose eigenvalues are all distinct. It suffices to verify this statement for a regular diagonal matrix since any regular symmetric matrix in P is conjugate by an element of $SO(n,\mathbb{R})$ to a regular diagonal matrix in P. By (22) a diagonal matrix $A=\mathrm{diag}(\lambda_1,\lambda_2,\ldots,\lambda_n)\in A$ is regular if and only if $\alpha_{ij}(A)=\lambda_i-\lambda_j\neq 0$ for every $i\neq j$. The regularity of A is thus equivalent to the distinctness of the $\{\lambda_i\}$, but these are the eigenvalues of A.

Weyl chambers of A

We show that there is a one-one correspondence between the Weyl chambers of A and the elements of the permutation group on n letters. Let $A=\mathrm{diag}(\lambda_1,\lambda_2,\ldots,\lambda_n)$ and $B=\mathrm{diag}(\mu_1,\mu_2,\ldots,\mu_n)$ be regular elements of A. Since A is regular the $\{\lambda_i\}$ are distinct and there exists a unique permutation σ of n letters such that

$$\lambda_{\sigma(1)} > \lambda_{\sigma(2)} > \ldots > \lambda_{\sigma(n)} .$$

Similarly there exists a unique permutation σ^* of n letters such that

$$\mu_{\sigma^*(1)} > \mu_{\sigma^*(2)} > \ldots > \mu_{\sigma^*(n)} .$$

By definition (23) the condition that A,B determine the same Weyl chamber is precisely the condition that

$$(\lambda_i-\lambda_k)\cdot(\mu_i-\mu_k) > 0$$

for all $i\neq k$. It is not difficult to see that this condition holds if and only if $\sigma=\sigma^*$.

k-flats in $P(n,\mathbb{R})$ $(k=n-1)$

The abelian subspace $A\subseteq P$ consisting of diagonal $n\times n$ matrices with trace zero determines the complete, flat, totally geodesic submanifold $F\subseteq P(n,\mathbb{R})$ of dimension n-1 consisting of all diagonal matrices of determinant 1. All other flats F^* through I_n have the form

$F^*=\phi F\phi^t=\phi F\phi^{-1}$ for some ϕ in $SO(n,\mathbb{R})$, and hence F^* is an abelian group contained in $P(n,\mathbb{R})$. All k-flats in $P(n,\mathbb{R})$ have the form gFg^t for some $g \in SL(n,\mathbb{R})$.

Decomposition of the space of Jacobi vector fields on a geodesic

In a symmetric space one can say a great deal about the structure of the Jacobi vector fields along a given geodesic γ. We shall look at Jacobi vector fields on a geodesic γ with a view to determining explicitly the spaces $J^P(\gamma)$, consisting of parallel Jacobi vector fields on γ, and $J^S(\gamma)$, consisting of "strong stable" Jacobi vector fields on (see section 4 and in particular the construction of the strong stable manifolds). This discussion will also show that regular unit vectors v in T_pM (i.e. $r(v)=r(M)$) correspond to regular elements X in $\mathcal{P}$ under the isomorphism $dp: \mathcal{P}\to T_pM$, where $\mathfrak{g}=\mathfrak{k}+\mathcal{P}$ is the Cartan decomposition determined by p. An alternate approach to the problem of describing Jacobi vector fields on a geodesic γ in a symmetric space M may be found in [Mi: pp. 109-111]. In conjunction with the curvature tensor formula (6) one can also use this approach to describe the spaces of parallel and strong stable Jacobi vector fields.

Our starting point is a description of the Jacobi vector fields on a geodesic γ that arise as a variation vector field of a geodesic variation of γ determined by a 1-parameter subgroup of isometries (i.e. the restriction of a Killing vector field on M to γ). Specifically, given a geodesic γ in M with $\gamma(0)=p$ we write $\gamma(t)=e^{tX}(p)$ for a suitable $X \in \mathcal{P}$, where $\mathfrak{g}=\mathfrak{k}+\mathcal{P}$ is the Cartan decomposition determined by p. Given $X^* \in \mathfrak{g}$ let $J_{X^*}(t)$ be the variation vector field $(\alpha)_*(\partial/\partial s)(t,0)$ on $\gamma(t)=\alpha(t,0)$, where $\alpha:\mathbb{R}\times\mathbb{R}\to M$ is the variation of γ given by

$$\alpha(t,s) = e^{sX^*} \circ e^{tX}(p)$$

It is not difficult to show the following

(27) Proposition a) Given $X^* \in \mathfrak{g}$ the vector field $J_{X^*}(t)$ is the unique Jacobi vector field on γ such that $J_{X^*}(0)=dp(X^*)$ and $J_{X^*}'(0)=dp([X^*,X])$, where ' denotes the covariant derivative along γ and $p:G\to M$ is the map $p(g)=g(p)$.
b) If $J(\gamma)$ denotes the 2n-dimensional space of Jacobi vector fields on γ, then the map

$J: \mathfrak{g} \to J(\gamma)$

given by $J(X^*) = J_{X^*}$ is a linear map whose kernel is $k \cap Z(X)$, where $Z(X) = \{Y \in \mathfrak{g} : adX(Y) = 0\}$.

c) $J(\mathfrak{g})$ is a proper subspace of $J(\gamma)$ of dimension $\leq 2n-k$, where $k = \text{rank}(\mathfrak{g})$, and equality holds if and only if X is a regular element of P.

First decomposition of $J(\mathfrak{g})$

For any element $X \in P$ the transformation $adX: \mathfrak{g} \to \mathfrak{g}$ is symmetric relative to the canonical inner product ϕ_p on $\mathfrak{g} \times \mathfrak{g}$ (see (17)) and $\mathfrak{g}$ is the direct sum of eigenspaces of adX that are orthogonal relative to ϕ_p (and hence to B). Let $\mathfrak{g}_X^+$ (respectively $\mathfrak{g}_X^-$) be the direct sum of all eigenspaces of adX corresponding to positive (respectively negative) eigenvalues. Then

(28) $\mathfrak{g} = Z(X) + \mathfrak{g}_X^+ + \mathfrak{g}_X^-$ (orthogonal direct sum)
and $J(\mathfrak{g}) = J(Z(X) + J(\mathfrak{g}_X^+) + J(\mathfrak{g}_X^-))$.

Second decomposition of $J(\mathfrak{g})$

Let $A \subseteq P$ be a maximal abelian subspace containing the element X such that $e^{tX}(p) = \gamma(t)$. Let $\mathfrak{g} = \mathfrak{g}_0 + \sum_{\alpha \in \Lambda} \mathfrak{g}_\alpha$ be the root space decomposition of $\mathfrak{g}$ determined by A. From elementary properties of the root space decomposition we obtain

(29) Proposition

1) $Z(X) = \mathfrak{g}_0 + \sum_{\alpha(X)=0} \mathfrak{g}_\alpha$

2) $\mathfrak{g}_X^+ = \sum_{\alpha(X)>0} \mathfrak{g}_\alpha$

3) $\mathfrak{g}_X^- = \sum_{\alpha(X)<0} \mathfrak{g}_\alpha$

4) If $X^* \in \mathfrak{g}_\alpha$, then
$J_{X^*}(t) = e^{-t\alpha(X)} E(t)$, where $E(t)$ is parallel along γ.

(30) Corollary

1) The subspace $J(Z(X))$ is precisely the subspace of parallel Jacobi vector fields on γ. Moreover $J(Z(X))$ has minimal dimension if and only if X is a regular element.

2) If $0 \neq X^* \in g_X{}^+$, then $\|J_{X^*}(t)\| \leq ce^{-\lambda t}$ for all $t \geq 0$ and suitable positive constants c, λ.

3) If $0 \neq X^* \in g_X{}^-$, then $\|J_X{}^*(t)\| \geq ce^{\lambda t}$ for all $t \geq 0$ and suitable positive constants c, λ.

4) $J(Z(X)) + J(g_X{}^+) = \{Y \in J(\gamma) : \|Y(t)\| \leq c$ for all $t \geq 0$ and some $c > 0$ depending on $Y\}$

5) $J(Z(X)) + J(g_X{}^-) = \{Y \in J(\gamma) : \|Y(t)\| \leq c$ for all $t \leq 0$ and some $c > 0$ depending on $Y\}$.

The Jacobi vector fields defined in 4) and 5) are called respectively the stable and unstable Jacobi vector fields on γ (compare with the discussion in section 4).

Stability group of a point at infinity

For each $x \in M(\infty)$, where M is a symmetric space of noncompact type and $M(\infty)$ denotes the boundary sphere defined in section 1, we consider the group

$$G_x = \{g \in G = I_0(M) : gx = x\}.$$

The groups G_x are precisely the parabolic subgroups of the semisimple Lie group G and are the elements of the Tits geometry associated to G[T], [Mos, section 16]. We conclude this appendix by using the root space decomposition to describe each G_x and to derive some useful facts about it. See also [Mos, section 2] and [Ka, pp. 77-84].

(31) Proposition Let M be a symmetric space of noncompact type, and let $x \in M(\infty)$ be given. Let $p \in M$ be a point, and let $g = k + P$ be the corresponding Cartan decomposition (see (1), (2)), where g is the Lie algebra of $G = I_0(M)$. Let $X \in P$ be that element such that $\gamma_{px}(t) = e^{tX}(p)$, and let $A \subseteq P$ be a maximal abelian subspace that contains X. Let g_x denote the Lie subalgebra of g corresponding to G_x. Then

$$g_X = g_0 + \sum_{\alpha(X)\geq 0} g_\alpha$$

where $g=g_0+\sum_{\alpha\in\Lambda} g_\alpha$ is the root space decomposition determined by A.

Before proving this result we discuss two corollaries. If k is the rank of M (=the rank of g) and if F is a k-flat in M that contains a point p, then $F=\exp(A)(p)$, where $A\subseteq P$ is some maximal abelian subspace of P and $g=k+P$ is the Cartan decomposition determined by p. Since F is a totally geodesic complete submanifold of M the set $F(\infty)$ of points at infinity for F becomes a subset of $M(\infty)$ in an obvious way. With this discussion in mind we now obtain as an immediate corollary of (31) the following:

(32) Proposition Let F be a k-flat in a symmetric space M of noncompact type and rank $k\geq 2$. Then there are only finitely many distinct groups $(G_x)_0$, where $x\in F(\infty)$. Here $(G_x)_0$ denotes the connected component of G_x that contains the identity.

Next consider two regular unit vectors v_1, v_2 in T_pM that determine the same Weyl chamber in T_pM as defined above in section 4. Let $x_1, x_2 \in M(\infty)$ be the points determined by the geodesics γ_{v_1} and γ_{v_2}. The geodesics γ_{v_1} and γ_{v_2} both lie in some k-flat $F=\exp(A)(p)$ and in fact if X_1, X_2 denote the vectors in A corresponding to v_1, v_2 then X_1, X_2 determine the same Weyl chamber in A as defined above in (23). From (31) we now obtain

(33) Proposition Let $v_1, v_2 \in T_pM$ define the same Weyl chamber in T_pM (in the sense of section 4), where M is a symmetric space of noncompact type. Let $x_1, x_2 \in M(\infty)$ be the asymptote classes of the geodesics $\gamma_{v_1}, \gamma_{v_2}$ respectively. Then $(G_{x_1})_0=(G_{x_2})_0$.

We now prove (31). We need the following:

(34) Lemma Using the notation of (31) set $\gamma(t)=\gamma_{px}(t)$, and let $Z\in g$ be given arbitrarily. Relative to the root space decomposition of g determined by A write $Z=Z_0+\sum_{\alpha\in\Lambda} Z_\alpha$, where $Z_0\in g_0$ and $Z_\alpha\in g_\alpha$ for each $\alpha\in\Lambda$. Then for all $s,t\in\mathbb{R}$ we have

$$d(e^{sZ}\gamma(t),\ \gamma(t)) = d(e^{sZ(t)}p,p)$$

where $Z(t)=\mathrm{Ad}(e^{-tX})(Z)=Z_0+\sum_{\alpha\in\Lambda} e^{-t\alpha(X)}Z_\alpha$.

Proof of the lemma $d(e^{sZ}\ (t),\ \ (t))=d(e^{sZ}e^{tX}(p),e^{tX}(p))=$
$=d(e^{-tX}e^{sZ}e^{tX}(p),p)=d(e^{sZ(t)}(p),p)$ by the definition of $\mathrm{Ad}(e^{-tX})$. The root space formula for $Z(t)$ follows from (21) above.

Proof of (31) If $Z\in \mathfrak{g}_0+\sum_{\alpha(X)\geq 0}\mathfrak{g}_\alpha$, $Z=Z_0+\sum_{\alpha(X)\geq 0}Z_\alpha$, then we see from the lemma that for each fixed $s\in\mathbb{R}$ the convex function

$$t\to d^2(e^{sZ}\gamma(t),\gamma(t))$$

converges to $d(e^{sZ^*}(p),p)$ as $t\to+\infty$, where $Z_*=Z_0+\sum_{\alpha(X)=0}Z_\alpha$. Hence $\{e^{sZ}\}\subseteq G_X$ and $Z\in\mathfrak{g}_X$.

Conversely let $Z\in\mathfrak{g}_X$ and write $Z=Z_1+Z_2$, where $Z_1\in\mathfrak{g}_0+\sum_{\alpha(X)\geq 0}\mathfrak{g}_\alpha$ and $Z_2\in\sum_{\alpha(X)<0}\mathfrak{g}_\alpha$. By the work above $Z_1\in\mathfrak{g}_X$ and hence $Z_2=Z-Z_1\in\mathfrak{g}_X$. Therefore the function $f(t)=d^2(e^{sZ_2}\gamma(t),\ \gamma(t))$ is bounded above for $t\geq 0$ and each fixed number $s\in\mathbb{R}$. On the other hand $f(t)\to 0$ as $t\to-\infty$ by the lemma above. Since $f(t)$ is bounded on $\mathbb{R}$ and convex it must be constant and hence identically zero. By the lemma above $Z_2(t)=\mathrm{Ad}(e^{-tX})(Z_2)\in\mathfrak{k}$ for all t and hence $Z_2=Z_2(0)\in\mathfrak{k}$. Thus $Z_2=\Theta_p(Z_2)\in\Theta_p(\sum_{\alpha(X)<0}\mathfrak{g}_\alpha)=\sum_{\alpha(X)>0}\mathfrak{g}_\alpha$ since $\Theta_p:\mathfrak{g}_\alpha\to\mathfrak{g}_{-\alpha}$ is an isomorphism for all roots α (here $\Theta_p:\mathfrak{g}\to\mathfrak{g}$ is the Cartan involution determined by p - see the discussion preceding (1)). Finally $Z_2\in(\sum_{\alpha(X)<0}\mathfrak{g}_\alpha)\cap(\sum_{\alpha(X)>0}\mathfrak{g}_\alpha)=\{0\}$ and $Z=Z_1\in\mathfrak{g}_0+\sum_{\alpha(X)\geq 0}\mathfrak{g}_\alpha$. This completes the proof of (31).

References

[A 1] D.V. Anosov, "Ergodic properties of geodesic flows on closed Riemannian manifolds", Soviet Math. Dokl. 4 (1963), 1153-1156.

[A 2] D.V. Anosov, "Roughness of geodesic flows on compact Riemannian manifolds of negative curvature", Soviet Math. Dokl. 3(1962), 1068-1070.

[A 3] D.V. Anosov, "Geodesic flows on Riemannian manifolds of negative curvature", Proc. Steklov Instit. Math 90 (1969).

[AS] D.V. Anosov and Ja. I. Sinai, "Certain smooth ergodic systems", Russian Math Surveys 22:5 (1967), 109-167.

[Ba 1] W. Ballmann, "Einige neue Resultate über Mannigfaltigkeiten nicht positiver Krümmung",Dissertation, Univ. Bonn, 1978 and "Axial isometries of manifolds of nonpositive curvature", Math. Ann. 259 (1982), 131-144.

[Ba 2] W. Ballmann, "Nonpositively curved manifolds of higher rank", to appear.

[BB] W. Ballmann and M. Brin, "On the ergodicity of geodesic flows", Erg. Th. Dyn. Syst. 2 (1982), 311-315.

[BBE] W. Ballmann, M. Brin and P. Eberlein, "Structure of manifolds of nonpositive curvature, I", preprint.

[BBS] W. Ballmann, M. Brin and R. Spatzier, "Structure of manifolds of nonpositive curvature, II", preprint.

[BE] W. Ballmann and P. Eberlein, "Fundamental group of manifolds of nonpositive curvature", in preparation.

[Be] M. Berger, "Sur les groupes d'holonomie des variétés a connexion affine et des variétés riemanniennes", Bull. Soc. Math. France 83 (1953), 279-330.

[BO] R. Bishop and B. O'Neill, "Manifolds of negative curvature", Trans. AMS 145 (1969), 1-49.

[BP] M. Brin and Ja. B. Pesin, "Partially hyperbolic dynamical systems," Math. USSR Izv. 8 (1974), 177-218.

[Bu] K. Burns, "Hyperbolic behavior of geodesic flows on manifolds with no focal points", dissertation, Univ. of Warwick, 1982 and Erg. Th. Dyn. Syst. 3 (1983), 1-12.

[BuS] K. Burns and R. Spatzier, to appear.

[CE] S. Chen and P. Eberlein, "Isometry groups of simply connected manifolds of nonpositive curvature", Ill. J. Math. 24 (1980), 73-103.

[ChE] J. Cheeger and D. Ebin, "Comparison Theorems in Riemannian Geometry",North-Holland, Amsterdam 1975.

[CG] J. Cheeger and D. Gromoll, "On the structure of complete manifolds of nonnegative curvature", Annals Math. 96 (1972),413-443.

[E 1] P. Eberlein, "A canonical form for compact nonpositively curved manifolds whose fundamental groups have nontrivial center", Math. Ann. 260 (1982), 23-29.

[E 2] P. Eberlein, "Euclidean de Rham factor of a lattice of nonpositive curvature", J. Diff. Geom. 18 (1983), 209-220.

[E 3] P. Eberlein, "Geodesic flows on negatively curved manifolds I", Annals Math. 95 (1972), 492-510.

[E 4] P. Eberlein, "Geodesic rigidity in compact nonpositively curved manifolds", Trans. Amer. Math. Soc. 268 (1981), 411-443.

[E 5] P. Eberlein, "Isometry groups of simply connected manifolds of nonpositive curvature II", Acta Math. 149 (1982), 41-69.

[E 6] P. Eberlein, "Rigidity of lattices of nonpositive curvature", Erg. Th. Dyn. Syst. 3 (1983), 47-85.

[E 7] P. Eberlein, "When is a geodesic flow of Anosov type?, I", J. Diff. Geom. 8 (1973), 437-463.

[E 8] P. Eberlein, "Lattices in spaces of nonpositive curvature", Annals of Math. 111 (1980), 435-476.

[EO] P. Eberlein and B. O'Neill, "Visibility manifolds", Pac. J. Math. 46 (1973), 45-109.

[Gra] A. Grant, "Surfaces of negative curvature and permanent regional transitivity", Duke Math. J. 5 (1939), 207-229.

[GKM] D. Gromoll, W. Klingenberg and W. Meyer, Riemannsche Geometrie im Großen, Lecture Notes in Mathematics, Vol. 55, Springer, Heidelberg, 1968.

[GW] D. Gromoll and J. Wolf, "Some relations between the metric structure and the algebraic structure of the fundamental group in manifolds of nonpositive curvature", Bull. Amer. Math. Soc. 77 (1971), 545-552.

[Gr 1] M. Gromov, "Lectures at College de France", Spring 1981.

[Gr 2] M. Gromov, "Three remarks on geodesic dynamic and fundamental group", preprint.

[GS] M. Gromov and V. Schroeder, book, to appear.

[Gu] R. Gulliver, "On the variety of manifolds without conjugate points", Trans. Amer. Math. Soc. 210 (1975), 185-201.

[Ha] J. Hadamard, "Les surfaces à courbures opposées et leur lignes géodésiques, Jour. de Math. Pures et Appliq. 5 (4) (1898), 27-74.

[Hed 1] G. A. Hedlund, "On the metrical transitivity of the geodesics on closed surfaces of constant negative curvature", Annals of Math., 35(2) (1934), 787-808.

[Hed 2] G.A. Hedlund, "A metrically transitive group defined by the modular group", Amer. J. Math. 57 (1935), 668-678.

[Hed 3] G.A. Hedlund, "Two dimensional manifolds and transitivity", Annals of Math. 37(2) (1936), 534-542.

[Hed 4] G. A. Hedlund, "Fuchsian groups and transitive horocycles", Duke Math. J., 2 (1936), 530-542.

[Hed 5] G.A. Hedlund, "On the measure of geodesic types on surfaces of negative curvature", Duke Math. J. 5 (1939), 230-248.

[Hed 6] G.A. Hedlund, "Fuchsian groups and mixtures", Ann. of Math. 40(2) (1939), 370-383.

[Hed 7] G.A. Hedlund, "The dynamics of geodesic flows", Bull. Amer. Math. Soc. 45 (1939), 241-260.

[Hel] S. Helgason, "Differential geometry and symmetric spaces", Academic Press, New York, 1962.

[Ho 1] E. Hopf, "Fuchsian groups and ergodic theory", Trans. Amer. Math. Soc. 39 (1936), 299-314.

[Ho 2] E. Hopf, "Ergodentheorie", Ergebnisse der Math., 5, Springer, Berlin, 1937 and Chelsea, New York, 1948.

[Ho 3] E. Hopf, "Statistik der geodätischen Linien in Mannigfaltigkeiten negativer Krümmung", Ber. Verh. Sächs. Akad. Wiss. Leipzig 91 (1939), 261-304.

[Ho 4] E. Hopf, "Statistik der Lösungen geodätischer Probleme vom unstabilen Typus, II", Math. Annal. 117 (1940), 590-608.

[Ka] F.I. Karpelevic, "The geometry of geodesics and the eigenfunctions of the Beltrami-Laplace operator on symmetric spaces", Trans. Moscow Math. Soc. (AMS Translation) Tom 14 (1965), 51-199.

[LY] H.B. Lawson and S.-T. Yau, "Compact manifolds of nonpositive curvature", J. Diff. Geom. 7 (1972), 211-228.

[Ma] F.I. Mautner, "Geodesic flows on symmetric Riemann spaces", Annals of Math. 65 (1957), 416-431.

[Mi] J. Milnor, Morse Theory, Annals of Math. Studies Number 51, Princeton University Press, Princeton, New Jersey 1963.

[Mor 1] M. Morse, "Recurrent geodesics on a surface of negative curvature", Trans. Amer. Math. Soc. 22 (1921), 84-100.

[Mor 2] M. Morse, "A fundamental class of geodesics on any closed surface of genus greater than one", Trans. Amer. Math. Soc. 26 (1924), 25-61.

[Mor 3] M. Morse, "Instability and transitivity", Journ. de Math. Pures et Appliq. 14(9), (1935), 49-71.

[Mos] G.D. Mostow, Strong rigidity of locally symmetric spaces, Annals of Math. Studies Number 78, Princeton University Press, Princeton, New Jersey, 1973.

[O] B. O'Neill, Semi-Riemannian Geometry with Applications to Relativity, Academic Press, New York, 1983.

[Pe 1] Ja. B. Pesin, "Characteristic Lyapunov indicators and smooth ergodic theory", Russian Math. Surveys 32:4 (1977), 55-114.

[Pe 2] Ja. B. Pesin, "Families of invariant manifolds of dynamical systems with nonzero characteristic Lyapunov indicators", Math. USSR-Izv. 10 (1976), 1261-1305.

[Pe 3] Ja. B. Pesin, "Geodesic flows on closed Riemannian manifolds without focal points", Math. USSR-Izv. 11 (1977), 1195-1228.

[Pe 4] Ja. B. Pesin, "Geodesic flows with hyperbolic behavior of the trajectories and objects connected with them", Russian Math. Surveys, 36:4 (1981), 1-59.

[Po] H. Poincaré, "Sur les lignes géodésiques des surfaces convexes", Trans. Amer. Math. Soc. 6 (1905), 237-274.

[Pr] A. Preissmann, "Quelques proprietes des espaces de Riemann", Comm. Math. Helv. 15 (1942-43), 175-216.

[PR] G. Prasad and M.A. Raghunathan, "Cartan subgroups and lattices in semisimple groups", Annals of Math. 96 (1972), 296-317.

[Sc 1] V. Schroeder, "A splitting theorem for spaces of nonpositive curvature", to appear.

[Sc 2] V. Schroeder, "Über die Fundamentalgruppe von Räumen nichtpositiver Krümmung mit endlichem Volumen", Dissertation, Universität Münster, West Germany, 1984.

[Sim] J. Simons, "On transitivity of holonomy systems", Annals of Math. 76 (1962), 213-234.

[Sin 1] Ja. G. Sinai, "Classical dynamical systems with a countably-multiple Lebesgue spectrum, II", Izv. Akad. Nauk. SSSR Ser. Mat. 30 (1966), 15-68.

[Sin 2] Ja. G. Sinai, "Geodesic flows on compact surfaces of negative curvature", Soviet Math. Dokl. 2 (1961), 106-109.

[T] J. Tits, <u>Buildings of Spherical Type and Finite BN-pairs</u>, Lecture Notes in Math., Vol. 386, Springer, Berlin-Heidelberg-New York, 1974

[W 1] J. Wolf, <u>Spaces of constant curvature</u>, 2nd edition, published by the author, Berkeley, California, 1972.

[W 2] J. Wolf, "Homogeneity and bounded isometries in manifolds of negative curvature", Ill. J. Math. 8 (1964), 14-18.

EQUIVALENCE OF ONE DIMENSIONAL LAGRANGIAN FIELD THEORIES IN THE PLANE I. *

ROBERT B. GARDNER and WILLIAM F. SHADWICK

1. Introduction.

In this paper we consider the problem of local equivalence of first order Lagrangian field theories in one dependent and two independent variables and provide a complete solution for regular Lagrangians. Our calculation follows the algorithmic approach to the equivalence problem which was presented in reference [8]. Indeed the solution of our problem is in many respects similar to that of the first order Lagrangian in one independent and one dependent variable which was given in Chapter 6 of reference [8]. As one might expect, the various normalizations and reductions which arise in that case are also present here. The remarkable difference in the field case is the existence of Lagrangians which admit infinite continuous pseudogroups. There are also other phenomena which divide the Lagrangians into classes controled by algebraic objects which were not present in the particle case. The calculation splits into two branches, according to the vanishing or non-vanishing of a vector z in R^2. Both cases are of interest: $z = 0$ includes the minimal surface Lagrangians

$$L = \sqrt{1 + (\partial z/\partial x)^2 + e\,(\partial z/\partial y)^2} \qquad e = \pm 1,$$

the case $z \neq 0$ includes the Lagrangians

$$L = \frac{1}{2}\left((\partial z/\partial x)^2 + e\,(\partial z/\partial y)^2\right) \quad e = \pm 1,$$

for the Laplace and wave equations.

* Research supported in part by NATO Grant 0546/82 and NSERC Grants T2100 and U0172.

In the case $z = 0$ we show in section (6) that the equivalence problem always determines an e-structure. The structure equations we obtain are the equations for a connection in a circle bundle over R^5, and in case certain invariants vanish the structure equations become the equations of a metric connection on R^3 with metric

$$(\tau^1)^2 + (\tau^2)^2 + e\,(\tau^3)^2, \quad \text{where } e = \pm 1$$

according to the case where the Euler-Lagrange equations are elliptic or hyperbolic. The only non-constant invariants in this case are two components of curvature for this connecion.

The branch given by $z \neq 0$ contains the case where the solution to the equivalence problem depends on arbitrary functions. We consider this branch of the problem in section (7) and show that unless all of the invariants take the constant values which they assume in the cases $L = \frac{1}{2}\,((\partial z/\partial x)^2 + e\,(\partial z/\partial y)^2)$, we can always reduce to an e-structure, either on R^6, or on R^5. When no reduction is possible the equivalence problem is determined by an exterior differential system in involution and the solution depends on two arbitrary functions of one variable and one arbitrary constant, as the Cartan's test indicates. Cartan devoted a large part of his book on Areal Geometries to discussion of this case and mentioned the example

$$L = \frac{1}{2}\,((\partial z/\partial x)^2 + e\,(\partial z/\partial y)^2),$$

many times in his works (see [1], [2], [3]). It is remarkable that he never seems to have pointed out that the only class of Lagrangians admitting pseudogroups is given by the equivalence class of

$$L = \frac{1}{2}\,((\partial z/\partial x)^2 + e\,(\partial z/\partial y)^2).$$

We present the intrinsic calculation of the invariants here, including only as much of the parametric calculation as is necessary to identify some interesting cases and eliminate certain degenerate ones. We leave to part II the complete discussion of the parametric form of the invariants and of sufficient conditions for equivalence in some cases of interest. As a by-product of our calculation we obtain the so called "Poincare-Cartan" form associated to the Lagrangian. This 2-form, which Cartan attributes to De Donder [7], contains all of the essentials of the variational problem. It is a product of the first normalization in the equivalence problem. We also obtain the contact invariance of the Euler-Lagrange equations for L and of rank and signature of Hessian (L), results whose classical proofs involve tedious calculation.

We wish to point out there is some history to this problem. Élie Cartan indicated a solution to this equivalence problem in 1933 in his book [1]. The solution, however, is not based on the usual method of equivalence, but on the identification of the problem with his analysis of areal geometry studied from the viewpoint of "the repère mobile". This approach to equivalence is quite difficult for the uninitatied, as the review [9] by Herman Weyl dramatically indicates, but even more important the formulation of the invariants in the areal geometry book is not in a form amenable to application to the study of Lagrangian field theories. There was also a short note in 1945 by Debever [5] which announced the solution of the equivalence problem for the $z = 0$ branch, but the details were not published [6].

Part of this research was performed while we were members of the Mathematical Sciences Research Institute at Berkeley and we would like to thank them for their hospitality. We would also like to thank Robert Bryant for helpful discussions on Cartan's book on Areal Geometry and some incisive comments on the work while it was in progress. In particular we are indebted to him for the observation that there is only one model in which a pseudogroup can arise.

2. The equivalence problem.

The origin of the problem we are considering is the variational problem for

$$(1) \qquad L(x,y,z,\partial z/\partial x,\partial z/\partial y)\ dx\ dy.$$

(In (1) and for the rest of the paper, all products of exterior forms are to be read as exterior products.) We consider L as a function on the jet space $J^1(R^2,R)$, with local coordinates x,y,z,p,q (see [8]), and want to determine the invariants of (1) thought of as a functional on 1-graphs

$$j^1(f) : R \longrightarrow J^1(R^2,R).$$

Thus one is led to study the equivalence of Lagrangians under diffeomorphisms

$$\rho : J^1(R^2,R) \longrightarrow J^1(R^2,R),$$

which for all $f : R^2 \longrightarrow R$ satisfy

$$(2) \qquad \int_{j^1(f)} \rho^*\bar{\Omega} = \int_{j^1(f)} \Omega$$

where $\Omega = L\ dx\ dy$ and $\bar{\Omega} = \bar{L}\ \overline{dx}\ \overline{dy}$.

Since we can choose f with arbitrarily small support this is equivalent to

$$(3) \qquad j^1(f)^*(\rho^*\bar{\Omega} - \Omega) = 0 \quad \text{for all} \quad f : R^2 \longrightarrow R.$$

We begin our analysis by studying 2-forms with property (3), that is 2-forms ϕ such that

$$(4) \qquad j^1(f)^*\phi = 0 \quad \text{for all} \quad f : R^2 \longrightarrow R.$$

We will analyze this property by lifting to the jet space $J^2(R^2,R)$ with local coordinates (x,y,z,p,q,r,s,t) via the canonical projection π^2_1 (see [8]). Let us set

$$\theta = dz - pdx - qdy$$

$$\theta_1 = dp - rdx - sdy \quad \text{and} \quad \theta_2 = dq - sdx - tdy.$$

We claim that any 2-form ϕ on $J^1(R^2,R)$ which satisfies (4) has the property

(5) $$\pi_1^{2*}\phi \in \text{ ideal } (\theta,\theta_1,\theta_2).$$

The 1-forms in $\pi_1^{2*}T^*J^1(R^2,R)$ are spanned by θ, θ_1, θ_2, dx, dy, hence

$$\pi_1^{2*}\phi \equiv A\ dx\ dy \bmod (\theta,\ \theta_1,\ \theta_2).$$

But by property (1)

$$j^2(f)^*\ \pi_1^{2*}\phi = j^1(f)^*\phi = 0, \quad \text{for all } f : R^2 \longrightarrow R$$

and this implies

$$A \circ j^2(f) = 0 \quad \text{for all } f : R^2 \longrightarrow R,$$

and hence A = 0, which verifies (5).

We now want to verify that any 2-form which satisfies (5) also satisfies

(6) $$\phi \in \text{ ideal } (\theta, d\theta).$$

We know we can write

$$\pi_1^{2*}\phi = \xi\theta + \xi^1\theta_1 + \xi^2\theta_2 .$$

Since

$$d(\pi_1^{2*}\phi) = \pi_1^{2*}\, d\,\phi$$

this three form involves at most differentials on $J^1(R^2,R)$, and hence

$$d(\pi_1^{2*}\phi)\ \theta\ \theta_1\theta_2 = 0,$$

since it is a six form involving at most the five independent differentials on $J^1(R^2,R)$. This implies

$$\xi^1 d\theta + \xi^2 d\theta \in \text{ ideal}(\theta,\theta_1,\theta_2)$$

and as a result

$$\xi^1\, dx = 0,\ \xi^1\, dy + \xi^2\, dx = 0,\ \xi^2\, dy = 0.$$

Thus by Cartan's lemma

$$\pi_1^{2*}\phi = \xi\theta + f(dx\ \theta_1 + dy\ \theta_2)$$

or equivalently

$$\phi = \xi\ \theta + f\ d\theta$$

which establishes (6).

With this preparation we see that our condition (3) for equivalence is that

$$(7) \qquad \rho^*\bar{\Omega} \equiv \Omega \quad \text{mod ideal } (\theta, d\theta).$$

We will call $\bar{\Omega}$ and Ω <u>ρ-equivalent</u> if there is a diffeomorphism satisfying (7). In order for this to be an equivalence relation we require transitivity, and thus if Ω'' and $\bar{\Omega}$ are ψ-equivalent and $\bar{\Omega}$ and Ω are ρ-equivalent we require that Ω'' and Ω be $(\psi \circ \rho)$-equivalent. Thus if

$$\psi * \Omega'' \equiv \Omega$$

and $\qquad\qquad$ mod ideal $(\theta, d\theta)$

$$\rho * \bar{\Omega} \equiv \Omega$$

we require

$$(\psi \circ \rho)^*\Omega'' \equiv \Omega \quad \text{mod ideal } (\theta, d\theta).$$

It is a consequence of this requirement that the diffeomorphisms involved must be contact transformations. To see this we note that for any 1-form π , $\bar{\Omega} + \pi\bar{\theta}$ and $\bar{\Omega}$ are identity-equivalent. If $\bar{\Omega}$ and Ω are ρ-equivalent it follows that

$$(\text{id} \circ \rho)^*(\bar{\Omega} + \pi\bar{\theta}) \equiv \Omega \quad \text{mod ideal } (\theta, d\theta)$$

and

$$\rho * \bar{\Omega} \equiv \Omega \qquad \text{mod ideal } (\theta, d\theta)$$

so

$$\rho^*(\pi\bar{\theta}) \in \text{ ideal } (\theta, d\theta)$$

and we have

$$\rho^*(\pi\bar{\theta}) = \xi\theta + f\, d\theta$$

for some 1-form ξ and function f. But squareing both sides and wedgeing with θ gives

$$0 = (\rho^*(\pi\bar{\theta})^2)\,\theta = f^2(d\theta)^2\theta,$$

and it follows that $f = 0$. As a result we see that if $\bar{\Omega}$ and Ω are ρ-equivalent then the diffeomorphism satisfies

$$\rho^*(\pi\bar{\theta}) \in \text{ideal } (\theta)$$

for arbitrary 1-forms π. It follows from this that

$$\rho * \theta \in \text{ideal } (\theta)$$

so ρ is a contact transformation. It now follows as well that if $\bar{\Omega}$ and Ω are ρ-equivalent then Ω and $\bar{\Omega}$ are ρ^{-1}-equivalent. By exactly the same sort of argument as above, we see that if

$$\rho * \bar{\Omega} = \Omega + \xi\theta + g\, d\theta,$$

then $g = 0$. Thus we are led to consider diffeomorphisms ρ whose Jacobians are given by

$$\rho * \begin{pmatrix} d\bar{z} - \bar{p}d\bar{x} - \bar{q}d\bar{y} \\ \bar{L}^{1/2}d\bar{x} \\ \bar{L}^{1/2}d\bar{y} \\ d\bar{p} \\ d\bar{q} \end{pmatrix} = \begin{pmatrix} A & 0 & 0 \\ & & \\ B & D & M \\ C & E & F \end{pmatrix} \begin{pmatrix} dz - pdx - qdy \\ L^{1/2}dx \\ L^{1/2}dy \\ dp \\ dq \end{pmatrix}$$

where $0 \neq A \in R$, $B, C \in R^2$, $D \in SL(2,R)$, $E, M \in M(2,2)$ and $F \in GL(2R)$. A straight-forward calculation shows that M must vanish and the diffeomorphisms involved are actually <u>prolonged point transformations</u> as in the first order Lagrangian problem in a single dependent variable (see [8]).

We now proceed to apply the equivalence algorithm essentially as described in [8], by lifting the problem to $U \times G$ where G is the group of all matricies of the form

$$\begin{pmatrix} A & O & O \\ B & D & O \\ C & E & F \end{pmatrix}$$

and applying Lie algebra compatible absorptions and the technique of group reduction. We say essentially in that we found it convenient in this paper to modify the description in [8] by interchanging rows with columns and left actions with right actions.

3. Application of the equivalence algorithm.

After lifting the problem to $U \times G$ we will exterior equations

$$(8) \qquad d\begin{pmatrix}\theta\\ w\\ \eta\end{pmatrix} = \begin{pmatrix}\alpha & \alpha_1 & \alpha_2\\ \beta & \delta & \alpha_3\\ \gamma & \varepsilon & \phi\end{pmatrix}\begin{pmatrix}\theta\\ w\\ \eta\end{pmatrix}$$

Since $S \in G$ satisfies

$$dS\,S^{-1} = \begin{pmatrix}* & 0 & 0\\ * & \Delta & 0\\ * & * & *\end{pmatrix} \qquad \text{with} \quad \operatorname{tr}\Delta = 0$$

we see

$$\alpha_1,\ \alpha_2,\ \alpha_3,\ \operatorname{tr}\delta \equiv 0 \mod (\theta, w, \eta),$$

and hence these are the principal components of order zero.

Inspection of the definition of θ and w shows that $d\theta$ and dw contain no terms quadratic in η. The term in $d\theta$ which is a multiple of θ can be absorbed in α, and the term δ in dw can be replaced by

$$\delta - \frac{1}{2}\operatorname{tr}\delta$$

giving rise to torsion terms of the form

$$\frac{1}{2}\operatorname{tr}\delta w = \sigma\theta + {}^t w P w + {}^t \eta v w.$$

Since we may assume ${}^tP = -P$, and hence that $\operatorname{tr} P = 0$, the first two terms in this torsion can be absorbed, by modifying β and δ respectively.

This leads us to

$$(9) \qquad d\begin{pmatrix}\theta\\ w\\ \eta\end{pmatrix} = \begin{pmatrix}\alpha & 0 & 0\\ \beta & \delta & 0\\ \gamma & \varepsilon & \phi\end{pmatrix}\begin{pmatrix}\theta\\ w\\ \eta\end{pmatrix} + \begin{pmatrix}{}^t wJw + {}^t\eta Kw\\ {}^t\eta\, v\, w\\ 0\end{pmatrix}$$

where J is antisymmetric, $\operatorname{tr}\delta = 0$, and no further absorptions are possible.

4. The first group reductions.

The infinitesimal group action on J, K and v, is given by the fiber variation in their exterior derivatives. We compute this by calculating the exterior derivatives modulo (Θ, w, η).

We will compute so many times modulo the base, that is modulo (Θ, w, η), that we fix the convention that any congruence is modulo base unless otherwise denoted.

The equations $0 = d^2\Theta$ and $0 = d^2w$ yield

$$2dJ + (\alpha + {}^t\delta)J + J\delta + \varepsilon K - {}^t\{(\alpha 1 + {}^t\delta)J$$
$$+ J\delta + \varepsilon J\} \equiv 0$$

and

(10) $$dK - K\phi + (\alpha - {}^t\delta)K \equiv 0$$

$$dv + \beta + {}^t\phi v \equiv 0.$$

A direct calculation from the defining equation for the initial coframe shows that $K = A^t D^{-1} F^{-1}/L^{1/2}$ and hence is non-singular.

As such we see from the infinitesimal action that we can effect a group reduction by translating J to the 0 matrix, and right or left multiplying K by its inverse to fix $K = 1$. We can also translate the vector v to zero.

At this point we can observe that if

$$\tilde{w} := w^1 w^2$$

then

$$\tilde{w} \equiv L \, dx \, dy$$

and

$$\text{mod } \Theta$$

$$d\tilde{w} \equiv 0$$

and hence $\tilde{w}$ is the Cartan 2-form for the Lagrangian L. We note that in [1] Cartan attributes the discovery of this 2-form to De Donder [7].

We deduce from the congruences (10), that the reduced group has new principal components given by

$$\varepsilon - {}^t\varepsilon \equiv 0$$

$$\alpha 1 - {}^t\delta - \phi \equiv 0$$

$$\beta \equiv 0$$

If we set

$$\beta = Gw + H\eta$$

then
$$dw = H\,\eta\theta + (G - \tfrac{1}{2}\operatorname{tr} G\,1)\,w\theta + \delta w + \tfrac{1}{2}\operatorname{tr} G\,\theta\,w,$$

and if we modify δ by subtracting $(G - \frac{1}{2}\operatorname{tr} G\,1)\theta$, we find

$$dw = H\,\eta\theta + \delta w + f\,w\,\theta,$$

where $f = -\frac{1}{2}\operatorname{tr} G$.

If we set

$$\left.\begin{aligned} \tfrac{1}{2}(\varepsilon - {}^t\varepsilon) &\equiv {}^t w\,M + {}^t\eta\,N \\ \phi - \alpha\,1 - {}^t\delta &\equiv {}^t w\,S + {}^t\eta\,T \end{aligned}\right\} \bmod \theta$$

and substitute into the equation for $0 = d^2\theta \bmod \theta$, we find the scalar relations,

$${}^t\eta\,T\,\eta = 0 \quad \text{and} \quad {}^t w\,{}^t w\,S\eta - {}^t w\,{}^t\eta\,N\,w = 0.$$

Now by suitably modifying γ and ε, we can absorb all the torsion terms in $d\eta$ so that

$$d\eta = \gamma\theta + \varepsilon w + (\alpha 1 - {}^t\delta)\,\eta$$

with $\varepsilon = {}^t\varepsilon$.

Thus we arrive at

$$\begin{aligned} d\theta &= \alpha\theta + {}^t w\eta \\ (11) \qquad dw &= H\eta\theta + fw\theta + \delta w \\ d\eta &= \gamma\theta + \varepsilon w + (\alpha 1 - {}^t\delta)\eta, \end{aligned}$$

and no further absorption is possible.

5. The second group reductions.

Now we compute the infinitesimal group action on H and f by computing the exterior derivatives modulo base from the integrability conditions resulting from (11).

The equation $0 = d^2 w$ yields

$$(12) \qquad \begin{aligned} dH + 2\alpha 1 H - (H{}^t\delta + \delta H) &\equiv 0 \\ \mathrm{tr}(df + f\alpha 1 + H\varepsilon) &\equiv 0 \end{aligned}$$

and the scalar relation that the matrix H is symmetric.

We see from the infinitesimal action that we can effect a group reduction by utilizing conjugation on H by GL(2,R). Since the only invariants of a symmetric matrix under conjugation are the rank and signature, these are the only invariants of H. A direct calculation from the defining equations for the initial coframe shows that

$$H = D\,\mathrm{Hess}(L)\,{}^tD/A^2$$

where

$$\mathrm{Hess}(L) = \begin{pmatrix} L_{pp} & L_{pq} \\ L_{qp} & L_{qq} \end{pmatrix}$$

Thus for the regular problem of the calculus of variations, which is characterized by Hess(L) being non-singular we have H non-singular. We restrict to this case for the rest of the paper.

As such we can effect a group reduction by normalizing H to the constant matrix Q, where e_1, e_2 are ±1

$$Q = \begin{pmatrix} e_1 & 0 \\ 0 & e_2 \end{pmatrix} \quad \text{and we set} \quad e = e_1 e_2$$

The Euler-Lagrange equations for L have symbol Q and are thus elliptic for $e = 1$, and hyperbolic for $e = -1$.

We also see from the infinitesmal action and the non-singularity of H that the function f may be translated to 0.

We deduce from the congruences (12), that the new reduced group has new principal components given by

$$\operatorname{tr} Q\,\varepsilon \equiv 0$$

$$2\alpha Q - ({}^tQ\delta + \delta Q) \equiv 0.$$

If we multiply the last congruence by Q^{-1} and then take the trace we find

$$\alpha \equiv 0$$

and

$$Q^t\delta + \delta Q \equiv 0$$

Thus δ lies in SO(1,1) or in SO(2) depending upon whether $e = \pm 1$.

By the fundamental theorem of Riemannian or pseudo Riemannian geometry, depending on the sign of e, we can find a matrix ${}^twM$ such that $\overline{\delta} = \delta + wM$ satisfies

$$dw = Q\eta\theta + \overline{\delta}w$$

and

$$Q^t\overline{\delta} + \overline{\delta}Q \equiv 0 \mod(\theta,\eta).$$

If we let

$$\alpha = {}^tz\eta + {}^tvw$$

and

$$\left.\begin{aligned} \operatorname{tr} Q\,\varepsilon &\equiv {}^taw + {}^tb\eta \\ Q^t\delta + \delta Q &\equiv {}^t\eta C \end{aligned}\right\} \mod \theta$$

then, by making use of identies on these torsion terms given by $0 = d^2 w$, we can absorb in $dQ\eta$ by modifying γ and ε

$$dQ\eta = \gamma\theta + Q\varepsilon w + \delta Q\eta - {}^t z\eta\eta - \frac{1}{2}{}^t vQ\eta\, w$$

where

$${}^t\varepsilon = \varepsilon, \; \operatorname{tr} Q\varepsilon = 0.$$

Now no further absorptions are possible.

The relation $0 = d^2\theta$ implies

$$(13) \qquad {}^t dz + Q^{-1}\delta Qz \equiv 0$$

and

$$dv + \varepsilon z + \gamma + {}^t\delta v \equiv 0.$$

Thus we can translate v to zero. Now it follows from (13) and

$$Q{}^t\delta + \delta Q \equiv 0$$

that

$$i^2 := {}^t z\, Q\, z$$

is an invariant.

The congruense (13) shows that we can effect a group reduction by normalizing z, provided the vector z is non-zero. Thus we have two cases to consider.

6. The case $z = 0$, the third reductions.

When $z = 0$, the structure equations become

$$d\theta = {}^t w \eta$$

$$dw = Q\eta\theta + \delta w$$

$$dQ\eta = Q\gamma\theta + Q\varepsilon w + \delta Q\eta ,$$

and we have $\gamma \equiv 0$.

If we write

$$\gamma = M_S w + M_A w + N\eta$$

where M_S is symmetric and M_A is skew symmetric, then we can absorb the symmetric part of M in ε and adjust so that $\operatorname{tr} Q\,\varepsilon = 0$. We are left with

$$dQ\eta = QM_A w\theta + fw\,\theta + QN\eta\theta + Q\varepsilon w + \delta Q\eta$$

Now $0 = d^2\theta$ gives

$$0 = -{}^t w\,{}^t\delta\eta - {}^t wM_A w\theta - {}^t wN\eta\theta$$
$$-{}^t wQ^{-1}\delta Q\eta$$

The term ${}^t wMw\theta$ must vanish separately since

$${}^t\delta + Q^{-1}\delta Q \equiv 0 \mod \theta, \eta$$

and so M_A must also be symmetric and hence is zero. Thus

$$-{}^t w({}^t\delta - N\theta + Q^{-1}\delta Q) = 0$$

or

$$Q\,{}^t\delta + \delta Q = Q N \theta .$$

Now

$$dQ\eta = fw\theta + QN\eta\theta + Q\varepsilon w + \delta Q\eta$$

and no further absorbtion is possible.

Next from $0 = d^2 w$ we have

$$0 = Q\varepsilon w\Theta + \delta Q\eta\Theta + (d\delta - \delta\delta)w - \delta Q\eta\Theta$$

so

(14) $$(d\delta - \delta\delta - Q\varepsilon\Theta)w = 0.$$

Now by Cartan's lemma we have

$$d\delta = \delta\delta + Q\varepsilon\Theta + {}^t w\Psi$$

where

(15) $$\mathrm{tr}\ {}^t w\Psi = 0, \text{ and } {}^t w\Psi w = 0.$$

Now, $${}^t\delta + Q\,\delta\,Q = N\Theta \quad\text{and}\quad \mathrm{tr}\,N = 0,$$

$${}^t(Q\,{}^t\delta + \delta Q) = \delta Q + Q\,{}^t\delta,$$

so

$${}^t(Q\,N) = Q\,N$$

and

$$Q\,N = {}^t N\,Q\,.$$

This forces

$$N = \begin{pmatrix} N_1 & eN_2 \\ N_2 & -N_1 \end{pmatrix}.$$

We now have

$$2\varepsilon Q = 2\begin{pmatrix} \varepsilon_1 & e\varepsilon_2 \\ \varepsilon_2 & -\varepsilon_1 \end{pmatrix}.$$

Utilizing $d^2Q\eta = 0$, (14) and (15) we get

$$dN - 2\varepsilon Q + N\delta Q - Q^{-1}\delta\,Q\,N \equiv 0$$

and $df \equiv 0$.

Thus N is infinitesimally translated by $2\,\varepsilon\,Q$ which has exactly the same symmetries as N. As such we can translate N to 0, and this results in $\varepsilon \equiv 0$. The function f is an invariant which we will denote by i.

The structure equations now read

$$d\theta = {}^t w\eta$$

$$dw = Q\eta\theta + \delta w$$

$$dQ\eta = iw\theta + Q\varepsilon w + \delta Q\eta .$$

Now we have

$$Q^t\delta + \delta Q = {}^t\eta G \quad \text{where} \quad {}^t\eta G\eta = 0,$$

and thus

$$(16) \quad \delta = \Delta M + {}^t\eta GQ \quad \text{where } M = \begin{pmatrix} 0 & 1 \\ -e & 0 \end{pmatrix} .$$

If we let

$$\varepsilon = A\theta + {}^t wB + {}^t\eta C,$$

we have

$$\operatorname{tr} QA = \operatorname{tr} Q\,w\,B = \operatorname{tr} Q\,C = 0,$$

and

$${}^tA = A, \quad {}^t({}^t wB) = {}^t wB, \quad {}^t({}^t\eta C) = {}^t\eta C.$$

Then

$$(17) \quad dQ\eta = iw\theta + QA\theta w + Q^t wBw + Q^t\eta Cw + \Delta M\,Q\,\eta .$$

It follows that Δ is uniquely determined so we have an {e}-structure on R^6 with structure equations

$$d\theta = {}^t w\eta$$

$$(18) \quad dw = Q\eta\theta + \Delta Mw + {}^t\eta GQ^{-1}w$$

$$dQ\eta = Q\bar{A}\theta w + Q^t wBw + Q^t\eta Cw + MQ\eta$$

By considering $0 = d^2w \bmod(\eta,\theta)$ and the equation $0 = d^2Q \bmod(w,\theta)$ we see that $d\Delta$ is independent of Δ and hence we have the structure equations of a circle bundle over R^5.

In the case $G = 0$, and $C = 0$, these are the structure equations for a metric connection on R^3 with the metric $(\Theta)^2 + (w^1)^2 + e(w^2)^2$ as Debever noted for the case $e_1 = 1$, and $e_2 = 1$.

7. <u>The case $z \neq 0$, the third reductions.</u>

In the case that $z \neq 0$, equation (13) shows that we can rotate z. We will assume in the case $e = -1$ that z lies in the forward light cone. This allows us to rotate z to $I = {}^t(i, 0)$.

After this reduction given by rotating z we have

$$\delta \equiv 0, \quad \gamma + \varepsilon I \equiv 0$$

$$Q^t\delta + \delta Q \equiv 0 \mod(\Theta, \eta)$$

$$\mathrm{tr}\,\delta = 0, \quad \mathrm{tr}\,Q\varepsilon = 0, \quad {}^t\varepsilon = \varepsilon .$$

The structure equations become

$$d\Theta = {}^tI\eta\Theta + {}^tw\eta$$

$$dw = Q\eta\Theta + A\Theta w + {}^twBw + {}^t\eta Cw$$

$$\begin{aligned} dQ\eta = & -Q\varepsilon I\Theta + Q\varepsilon w + A\Theta Q\eta \\ & + {}^twBQ\eta + {}^t\eta CQ\eta - {}^tI\eta Q\eta \\ & + QM_s w\Theta + QM_A w\Theta + QN\eta\Theta \end{aligned}$$

where

$$\gamma = -\varepsilon I + M_s w + M_A w + N\eta$$

$$M_s = \frac{1}{2}(M + {}^tM), \quad M_A = \frac{1}{2}(M - {}^tM)$$

$$\mathrm{tr}\,A = \mathrm{tr}\,{}^twB = \mathrm{tr}\,{}^t\eta C = 0.$$

$$Q^t({}^twB) + ({}^twB)Q = 0.$$

Now we can absorb the symmetric matrix M in ε and adjust so that $\mathrm{tr}\, Q\,\varepsilon = 0$. We are left with

$$\begin{aligned} dQ\,\eta = & -Q\varepsilon I\theta + Q\varepsilon w + {}^t wBQ\eta \\ & + {}^t\eta CQ\eta + (QN - AQ)\eta\theta - {}^t I\eta Q\eta \\ & + QM_A w\theta + fw\theta. \end{aligned}$$

We note that the remaining freedom in modifying ε at this point is that we can replace ε_1 and ε_2 by

$$\varepsilon_1 + ih_1\theta + h_1 w^1 + h_2 w^2$$

$$\varepsilon_2 + ih_2\theta + h_2 w^1 - eh_1 w^2.$$

It follows that if either of ε_1 or ε_2 becomes congruent to zero the remaining one is uniquely determined and we have an $\{e\}$-structure.

Thus we may proceed at this point by calculating the group action on A, B, C, N, and f and obtaining the conditions under which we will be unable to achieve any group reduction.

Now $0 = d^2\theta$ shows that the antisymmetric matrix M_A must also be symmetric and hence vanishes. The same equation shows that N is expressible in terms of A, B, and P where

$$dI \equiv Pw \mod \theta,\eta$$

and hence whenever A and B are invariants, so is N. Finally from $0 = d^2\theta$ we get

$$2\,{}^t I\eta Q\eta = Q\,{}^t({}^t\eta C)\eta + {}^t\eta CQ\eta,$$

and if

$${}^t\eta C = \begin{pmatrix} C_1^a\eta_a & C_2^a\eta_a \\ C_3^a\eta_a & -C_1^a\eta_a \end{pmatrix}$$

this equation gives

$$C_3^1 - 2C_1^2 + eC_2^1 = 0$$

and

$$-eC_2^2 - 2eC_1^1 - C_3^2 = 2ei.$$

Now the equation $0 = d^2 w$ shows that

$$dC \equiv 0$$

and the action on A and B is given by

$$dA - (Q\varepsilon + {}^{t}I\varepsilon QC) \equiv 0$$

$${}^{t}w(dB + \varepsilon C)w \equiv 0.$$

If we compute the action in components we get

$$(19) \quad dA + \left(\begin{array}{c|c} \varepsilon_1(1+iC_1^1) + C_1^2 i\varepsilon_2 & \varepsilon_1 iC_2^1 + \varepsilon_2(1+iC_2^2) \\ \hline \varepsilon_1(iC_3^1) + \varepsilon_2(e+iC_3^2) & -(\varepsilon_1(1+iC_1^1) + C_1^3 i\varepsilon_2) \end{array}\right)$$

congruent to zero modulo w,θ,η, and since

$${}^{t}w\,B\,w = -\begin{pmatrix} eb_1 \\ \\ b_2 \end{pmatrix} w^1w^2 ,$$

the action on B is given by

$$(20) \begin{pmatrix} e\,db_1 \\ \\ db_2 \end{pmatrix} + \begin{pmatrix} \varepsilon_1(C_1^2 + eC_1^2) + \varepsilon_2(C_2^2 - C_1^1) \\ \\ \varepsilon_1(eC_3^2 - C_1^1) - \varepsilon_2(C_3^1 + C_1^2) \end{pmatrix} \equiv 0 .$$

Thus unless $dA \equiv 0$ and $dB \equiv 0$ we can reduce to an {e}-structure. In order for dA to be invariant we see from equation (19) that we must have

$$C_1^1 = -1/i,\ C_1^2 = 0,\ C_2^1 = 0,\ C_2^2 = -1/i$$

$$C_3^1 = 0,\ C_3^2 = -e/i.$$

In this case (19) gives $i = \sqrt{2}$, and (20) shows that

$$dB \equiv 0$$

as well.

We continue on the assumption that $dA \equiv 0$. In this case the structure equation reduce to

$$d\theta = {}^t I\eta\theta + {}^t w\eta$$

$$dw = Q\eta\theta + A\theta w + {}^t wBw + {}^t\eta Cw$$

$$dQ\eta = -Q\varepsilon I\theta + Q\varepsilon w + (QN - AQ)\,\eta\theta + {}^t wBQ + fw\theta,$$

where

$${}^t wB = \begin{pmatrix} 0 & -e\,{}^t wb \\ {}^t wb & 0 \end{pmatrix} \quad \text{and } b = {}^t(b_1, b_2).$$

Now $0 = d^2\theta$ shows that

$$QN - AQ = Q\,{}^t A - \sqrt{2}\begin{pmatrix} 0 & b_1 \\ 0 & eb_2 \end{pmatrix}$$

It follows that $dN \equiv 0$ since we have $dA \equiv 0$ and $dB \equiv 0$. Thus the only case in which reduction to an {e}-structure is possible is the case in which we can normalize the torsion term f.

We calculate $d^2Q\eta$ mod η to obtain the infinitesimal group action on f. This yields

$$0 \equiv (d(Q\varepsilon) - Q\varepsilon\,{}^t wB - {}^t wBQ\varepsilon)w$$
$$- (d(Q\varepsilon) - {}^t wBQ\varepsilon)I\theta$$
$$+ \left(Q\,{}^t A\varepsilon + Q\varepsilon A - \sqrt{2}\begin{pmatrix} 0 & b_1 \\ 0 & eb_2 \end{pmatrix}\varepsilon + df_1\right)w\theta$$

mod η. It follows from the Cartan lemma that

$$d(Q\varepsilon) = Q\varepsilon\,{}^t wB + {}^t wBQ\varepsilon + {}^t w\Psi + \theta\Phi$$

where, ${}^t w\Psi$ and Φ have the same symmetries as $Q\varepsilon$ modula base. If we substitute the expression for $d(Q\varepsilon)$ in the congruence above and wedge with $(w^2, -w^1)$ we get

$$df + (2a_1 + \sqrt{2}b_2)\varepsilon_1 + (a_3 + ea_2 - \sqrt{2}b_1)\varepsilon_2 \equiv 0,$$

where

$$A = \begin{pmatrix} a_1 & a_2 \\ a_3 & -a_1 \end{pmatrix} .$$

If we write A as the sum of its "Q-symmetric" and "Q-antisymmetric" parts,

$$A = \frac{1}{2}(A + Q^{-1t}AQ) + \frac{1}{2}(A - Q^{-1t}AQ)$$

then

$$A = \frac{1}{2}\begin{pmatrix} 2a_1 & a_2 + ea_3 \\ a_3 + ea_2 & -2a_2 \end{pmatrix} + \frac{1}{2}\begin{pmatrix} 0 & a_2 - ea_3 \\ a_3 - ea_2 & 0 \end{pmatrix}.$$

Thus unless the Q-symmetric part of A is equal to

$$1/\sqrt{2}\begin{pmatrix} -b_2 & eb_1 \\ b_1 & b_2 \end{pmatrix}$$

we can normalize f.

If we let $a_2 - ea_2 = -2\,ea$ then we obtain as necessary conditions for no further reduction to be possible that,

$$A = 1/\sqrt{2}\begin{pmatrix} -b_2 & eb_1 \\ b_1 & b_2 \end{pmatrix} + \begin{pmatrix} 0 & -ea \\ a & 0 \end{pmatrix}.$$

We can obtain further relations on the invariants a, b_1, b_2 and their covariant derivatives from the equation $0 = d^2w$.
By calculating dA and dB in this way we find two equations for the derivatives of b_2 in the η_1 and η_2 directions:

$$(21)\qquad \begin{matrix} db_2 - 1/\sqrt{2}b_2\eta_1 + (1/\sqrt{2}b_1 + a)\eta_2 \\ db_2 - 1/\sqrt{2}b_2\eta_1 + (1/\sqrt{2}b_1 - a)\eta_2 \end{matrix} \qquad \text{mod } \theta, w$$

Thus the integrability conditions force $a = 0$. Similarly we find

$$db_1 - 1/\sqrt{2}b_2\eta_1 + 1/\sqrt{2}b_1\eta_2 \mod \theta, w,$$

and we also obtain the relation that

$$f = \frac{1}{2}\{(b_1)_2 - (b_2)_1\} - 1/\sqrt{2}(b_2)_\theta$$

where $(b_1)_2$, $(b_2)_1$, and $(b_2)_\theta$ are covariant derivatives in the w^1, w^2, and θ directions respectively.

The structure equations are now

$$d\theta = {}^tI\eta\theta + {}^tw\eta$$

$$dw = Q\eta\theta + {}^twBw + A\theta w + {}^t\eta Cw$$

$$dQ\eta = -Q\varepsilon I\theta + Q\varepsilon w + {}^twBQ\eta + fw\theta$$
$$+ M\eta\theta$$

where

$$M = 1/\sqrt{2}\begin{pmatrix} -b_2 & -b_1 \\ b_1 & -eb_2 \end{pmatrix}$$

We can obtain more equations for the covariant derivatives of b_1 and b_2 from $d^2Q\eta$ as they appear in dM. If we now calculate $0 = d^2Q\eta \mod \theta$ we find, by using Cartan's lemma that

$$d(Q\varepsilon) = Q\varepsilon{}^twB + {}^twBQ\varepsilon + {}^tw\Psi + {}^t\eta\Pi \mod \theta$$

If we substitute in $0 \equiv d(Q\eta) \mod \theta$ we get, as a necessary condition for dB to be invariant that

$${}^t\eta\Pi = 1/\sqrt{2}\begin{pmatrix} \varepsilon_1\eta_1 - e\varepsilon_2\eta_2 & \varepsilon_1\eta_2 + \varepsilon_2\eta_1 \\ e\varepsilon_2\eta_1 + \varepsilon_2\eta_2 & e\,\varepsilon_2\eta_2 - \varepsilon_1\eta_1 \end{pmatrix} \mod \theta, w$$

It follows that when we compute $d^2(Q\eta) \mod w$ all of the fibre terms vanish and we obtain the congruence

$$0 \equiv dM\eta\theta - {}^t(Q\eta)\,BQ\eta\theta - M\eta{}^tI\eta\theta \mod w\,.$$

This yields

$$db_2\eta_1 + db_1\eta_1 \equiv 0$$
$$\mod(\theta,w)$$
$$db_1\eta_1 - edb_2\eta_2 \equiv 0.$$

But we already have (21), and thus we find

$$b_1\eta_1\eta_2 \equiv 0$$
$$\mod(\theta,w)$$
$$b_2\eta_1\eta_2 \equiv 0$$

so $b_1 = b_2 = 0$, and hence $f = 0$ as well.

Thus we have precisely one case in which there is no reduction. It is the case $I = {}^t(\sqrt{2},0)$, $A = B = f = 0$,

$$C = -1/\sqrt{2}\begin{pmatrix} \eta & \eta_2 \\ e\eta_2 & -\eta_2 \end{pmatrix}.$$

These conditions are satisfied by the Lagrangians

$$L = \frac{1}{2}(p^2 + eq^2)$$

so we see that these represent the only equivalence class for which we can not obtain an identity structure.

The structure equations in this case are

$$d\theta = {}^tI\eta\theta + {}^tw\eta$$

$$dw = Q\eta\theta + {}^t\eta Cw$$

$$d\eta = -\varepsilon I\theta + \varepsilon w.$$

If we re-write these equations in components with

$$\tau^1 = \theta,\ \tau^2 = w^1,\ \tau^3 = w^2$$

$$\tau^4 = \eta_1,\ \tau^5 = \eta_2$$

we have

$$d\tau^i = b_j^{i\rho}\varepsilon_\rho\tau^j + \gamma^i_{jk}\tau^j\tau^k$$

Following Cartan [4], we compute σ_1 which is the rank of the 5 X 3 matrix

$$\begin{pmatrix} b_1^{i1}\, t_i & b_1^{i2}\, t_i \\ b_2^{i1}\, t_i & b_2^{i2}\, t_i \\ b_3^{i1}\, t_i & b_3^{i2}\, t_i \\ b_4^{i1}\, t_i & b_4^{i2}\, t_i \\ b_5^{i1}\, t_i & b_5^{i2}\, t_i \end{pmatrix} \quad \begin{pmatrix} -\sqrt{2}\, t_4 & -\sqrt{2}\, t_5 \\ t_4 & t_5 \\ -e\, t_5 & t_4 \\ 0 & 0 \\ 0 & 0 \end{pmatrix}$$

where $t_1,\ldots,t_5$ are independent parameters. Thus $\sigma_1 = 2$, the higher σ_i are all zero, and

$$\sigma_0 + 2 = 5 .$$

Since there are two arbitrary parameters which give the ambiguity in i_1 and ε_2, Cartan's test ([4] p. 73) shows that the system is in involution and the general solution depends on $\sigma_1 = 2$ functions of 1 variable and $\sigma_0 = 1$ constant.

If we consider the example of the Lagrangian $L = \frac{1}{2} p\, q$, which is equivalent to $\frac{1}{2}(p^2 - q^2)$, it is clear that the 2-form

$$L\, dx\, dy = \frac{1}{2} p\, q\; dx\, dy$$

is invariant under the prolongation of the change of variables given by

$$\overline{x} = F(x) \qquad \overline{y} = G(y) \qquad \overline{z} = z + c.$$

This yiels

$$\overline{p} = F'(x)p \qquad \overline{q} = G'(y)q$$

so

$$\frac{1}{2}\, \overline{p}\, \overline{q}\, \overline{dx}\, \overline{dy} = \frac{1}{2} p\, q\, dx\, dy.$$

Similarly one may demonstrate the invariance of $L = \frac{1}{2}(p^2 + q^2)$ under a pseudogroup of coordinate transformations [1].

BIBLIOGRAPHY

1. Cartan, Élie Les Espaces Mêtriques Fondés sur la Notion d'Aire Hermann, Paris, 1933.

2. Notice sur les travaux scientifiques Gauthier-Villars, Paris, 1974.

3. "Espaces Generalisés" Actualites Scientifques et Industrielles No. 194, Hermann, Paris, 1935

"les sous-groupes des groupes continus de transformations" Ann. Éc. Normale, t.25, 1908, p.57-194

5. Debever, Robert "Sur un probleme d'équivalence et son application a la théorie des espaces métriques fondés sur la notion d'aire", Comptes Rendus de l'A.F.A.S. October 1945, Paris, pp. 83-84.

6. personal communication November, 1984.

7. De Donder, Th. "Théorie invariantive du calcul des variations", Bull. Acad. de Belg. 1929, chapter 1.

8. Gardner, Robert "Differential Geometric Methods Interfacing Control Theory" in Differential Geometric Control Theory, Ed. Brockett et al., Birkhäuser, N.Y., (1983) pp. 117-180.

9. Weyl, Hermann "Cartan on Groups and Differential Geometry" Bull. A.M.S. (1938), pp. 598-601.

University of North Carolina
Chapel Hill, N.C. 27514

University of Waterloo
Waterloo, Ontario N2L 3G1

Applications of the Gauss mapping for hypersurfaces of the sphere

by

Th.Hasanis and D.Koutroufiotis

Let M^n be an n-dimensional smooth oriented Riemannian manifold, and $\varphi : M^n \to S^{n+1}$ an isometric immersion of M^n into the unit hypersphere of R^{n+2}. The unit normal vectorfield of M^n in S^{n+1} induces a mapping $\nu : M^n \to S^{n+1}$, the Gauss mapping of the immersion φ. If we pull back onto M^n through ν the standard metric of S^{n+1}, we obtain on M^n a 2-covariant tensorfield, which is a new metric on M^n provided the Weingarten operator of φ is non-singular everywhere. The analogous construction for immersions in R^3 (more generally, R^{n+1}) is classical and rich in applications. In the sequel we shall use this new metric on M^n to obtain information on φ and M^n, under appropriate assumptions.

We denote by e the unit normal vectorfield of the immersion φ. The Gauss mapping ν is defined as follows: for any $p \in M^n$, $\nu(p)$ is the endpoint of the vector obtained by tranlating $e(p)$ parallel to itself in R^{n+2} so as to have initial point the center of S^{n+1}. If we identify M^n and $\varphi(M^n)$ locally, then, for any tangent vector X of M^n, we have $\nabla_X e = -AX$, where ∇ is the connection of S^{n+1} and A is the Weingarten operator of φ. Now we have in R^{n+2},

$$d\nu(X) = -AX \quad ; \tag{1}$$

the proof of this identity is exactly the same as in the case where φ immerses M^n in R^{n+1}. From (1) we deduce that if $\det A \neq o$ everywhere, the mapping ν is an immersion of M^n in S^{n+1}, and in fact an isometric immersion if we endow M^n with the new metric $< >_*$ defined by

(2) $$<X, Y>_* = <AX, AY> \quad .$$

Lemma 1. *If $\det A \neq o$, the immersion $\nu : M^n \to S^{n+1}$ has Weingarten operator A^{-1}.*

Proof. Observe first that the unit normal vectorfield e_* of ν at $p \in M^n$ is $e_*(p) = \pm x(p)$, where x denotes the position vector of $\varphi(p)$; for the tangent n-planes of $\varphi(M^n)$ at $\varphi(p)$ and $\nu(M^n)$ at $\nu(p)$ are parallel by (1), and e_* is perpendicular to e. We may assume $e_* = x$. Let A_* be the Weingarten operator of ν, and D the standard connection of R^{n+2}. We have for any tangent vector X of M^n:

$$\nabla_X e_* = -d\nu(A_* X) = A A_* X \quad ,$$

and the Gauss equation,

$$\nabla_X e_* = D_X x + <X, x> x = X \quad .$$

So $A A_* X = X$, and $A_* = A^{-1}$.

By virtue of Lemma 1 and the identity $<A^{-1}X, Y>_* = <AX, Y>$, the principal directions of ν are the same as those of φ, and the principal curvatures of ν are the reciprocals of those of φ. Therefore, the sectional curvature K_* of M^n in the metric $< >_*$ for the plane spanned by the orthogonal principal directions X_1, X_2, is

(3) $$K_*(X_1 \wedge X_2) = 1 + \frac{1}{k_1 k_2} \quad ,$$

where k_1, k_2 are the corresponding principal curvatures . Actually , if A is an arbitrary Codazzi tensor with $\det A \neq o$ on the Riemannian manifold M^n, then Wegner ([10],p.65) showed more generally that the sectional curvatures of $\langle\ \rangle_*$ and the original metric for the plane $X_1 \wedge X_2$, where $AX_i = k_i X_i$, are related by

(4) $$K_*(X_1 \wedge X_2) = \frac{K(X_1 \wedge X_2)}{k_1 k_2} \quad ;$$

and used this formula to obtain results on submanifolds of space forms .

Eisenhart ([1],p.159) ascribes the following lemma to Bianchi in the case of a hypersurface in a 4-dimensional space form . We owe to J.-P.Bourguignon the observation that it is valid in any dimension .

Lemma 2. *Let M^n be a hypersurface in the space form $N^{n+1}(c)$. Consider the function "sectional curvature of M^n at $p \in M^n$" : it attains its absolute extrema at 2-planes spanned by principal directions of M^n .*

Proof. Let $E_1, \ldots, E_n$ be an orthonormal basis of T_pM^n , consisting of eigenvectors of the Weingarten operator A at p ; that is , $AE_i = k_i E_i$, $i = 1,\ldots,n$. Set $M = \max\{ k_i k_j \mid i, j = 1,\ldots,n$ and $i \neq j\}$ and $m = \min\{ k_i k_j \mid i, j = 1,\ldots,n$ and $i \neq j\}$. Let X, Y be a pair of orthonormal vectors at p . Writing $X = \sum_i x_i E_i$ and $Y = \sum_i y_i E_i$, we have for the sectional curvature at the 2-plane $X \wedge Y$:

$$K(X \wedge Y) = c + \langle AX, X\rangle \langle AY, Y\rangle - \langle AX, Y\rangle^2$$

$$= c + (\sum_i k_i x_i^2)(\sum_j k_j y_j^2) - (\sum_i k_i x_i y_i)^2$$

$$= c + \sum_{i \neq j} k_i k_j x_i^2 y_j^2 - 2 \sum_{i<j} k_i k_j x_i y_i x_j y_j$$

$$= c + \sum_{i<j} (k_i k_j x_i^2 y_j^2 + k_j k_i x_j^2 y_i^2) - 2 \sum_{i<j} k_i k_j x_i y_i x_j y_j$$

$$= c + \sum_{i<j} k_i k_j (x_i y_j - x_j y_i)^2 \leq c + M \sum_{i<j} (x_i y_j - x_j y_i)^2 \quad .$$

Using the identity of Lagrange , $(\sum_i x_i^2)(\sum_i y_i^2) - (\sum_i x_i y_i)^2 = \sum_{i<j} (x_i y_j - x_j y_i)^2$, and $\sum_i x_i^2 = \sum_i y_i^2 = 1$, $\sum_i x_i y_i = 0$, we obtain $K(X \wedge Y) \leq c + M$, and similarly $c + m \leq K(X \wedge Y)$. The values $c+m$ and $c+M$ are clearly assumed by the function K at 2-planes spanned by eigenvectors of A .

Lemma 3. *Let M^n be an oriented hypersurface in the space form $N^{n+1}(c)$. Suppose there exist constants H_0 and α so that the mean curvature H and the Ricci curvature Ric of M^n satisfy everywhere $|H| \leq H_0$ and $Ric(X, X) \leq \alpha |X|^2$ for all X, where $\alpha < (n-1)c$. Then all the principal curvatures of M^n satisfy $|k_i| \geq \beta$ everywhere for a certain positive constant β. It follows that, if M^n is complete in the induced metric, then $\langle X, Y \rangle_* = \langle AX, AY \rangle$ is a new complete metric on M^n.*

Proof. We have the well-known identity

$$Ric(X, X) = (n-1)c \langle X, X \rangle + \mathrm{trace}\, A \langle AX, X \rangle - \langle A^2 X, X \rangle \quad . \tag{5}$$

Setting $X = E_i$, where $\{E_i\}$ is an orthonormal basis with $A E_i = k_i E_i$, we obtain $(n-1)c + n H k_i - k_i^2 \leq \alpha$, which we rewrite as

$$(k_i - \frac{nH}{2})^2 \geq \frac{n^2H^2}{4} + (n-1)c - \alpha \quad ;$$

from this follows

$$|k_i| \geq \left[\frac{n^2H^2}{4} + (n-1)\,c - \alpha \right]^{\frac{1}{2}} - \frac{n|H|}{2} \quad .$$

We can estimate the right-hand side of this last inequality from below using $|H| \leq H_0$, and obtain immediately

$$|k_i| \geq \frac{(n-1)c-\alpha}{\left[\frac{n^2H_0^2}{4} + (n-1)c-\alpha\right]^{\frac{1}{2}} + \frac{nH_0}{2}} = \beta > o \quad .$$

Because $k_i \neq o$, $< \; >_*$ is indeed a metric on M^n , and setting $X = \Sigma x_i E_i$ we compute

$$< X , X >_* = \sum_i k_i^2 x_i^2 \geq \beta^2 < X , X > \quad ;$$

So $< \; >_*$ is complete if $< \; >$ is .

Let a be a unit vector in R^{n+2} . The isometric immersion φ determines the smooth functions $z = < x , a >$ and $\zeta = < e , a >$, defined on M^n ; it is well-known and easily verified that they satisfy the equations

$$\Delta z + n z = (trace\, A)\, \zeta \quad ,$$

$$\Delta \zeta + trace\, (A^2)\zeta = (trace\, A)z - < grad(trace\, A) , a > \quad ,$$

where Δ and *grad* denote respectively the Laplacian and the gradient on M^n with respect to the original metric . Because ν is an isometric immersion with respect to $\langle\ \rangle_*$, and $x_* = e$, $e_* = x$, we have the corresponding functions $z_* = \zeta$ and $\zeta_* = z$ which satisfy

(6) $$\Delta_* \zeta + n\zeta = (\mathrm{trace}\, A^{-1})\, z \quad ,$$

(7) $$\Delta_* z + (\mathrm{trace}\, A^{-2})\, z = (\mathrm{trace}\, A^{-1})\, \zeta - \langle \mathrm{grad}_*(\mathrm{trace}\, A^{-1}), a \rangle \quad ,$$

where Δ_* and $grad_*$ are respectively , the Laplacian and the gradient on M^n with respect to $\langle\ \rangle_*$.

Theorem 1. *Suppose* M^n $(n \geq 2)$ *is a compact orientable manifold , and* $\varphi : M^n \to S^{n+1}$ *an immersion with* $\det A \neq 0$ *and* $\mathrm{trace}\, A^{-1} = \mathrm{const.}$ *If* $\varphi(M^n)$ *lies in a closed hemisphere , then* φ *imbeds* M^n *as a small hypersphere of* S^{n+1} .

Proof. We assume that $\varphi(M^n)$ lies in the hemisphere whose north pole has position vector a with respect to the center of S^{n+1} , so that $z \geq 0$. We set $\sum_i k_i^{-1} = c$, and compute immediately from (6) and (7) :

$$\Delta_*(c\zeta + nz) = (c^2 - n \sum_i k_i^{-2})\, z \quad .$$

By the Cauchy-Schwarz inequality , $c^2 = (\sum_i k_i^{-1})^2 \leq n \sum_i k_i^{-2}$, with equality if and only if $c \neq 0$ and $k_1 = \ldots = k_n = n/c$. Thus , $\Delta_*(c\zeta + nz) \leq 0$ on M^n , so $c\zeta + nz = \mathrm{const.}$ by the maximum principle , consequently $(c^2 - n \sum_i k_i^{-2}) z \equiv 0$. We cannot have $c = 0$, for in that case $z \equiv 0$, which would imply that $\varphi(M^n)$ is a great hypersphere , hence $\det A = 0$. Let $W = \{ p \in M^n \mid z(p) > 0 \}$. The set W is non-void , open , and $k_1 = \ldots = k_n = n/c$ on it . It follows that all points of the closure $\bar{W}$ are also mapped by φ onto umbilics . Therefore , by continuity , the open set $M^n - \bar{W}$ must be empty , since $\varphi(M^n - \bar{W})$ is made up of flat points , being part of a great hypersphere . So $\varphi(M^n)$ is all-umbilical , that is , a small hypersphere Σ . Because M^n is compact , the immersion $\varphi : M^n \to \Sigma$ is a covering mapping , and so a diffeomorphism since Σ is simply-connected .

The example of the Clifford torus with $k_1 = 1$, $k_2 = -1$ in S^3 shows that the hypothesis that $\varphi(M^n)$ lies in a closed hemisphere cannot be omitted in Theorem 1 . We mention that our result can also be obtained from a theorem of Nomizu and Smyth ([7],p.490) in conjunction with Lemma 1 and the observations on the normal image of $\varphi(M^n)$. We may also formulate a slightly weaker form of Theorem 1 as a Christoffel-type theorem : if M^n is compact , and $\varphi : M^n \to S^{n+1}$ is an immersion with all $k_i > 0$ and $\sum_i k_i^{-1} = const.$, then φ imbeds M^n as a small hypersphere . Indeed , the condition $k_i > 0$ implies that all the sectional curvatures of M^n are greater than 1 , so $\varphi(M^n)$ lies in an open hemisphere according to do Carmo and Warner ([3],p.133) . Note that here we can formally weaken the assumption that M^n be compact , and merely assume M^n to be complete .

The duality we established between the immersion φ and the immersion ν , supposing $\det A \neq 0$, permits us to write down immediately theorems that are "dual" to the rigidity theorems of Reilly ([8],p.492) concerning φ . Thus , denoting with σ_r^* $(1 \leqq r \leqq n-1)$ the normalized elementary symmetric functions of the eigenvalues of A^{-1} , we have : if either (i) σ_r^* and σ_{r-1}^* are constant and $z \geqq 0$, or (ii) σ_r^* and σ_{r+1}^* are constant and $\zeta \geqq 0$, then φ imbeds M^n into S^{n+1} as a small hypersphere . Our Theorem 1 is the case (i) , $r = 1$.

Given the connected Riemannian manifold M^n , we denote by $\bar{M}^n$ its orientable 2-sheeted Riemannian covering space , and by $\pi : \bar{M}^n \to M^n$ the covering projection , which is a local isometry .

Theorem 2. *Let* M^2 *be complete , with Gauss curvature* $K \leq o$. *If* $\varphi : M^2 \to S^3$ *is an isometric immersion satisfying* $|H| \leq H_0 \leq \infty$, *then* $\sup K = o$; *furthermore ,* $\bar{M}^2$ *is diffeomorphic to* R^2 *in case* $K \not\equiv o$.

Proof. We assign an orientation to the complete $\bar{M}^2$, and consider the isometric immersion $\varphi\pi$. By Lemma 3 , we have a new complete metric on $\bar{M}^2$, given by (2) . Since $K = 1 + k_1 k_2$, the Gauss curvature of this metric is , according to (3) ,

$$K_* = \frac{K}{K-1} \geq o \quad . \tag{8}$$

If $\sup K = -c^2$, then

$$K = \frac{K_*}{K_* - 1} \leq -c^2 \quad ,$$

which is equivalent to

$$K_* \geq \frac{c^2}{1+c^2}$$

because $K_* \geq o$. If we had $c \neq o$, the Bonnet-Hopf-Rinow theorem applied to $(\bar{M}^2 , < >_*)$ would imply that $\bar{M}^2$ is compact , and in fact of genus zero by the Gauss-Bonnet formula ; that is impossible since $\bar{M}^2$ also carries a metric of non-positive Gauss curvature ; so $c = o$.

Assume now $\sup K = o$ and $K \not\equiv o$; then $K_* \geq o$ and $K_* \not\equiv o$ by (8) . It follows that $\bar{M}^2$ cannot be compact ; otherwise

$$\int_{\bar{M}^2} K \, dA = \int_{\bar{M}^2} K_* \, dA_* > o \quad .$$

By the Blanc-Fiala-Huber theorem , $(\bar{M}^2, < >_*)$ is parabolic ; in particuler , $\bar{M}^2$ is diffeomorphic to R^2 .

The Clifford flat torus in S^3 with $H=const.$, and its covering spaces , the cylinder and the plane , are trivial illustrations of Theorem 2 . We have no example showing that the alternative $\sup K = o$, $K \not\equiv o$ can actually occur .

The Clifford flat torus can also be characterized as the only isometric immersion in S^3 of a complete orientable M^2 with $K \leqq o$, which satisfies $trace\, A^{-1} = const.$ To prove this , we observe first that the conditions $k_1 k_2 \leqq -1$ and $k_1^{-1} + k_2^{-1} = const.$ imply the k_i are bounded away from zero , so $(M^2, < >_*)$ is complete . Further , we have $K_* \geqq o$ by (8) . We proceed now in a standard way . Using Simons' formula , we find $\Delta_*(trace\, A^{-2}) \geqq o$; therefore $trace\, A^{-2} = const.$, because $(M^2, < >_*)$ is parabolic and $trace\, A^{-2}$ is bounded above . Thus , the k_i are constant , and $K \equiv o$ by Theorem 2 , etc.

The proof of Theorem 2 does not use Lemma 1 , only Lemma 3 and the formula (4) for the Gauss curvature of the metric $< >_*$. We may thus apply the method of proof also to immersions in other space forms . In the case of euclidean 3-space , we obtain in this manner a weak form of Efimov's theorem which is due to Stoker ([9],p.380) . In fact , it is easy to show with this method that a complete M^n with $Ric \leqq o$, isometrically immersed in R^{n+1} with $|H| \leqq H_0$, must satisfy $\sup Ric = \sup \{ Ric(X_p, X_p) , p \in M^n , |X_p| = 1 \} = o$. In the case of hyperbolic 3-space , we have $K_* = K/(K+1)$, and we obtain using this formula : if M^2 is complete , with $K \leqq -1$, and $\varphi : M^2 \to H^3(-1)$ is an isometric immersion with $|H| \leqq H_0$, then $\sup K = -1$.

Let $S^n(r) = \{ x \in R^{n+1} , |x| = r \}$, $S^m(s) = \{ y \in R^{m+1} , |y| = s \}$, where r and s are positive numbers with $r^2 + s^2 = 1$; then $S^n(r) \times S^m(s) = \{ (x, y) \in R^{n+m+2} , x \in S^n(r) , y \in S^m(s) \}$ is a hypersurface of $S^{n+m+1}(1)$. As is well-known , this hypersurface has two distinct constant principal curvatures : one is s/r of multiplicity n, the other is $-r/s$ of multiplicity m; it is minimal only in the case $r=\sqrt{n/(n+m)}$, $s=\sqrt{m/(n+m)}$.

Theorem 3. *Let M^3 be a complete 3-dimensional Riemannian manifold, and $\varphi : M^3 \to S^4$ an isometric minimal immersion; then $\sup Ric \geq 3/2$. If M^3 is compact, this supremum is strictly larger than $3/2$, unless $\varphi(M^3) = S^1(1/\sqrt{3}) \times S^2(\sqrt{2/3})$.*

Proof. We shall show that

$$Ric(X, X) \leq \alpha < 3/2 \quad , \quad \forall X \text{ with } |X| = 1 \tag{9}$$

leads to a contradiction. Going over to the orientable two-sheeted covering space of M^3 if need be, we assume that M^3 is oriented.

Applying Lemma 3 $(\alpha < 2)$, we infer that all the principal curvatures k_i of φ are non-zero, and that $(M^3, <\ >_*)$ is complete. Since $k_1+k_2+k_3=0$, we may suppose after an eventual change of orientation, that

$$k_1 > 0 \quad , \quad k_2 < 0 \quad , \quad k_3 < 0 \quad .$$

Let $AX_i = k_i X_i$, $|X_i| = 1$; applying the identity (5) for $X = X_i$, we have from (9) :

$$k_1 \geq \sqrt{2-\alpha} \quad , \quad -k_2 \geq \sqrt{2-\alpha} \quad , \quad -k_3 \geq \sqrt{2-a} \quad .$$

We set $k_1 = \sqrt{2-\alpha}+\delta$, and obtain now :

$$\sqrt{2-\alpha}+\delta = -k_2 - k_3 \geq 2\sqrt{2-\alpha} \quad ,$$

so $\delta \geq \sqrt{2-a}$ and hence $k_1 \geq 2\sqrt{2-\alpha}$. Consequently

(10) $$-k_1k_2 \geqq 2(2-\alpha) \quad , \quad -k_1k_3 \geqq 2(2-\alpha) \quad , \quad k_2k_3 \geqq 2-\alpha \quad .$$

We consider next the isometric immersion ν of $(M^3, <\ >_*)$ onto S^4. The principal directions of ν are the same as those of φ, by Lemma 1, and so, using (3) and (10), we have :

(11) $$K_*(X_1 \wedge X_2) \geqq 1 - \frac{1}{2(2-\alpha)} = \frac{3-2\alpha}{2(2-\alpha)} > 0 \quad ,$$

$$K_*(X_1 \wedge X_3) \geqq \frac{3-2\alpha}{2(2-\alpha)} > 0 \quad , \quad K_*(X_2 \wedge X_3) > 1 \quad .$$

Therefore, by Lemma 2, all the sectional curvatures of $(M^3, <\ >_*)$ are bounded from below by a positive constant ; so M^3 is compact by the Bonnet-Hopf--Rinow theorem.

We compute now the scalar curvature Sc_* of $(M^3, <\ >_*)$:

$$Sc_* = Ric_*(X_1, X_1) + Ric_*(X_2, X_2) + Ric_*(X_3, X_3)$$

$$= 2\big(K_*(X_1 \wedge X_2) + (K_*(X_1 \wedge X_3) + K_*(X_2 \wedge X_3)\big)$$

$$= 2\Big(3 + \frac{1}{k_1k_2} + \frac{1}{k_1k_3} + \frac{1}{k_2k_3}\Big) = 2\Big(3 + \frac{k_1+k_2+k_3}{k_1k_2k_3}\Big) = 6 \quad .$$

Thus, $\nu(M^3)$ is a compact hypersurface of S^4, with positive sectional curvatures and constant normalized scalar curvature 1. According to the classification of Cheng and Yau ([2], Theorem 2), $\nu(M^3)$ is either totally umbilical, or the Riemannian product of two totally umbilical constantly curved submanifolds. The first possibility cannot occur because the k_i do not all have the same sign ; neither can the second, otherwise there would exist 2-planes where $K_* = 0$. This proves the first part of our theorem.

We suppose now that M^3 is compact , and that $\alpha \leq 3/2$ in (9) . We deduce from the inequalities (11) that all the sectional curvatures of $(M^3, < >_*)$ are non-negative . Since the $Sc_* = 6$, the theorem of Cheng and Yau is applicable . The first possibility cannot occur for the same reason as before , but the second can . The principal curvatures $1/k_i$ of $\nu(M^3)$ satisfy $1+(1/k_i k_j) = o$ if $k_i \neq k_j$; see [2] , equation (3.3) . Thus , setting $k_1 = c$, we have $k_2 = k_3 = -1/c$. In order to compute c , we first use the identity $Ric(X_1, X_1) = 2 + k_1 k_2 + k_1 k_3$ to compute $Ric_*(X_1, X_1) = o$, $Ric_*(X_2, X_2) = 1 + c^2$, $Ric_*(X_3, X_3) = 1 + c^2$, and then $6 = Sc_* = 2 + 2c^2$ to find $c = \sqrt{2}$. Now we have $k_1^2 + k_2^2 + k_3^2 = 3$, and appealing , for example , to a theorem of Nomizu-Smyth ([6],p.376) , we conclude that $\varphi(M^3) = S^1(r) \times S^2(s)$. Since φ is minimal , $r = \sqrt{1/3}$ and $s = \sqrt{2/3}$.

What can we say about $\sup Ric$ if the immersion $\varphi : M^3 \to S^4$ is not assumed to be minimal , all other assumptions in Theorem 3 remaining ? One computes easily that in the example $S^1(r) \times S^2(s) \subset S^4$ we have $\sup Ric = 1/s^2$. Taking this as a clue , we shall show that at least "in most cases" , we have $\sup Ric > 1$.

Proposition. *Let* M^3 *be a complete 3-manifold , and* $\varphi : M^3 \to S^4$ *an isometric immersion satisfying* $|H| \leq H_0$ *; then* $\sup Ric > 1$ *unless* M^3 *is topologically a space form of positive sectional curvature .*

Proof. Suppose $\sup Ric \leq 1$. Going over to the oriented $\bar{M}^3$ if necessary , we consider the isometric immersion $\varphi\pi : \bar{M}^3 \to S^4$. We apply Lemma 3 , and obtain the complete $(\bar{M}^3, < >_*)$. The principal curvatures k_i of $\varphi\pi$ are all different from zero , and not of the same sign ; otherwise we would have $Ric(X, X) > 1$ for all principal unit vector X . So we may assume $k_1 > o$, $k_2 < o$, $k_3 < o$. We have $k_2 k_3 > \beta^2$ for some positive constant β , by Lemma 3 .

If $AX_2 = k_2X_2$, $|X_2| = 1$, then

$$Ric(X_2, X_2) = 2 + k_1k_2 + k_3k_2 \leqq 1 \quad ;$$

hence , $k_1k_2 \leqq -1-k_2k_3 < -(1+\beta^2)$, and similarly $k_1k_3 < -(1+\beta^2)$. Using equation (3) , we obtain now :

$$K_*(X_2 \wedge X_3) > 1 \quad , \quad K_*(X_1 \wedge X_2) > \frac{\beta^2}{1+\beta^2} \quad , \quad K_*(X_1 \wedge X_3) > \frac{\beta^2}{1+\beta^2} \quad .$$

By Lemma 2 , the sectional curvature of $(\bar{M}^3, < \ >_*)$ is bounded from below by a positive constant . So $\bar{M}^3$ is a compact 3-manifold that admits a metric of positive sectional curvature , and the same holds for M^3 . By Hamilton's theorem [4] , M^3 admits a metric of constant positive sectional curvature .

Do there exist hypersurfaces in S^4 , which are topologically S^3 and carry metrics that satisfy $\sup Ric \leqq 1$? If such do exist , their Ricci forms must have a negative eigenvalue at every point ; this is an easy algebraic deduction from (5) and the well-known identity

$$2K(X_1 \wedge X_2) = Ric(X_1, X_1) + Ric(X_2, X_2) - Ric(X_3, X_3) \quad ,$$

$\{X_1, X_2, X_3\}$ an orthonormal frame of M^3 .

REFERENCES

[1] L.P.Eisenhart , Riemannian Geometry , Princeton University Press , 1966

[2] S.-Y.Cheng and S.-T.Yau , Hypersurfaces with constant scalar curvature , Math. Ann. 225(1977) 195-204 . (Zbl. 349.53041).

[3] M.do Carmo and F.W.Warner , Rigidity and convexity of hypersurfaces in spheres , J.Diff.Geom. 4(1970)133-144 . (Zbl. 201,237).

[4] R.S.Hamilton , Three-manifolds with positive Ricci curvature , J.Diff.Geom. 17(1982) 255-306 . (Zbl. 504.53034).

[5] A.Huber , On subharmonic functions and differential geometry in the large , Comm.Math.Helv. 32(1957) 13-72 . (Zbl. 80,150).

[6] K.Nomizu and B.Smyth , A formula of Simon's type and hypersurfaces with constant mean curvature , J.Diff.Geom. 3(1969) 367-377 . (Zbl. 196,251).

[7] K.Nomizu and B.Smyth , On the Gauss mapping for hypersurfaces of constant mean curvature in the sphere , Comm.Math.Helv. 44(1969) 484-490 . (Zbl. 184,470).

[8] R.Reilly , Extrinsic rigidity theorems for compact submanifolds of the sphere , J.Diff.Geom. 4(1970) 487-497 . (Zbl. 208,495).

[9] J.J.Stoker , On the form of complete surfaces in three-dimensional space for which $K \leq -c^2$ or $K \geq c^2$, Studies in Math.Analysis and Related Topics , pp.377-387 , Stanford University Press 1962 . (Zbl. 137,407).

[10] B.Wegner , Codazzi-Tensoren und Kennzeichnungen sphärischer Immersionen , J.Diff.Geom. 9(1974) 61-70 . (Zbl. 278.53041).

University of Ioannina , Greece .

SUBMANIFOLDS AND THE SECOND FUNDAMENTAL TENSOR

Stefana Hineva

Faculty of Mathematics, Sofia University, 1126 Sofia, Bulgaria

0. Introduction

Let M^n be an n-dimensional submanifold of an (n+p)-dimensional Riemannian manifold N^{n+p}.

In this paper we topologically characterize M^n in the case when the square S of the length of the second fundamental tensor of M^n and the mean curvature h of M^n satisfy certain inequalities. More precisely, we estimate the sectional curvature of M^n in terms of the length S of the second fundamental form and the mean curvature and when the sectional curvature keeps a constant sign we use the theorems of Hadamard - Cartan, Bonnet - Myers (see [4] p.44), Gromoll - Meyer [6] and Gheeger - Gromoll [5] in order to characterize M^n.

1. Preliminaries

Let M^n be an n-dimensional submanifold of an (n+p)-dimensional Riemannian manifold N^{n+p}. We choose a local frame of orthonormal fields $e_1, \ldots, e_{n+p}$ in N^{n+p} such that, restricted to M^n the vectors $e_1, e_2, \ldots, e_n$ are tangent to M^n and the remaining vectors $e_{n+1}, \ldots, e_{n+p}$ are normal to M^n.

We shall use the following convention on the ranges of indices:

$$1 \leqq i, j, k, \ldots \leqq n; \qquad 1 \leqq \lambda, \mu, \nu, \ldots \leqq p$$

We denote the second fundamental form $\alpha : T_x M \times T_x M \to T_x^{\perp} M$ of M^n for $x \in M^n$ where $T_x M$ is the tangent space of M^n at x and $T_x^{\perp} M$ is the normal space to M^n at x, by its components h_{ij}^{λ} with respect to the frame $e_1, \ldots, e_{n+p}$.

We call

(1.1) $H=\sum_{\lambda}\frac{1}{n}h^{\lambda}e_{\lambda}$, $h^2=\frac{1}{n^2}\sum(h^{\lambda})^2$, where $h^{\lambda}=\sum_{i}h_{ii}{}^{\lambda}$

the mean curvature vector of M^n. h is the mean curvature of M^n. The square S of the length of the second fundamental form is given by:

(1.2) $$S=\sum_{\lambda}\Big[\sum_{i,j}(h_{ij}{}^{\lambda})^2\Big]$$

In general for a matrix $A=(a_{ij})$ we denote by $N(A)$ the square of the norm of A, i.e. $N(A)=\operatorname{trace} A.A^t=\sum_{i,j}(a_{ij})^2$

We have

(1.3) $$|\operatorname{trace} A|\leqq\sqrt{n.N(A)}\ .$$

S and h^{λ} are independant of our choice of orthonormal basis. Let X and Y be a pair of orthonormal vectors tangent to M^n at a point $x\in M^n$ and let us suppose that the local frame $e_1,\dots,e_{n+p}$ is so chosen that X and Y coincide with two arbitrary vectors of that frame. Let $X=e_{n-1}$, $Y=e_n$. Then the sectional curvature $k_M(\sigma)$ of M^n at the point x for the plane σ spanned by X and Y is written as follows:

(1.4) $$k_M(\sigma)=k_N(\sigma)+\sum_{\lambda}\Big[h^{\lambda}_{n-1,n-1}h^{\lambda}_{nn}-(h^{\lambda}_{n-1,n})^2\Big],$$

where $k_N(\sigma)$ is the sectional curvature of N^{n+p}.

2. Lemmas

We need the following lemmas.

Lemma 2.1. Let $A=(a_{ij})$ be a symmetric $(n\times n)$-matrix $(n\geqq 2)$. Then

(2.1) $$a_{n-1,n-1}a_{nn}-a^2_{n-1,n}\leqq\frac{1}{2n^2}\Big[(4-n)a^2+n(n-2)b+2|a|\sqrt{2(n-2)(bn-a^2)}\Big]$$

(2.2) $$a_{n-1,n-1}a_{nn}-a^2_{n-1,n}\geqq\begin{cases}\dfrac{a^2}{2(n-1)}-\dfrac{b}{2}, & \text{when } \dfrac{a^2}{b}<n-1\\[2ex] 0, & \text{when } n-1\leqq\dfrac{a^2}{b}\end{cases}$$

where $a=\operatorname{trace} A$, $b=N(A)$. The equality in (2.1) holds if and only if either $n=2$, or if $n \geqq 3$, A is of the form

$$A = \begin{pmatrix} a_1 & \cdots & 0 & 0 & 0 \\ \cdots & \cdots & \cdots & \cdots & \cdots \\ 0 & \cdots & a_1 & 0 & 0 \\ 0 & \cdots & 0 & a_n & 0 \\ 0 & \cdots & 0 & 0 & a_n \end{pmatrix},$$

where $a_1 = \frac{a}{n} \mp \frac{1}{n}\sqrt{\frac{2(nb-a^2)}{n-2}}$, $a_n = \frac{a}{n} \pm \frac{1}{n}\sqrt{\frac{(n-2)(nb-a^2)}{2}}$.

The equality in (2.2) is fulfilled only when either $n=2$ or if $n \geqq 3$, A is of the form

$$A = \begin{pmatrix} a_1 & \cdots & 0 & 0 & 0 \\ \cdots & \cdots & \cdots & \cdots & \cdots \\ 0 & \cdots & a_1 & 0 & 0 \\ 0 & \cdots & 0 & \frac{a_1 \mp c}{2} & a_{n-1,n} \\ 0 & \cdots & 0 & a_{n-1,n} & \frac{a_1 \pm c}{2} \end{pmatrix}.$$

where $c = \frac{1}{n-1}\sqrt{a^2(3-2n) + 2(b-2a_{n-1,n}^2)(n-1)^2}$, $a_{n-1,n}^2 \leqq \frac{b}{2} + \frac{a^2(3-2n)}{4(n-1)^2}$,

$a_1 = \frac{a}{n-1}$ or

$$A = \begin{pmatrix} a_{11} & a_{12} & \cdots & a_{1,n-1} & a_{1n} \\ a_{21} & a_{22} & \cdots & a_{2,n-1} & a_{2n} \\ \cdots & \cdots & \cdots & \cdots & \cdots \\ a_{n-1,1} & a_{n-1,2} & \cdots & 0 & 0 \\ a_{n1} & a_{n2} & \cdots & 0 & 0 \end{pmatrix}$$

Proof. We consider all symmetric (n×n) matrices $C = (c_{ij})$ and apply Lagrange - multiplier method to determine the extreme values of the function

$$f = c_{n-1,n-1}\, c_{nn} - c_{n-1,n}^2 \tag{2.3}$$

when c_{ij} satisfy the complementary conditions:

$$\sum_i c_{ii} = a$$

$$\sum_i c_{ii}^2 + 2\sum_{i<j} c_{ij}^2 = b \tag{2.4}$$

We construct the function

$$\phi = c_{n-1,n-1}c_{nn} - c_{n-1,n}^2 + \lambda\left(\sum_i c_{ii} - a\right) + \mu\left(\sum_i c_{ii}^2 + 2\sum_{i<j} c_{ij}^2 - b\right), \tag{2.5}$$

where λ and μ are the Lagrange multipliers. Partially differentiating with respect to the elements of the matrix C and equating the derivatives to zero, we obtain the following system of equations

$$(2.6)\qquad \begin{aligned} &\lambda + 2\mu c_{ii} = 0 && i = 1, \ldots, n-1 \\ &c_{nn} + \lambda + 2\mu c_{n-1,n-1} = 0 \\ &c_{n-1,n-1} + \lambda + 2\mu c_{nn} = 0 \\ &(2\mu - 1) c_{n-1,n} = 0 \\ &\mu c_{ij} = 0 && 1 \leq i < j \leq n,\ i \neq n-1 \end{aligned}$$

We consider three cases.

i/ $\mu \neq 0, \frac{1}{2}$

From (2.6) and (2.4) we obtain

$$(2.7)\qquad \begin{aligned} &c_{n-1,n} = 0 \\ &c_{ij} = 0 \\ &c_{ii} = \frac{a}{n} \mp \frac{1}{n}\sqrt{\frac{2(nb-a^2)}{n-2}} \\ &c_{n-1,n-1} = c_{nn} = \frac{a}{n} \pm \frac{1}{n}\sqrt{\frac{(n-2)(nb-a^2)}{2}} \end{aligned}$$

The extremal values of the function f are

$$(2.8)\qquad f_1^{extr} = \frac{1}{2n^2}\left[(4-n)a^2 + n(n-2)b \pm 2a\sqrt{2(n-2)(nb-a^2)}\,\right]$$

The corresponding matrices to these extremal values are

$$C^{extr} = \begin{pmatrix} a_1 & \ldots & 0 & 0 & 0 \\ \ldots & \ldots & \ldots & \ldots & \ldots \\ 0 & \ldots & a_1 & 0 & 0 \\ 0 & \ldots & 0 & a_n & 0 \\ 0 & \ldots & 0 & 0 & a_n \end{pmatrix}, \qquad \text{where } \begin{aligned} a_1 &= c_{ii} \\ a_n &= c_{nn} \end{aligned} \qquad \text{(from (2.7))}$$

ii/ $\mu = 0$

From (2.6) we have

$$(2.9)\qquad \begin{aligned} &c_{ii},\ c_{ij} \ \text{- arbitrary numbers} \\ &c_{n-1,n-1} = c_{nn} = c_{n-1,n} = 0 \end{aligned}$$

and then $f_2^{extr} = 0$

iii/ $\mu = \frac{1}{2}$

In this case (2.6) reduces to

$$(2.10)\qquad \begin{aligned} &c_{nn} + c_{n-1,n-1} = c_{ii}\,, \qquad c_{ij} = 0 \\ &c_{n-1,n} \ \text{- arbitrary number} \end{aligned}$$

From (2.4) and (2.10) we determine

$$c_{11} = \frac{a}{n-1}$$

(2.11) $$c_{n-1,n-1} = \frac{a}{2(n-1)} \mp \frac{1}{2(n-1)} \sqrt{a^2(3-2n) + 2(b-2c_{n-1,n}^2)(n-1)^2}$$

$$c_{n,n} = \frac{a}{2(n-1)} \pm \frac{1}{2(n-1)} \sqrt{a^2(3-2n) + 2(b-2c_{n-1,n}^2)(n-1)^2}$$

where $c_{n-1,n}^2 \leqq \frac{b}{2} + \frac{a^2(3-2n)}{4(n-1)^2}$

(2.12) $$f_3^{extr} = \frac{a^2}{2(n-1)} - \frac{b}{2}$$

The corresponding matrix is of the form

$$C^{extr} = \begin{pmatrix} c_1 & \cdots & 0 & 0 & 0 \\ \vdots & \ddots & & & \vdots \\ 0 & \cdots & c_1 & 0 & 0 \\ 0 & \cdots & 0 & \frac{c_1 \mp d}{2} & c_{n-1,n} \\ 0 & \cdots & 0 & c_{n-1,n} & \frac{c_1 \pm d}{2} \end{pmatrix}$$

where $c_1 = c_{11} = \frac{a}{n-1}$, $d = \frac{1}{n-1}\sqrt{a^2(3-2n) + 2(b-2c_{n-1,n}^2)(n-1)^2}$

We shall compare the extremal values of f .

As the two parts of f_1^{extr} are symmetric with respect to a we shall investigate only the part

(2.13) $$f_{11}^{extr} = \frac{1}{2n^2}\left[(4-n)a^2 + n(n-2)b - 2a\sqrt{2(n-2)(nb-a^2)}\right], \quad |a| \leqq \sqrt{nb} .$$

We consider f_{11}^{extr} as a function of $\alpha = \frac{a}{\sqrt{b}}$, b and n being fixed and write it down in the form $f_{11}^{extr} = Fb$, where

(2.14) $$F = \frac{1}{2n^2}\left[(4-n)\alpha^2 + n(n-2) - 2\alpha\sqrt{2(n-2)(n-\alpha^2)}\right].$$

We have

(2.15) $$F^{extr} = \begin{cases} 0 & \text{when } \alpha = \sqrt{n-2} \\ \frac{1}{2} & \text{when } \alpha = -\sqrt{2} . \end{cases}$$

From (2.15) we conclude that

(2.16) $F \geqq 0,\ F = 0$ iff $\alpha = \sqrt{n-2}$

and consequantly $f_{11}^{extr} \geqq 0$, $f_{11}^{extr} = 0$ iff $\alpha = \sqrt{n-2}$.

Other remarkable points of F' are $F=\frac{n-2}{2n}$ for $\alpha=0$ and $F=\frac{1}{n}$ for $\alpha^2=n$.

We represent f_3^{extr} as $f_3^{extr}=\phi b$, where

(2.17) $$\phi=\frac{1}{2(n-1)}(\alpha^2-n+1), \quad \alpha^2=\frac{a^2}{b}.$$

We have

(2.18) $$\phi^{extr}=-\frac{1}{2} \qquad \text{for } \alpha=0.$$

Other interesting values of ϕ are $\phi=0$ for $\alpha^2=n-1$ and $\phi=\frac{1}{2n-2}$ for $\alpha^2=n$

Thus we have

(2.19) $$\begin{aligned} &\phi<0 \qquad \text{for } |\alpha|<\sqrt{n-1}, \\ &\phi\geqq 0 \qquad \text{for } |\alpha|\geqq\sqrt{n-1},\ \phi=0 \text{ iff } |\alpha|=\sqrt{n-1}. \end{aligned}$$

Next we shall find the common points of F and ϕ. From (2.16) and (2.19) we see that when $|\alpha|\leqq\sqrt{n-1}$, F and ϕ have no common points. The unique common point of F and ϕ is when $\alpha=(n-1)\sqrt{\frac{2}{2n-3}}$ and at this point $\phi=F=\frac{1}{2(2n-3)}$ and $\phi'=F'$. Consequently we always have $\phi\leqq F$.

Thus

(2.20) $$f^{max}=\frac{1}{2n^2}\left[(4-n)a^2+n(n-2)b+2|a|\sqrt{2(n-b)(bn-a^2)}\right],\ f^{max}>0,$$

$$f^{min}=\begin{cases}\frac{a^2}{2(n-1)}-\frac{b}{2} & \text{when } \frac{a^2}{b}<n-1\\ 0 & \text{when } n-1\leqq\frac{a^2}{b}\leqq n.\end{cases}$$

<u>Lemma 2.2.</u> Let $A=(a_{ij})$ be a symmetric $(n\times n)$ matrix $(n\geqq 2)$. Then

(2.21) $$|a_{n-1,n-1}a_{nn}-a_{n-1,n}^2|\leqq\frac{1}{2}b, \qquad b=N(A).$$

The equalities in (2.21) hold only when either $a=trace\,A=\pm\sqrt{2b}$ and the matrix is of the form

$$A=\begin{pmatrix}0 & \cdots & 0 & 0 & 0\\ \cdots & \cdots & \cdots & \cdots & \cdots\\ 0 & \cdots & 0 & 0 & 0\\ 0 & \cdots & 0 & a_n & 0\\ 0 & \cdots & 0 & 0 & a_n\end{pmatrix} \qquad a_n=\pm\sqrt{\frac{b}{2}}$$

or $a=0$ and A is of the form

$$A = \begin{pmatrix} 0 & \cdots & 0 & 0 & 0 \\ \cdots & \cdots & \cdots & \cdots & \cdots \\ 0 & \cdots & 0 & 0 & 0 \\ 0 & \cdots & 0 & c & a_{n-1,n} \\ 0 & \cdots & 0 & a_{n-1,n} & -c \end{pmatrix} .$$

where $c = \pm\frac{1}{2}\sqrt{2(b-2a^2_{n-1,n})}$, $a^2_{n-1,n} < \frac{b}{2}$.

Proof. The proof follows from that of lemma 2.1. Really from (2.15) and (2.18) we observe that when a runs the interval $[-\sqrt{nb}, +\sqrt{nb}]$, where b and n are fixed the biggest value of f is $b/2$, when $a^2 = 2b$ and the smallest value of f is $-b/2$, when $a^2 = 0$.

3. Estimates of the sectional curvature of M^n

We shall estimate the sectional curvature of M^n at a point $x \in M^n$ by terms of the square S of the length of the second fundamental form and the mean curvature of M^n at this point.

Theorem 3.1. Let M^n be an n-dimensional submanifold in an (n+p)-dimensional Riemanniam manifold N^{n+p}. Let X and Y be a pair of orthonormal vectors in $T_x M^n$, where $x \in M^n$. For the sectional curvature $k_M(\sigma)$ of the plane σ spanned by X and Y at a point $x \in M^n$ we have

(3.1) $$k_M(\sigma) \leqq k_N(\sigma) + \frac{1}{2n^2}\Big[(4-n)\sum_\lambda (h^\lambda)^2 + n(n-2)S + 2\big(\sum_\lambda |h^\lambda|\big)\sqrt{2(n-2)[nS - \sum_\lambda (h^\lambda)^2]}\ \Big].$$

(3.2) $$k_M(\sigma) \geqq k_N(\sigma) + \frac{1}{2(n-1)}\sum_\lambda (h^\lambda)^2 - \frac{1}{2}S , \quad \text{when } \frac{1}{n-1}\sum_\lambda (h^\lambda)^2 < S ,$$

$$k_M(\sigma) \geqq k_N(\sigma) , \qquad \text{when } S \leqq \frac{1}{n-1}\sum_\lambda (h^\lambda)^2 ,$$

where $k_N(\sigma)$ denotes the sectional curvature of N^{n+p}.

Proof. Let X and Y be a pair of orthonormal vectors in $T_x M^n$, $x \in M^n$. We choose a local frame of orthonormal fields $e_1, e_2, \ldots, e_{n+p}$ for N^{n+p} such that $e_1, \ldots, e_n$ are tangent to M^n and $X = e_{n-1}$ and $Y = e_n$. Then for the sectional curvature $k_M(\sigma)$ of M^n for the plane σ spanned by e_{n-1} and e_n at a point $x \in M^n$ we have (1.4).

Using lemma 2.1 for the matrices (h_{ij}^{λ}) from (2.1), (2.2), (1.1) and (1.2) we obtain the following estimates for the function $\psi = \sum_{\lambda}[h_{nn}^{\lambda}h_{n-1,n-1}^{\lambda} - (h_{n-1,n}^{\lambda})^2]$:

$$\psi \leqq \frac{1}{2n^2}\Big[(4-n)\sum_{\lambda}(h^{\lambda})^2 + n(n-2)S + 2\big(\sum_{\lambda}|h^{\lambda}|\big)\sqrt{2(n-2)\big[nS - \sum_{\lambda}(h^{\lambda})^2\big]}\Big] \tag{3.3}$$

$$\psi \geqq \begin{cases} \frac{1}{2(n-1)}\sum_{\lambda}(h^{\lambda})^2 - \frac{1}{2}S & \text{when } \frac{1}{n-1}\sum_{\lambda}(h^{\lambda})^2 < S, \\ 0 & \text{when } S \leqq \frac{1}{n-1}\sum_{\lambda}(h^{\lambda})^2 . \end{cases} \tag{3.4}$$

From (1.4), (3.3) and (3.4) follow (3.1) and (3.2).

We shall give some corollaries of this theorem.

Corollary 3.1. The sectional curvature $k_M(\sigma)$ of M^n at a point x for all planes $\sigma \in T_x M^n$ is non-negative $(k_M(\sigma) \geqq 0)$ if

$$k_N(\sigma) \geqq \frac{1}{2}S - \frac{1}{2(n-1)}\sum_{\lambda}(h^{\lambda})^2, \qquad \text{when } \frac{1}{n-1}\sum_{\lambda}(h^{\lambda})^2 < S, \tag{3.5}$$

or

$$k_N(\sigma) \geqq 0, \qquad \text{when } S \leqq \frac{1}{n-1}\sum_{\lambda}(h^{\lambda})^2 . \tag{3.6}$$

Corollary 3.2. $k_M(\sigma) \geqq k_N(\sigma)$ for all planes $\sigma \in T_x M^n$ at a point $x \in M^n$, when

$$S \leqq \frac{1}{n-1}\sum_{\lambda}(h^{\lambda})^2 . \tag{3.7}$$

Corollary 3.3. $k_M(\sigma) \leqq 0$ for all planes $\sigma \in T_x M^n$ at a point $x \in M^n$, when

$$k_N(\sigma) \leqq -\frac{1}{2n^2}\Big[\sum_{\lambda}(h^{\lambda})^2(4-n) + n(n-2)S + 2\big(\sum_{\lambda}|h^{\lambda}|\big)\sqrt{2(n-2)\big[nS - \sum_{\lambda}(h^{\lambda})^2\big]}\Big] . \tag{3.8}$$

(3.8) holds only when $k_N(\sigma)$ is negative as the right side of (3.8) is negative.

Theorem 3.2. Of all n-dimensional submanifolds in a Riemannian manifold N^{n+p}, which pass through a point $x \in N^{n+p}$ and have at x :

a/ the same tangent space b/ the same square of the length of the second fundamental form, the minimal submanifolds whose second fundamental tènsors are of the form

$$(h_{ij}^{\lambda}) = \begin{pmatrix} 0 & \dots & 0 & 0 & 0 \\ \dots & \dots & \dots & \dots & \dots \\ 0 & \dots & 0 & 0 & 0 \\ 0 & \dots & 0 & c^{\lambda} & h_{n-1,n}^{\lambda} \\ 0 & \dots & 0 & h_{n-1,n}^{\lambda} & -c^{\lambda} \end{pmatrix},$$

where $c^{\lambda} = \pm \frac{1}{2}\sqrt{2[S-2(h_{n-1,n}^{\lambda})^2]}$, $(h_{n-1,n}^{\lambda})^2 \leq \frac{1}{2}\sum_{i,j}(h_{ij}^{\lambda})^2$,

have the smallest sectional curvature at the point x for all σ ; the submanifolds whose second fundamental tensors are of the form

$$(h_{ij}^{\lambda}) = \begin{pmatrix} 0 & \dots & 0 & 0 & 0 \\ \dots & \dots & \dots & \dots & \dots \\ 0 & \dots & 0 & 0 & 0 \\ 0 & \dots & 0 & a_n & 0 \\ 0 & \dots & 0 & 0 & a_n \end{pmatrix} \qquad a_n = \pm\sqrt{\frac{b}{2}}$$

have the biggest sectional curvature at the point x for all planes σ.

We prove the theorem by the aid of lemma 2.2.

4. Characterization of the submanifolds in N^{n+p}

Theorem 4.1. A complete simply connected n-dimensional submanifold M^n in a Riemannian manifold N^{n+p} of negative sectional curvature is diffeomorphic to R^n if the second fundamental tensor of M^n satisfies (3.8).

The proof follows from corollary 3.3 and the theorem of Hadamard - Cartan (see [4] , p.44).

Corollary 4.1. If the second fundamental tensor of an n-dimensional complete simply connected submanifold M^n in an (n+p)-dimensional Riemannian manifold N^{n+p} of constant negative curvature $(c < 0)$ satisfies

$$\text{(4.1)} \qquad \frac{1}{2n^2}\Big[(4-n)\sum_{\lambda}(h^{\lambda})^2 + n(n-2)S + 2\big(\sum_{\lambda}|h^{\lambda}|\big)\sqrt{2(n-2)\big[nS - \sum_{\lambda}(h^{\lambda})^2\big]}\Big] \leq -c$$

then M^n is diffeomorphic to R^n.

Theorem 4.2. A complete connected n-dimensional submanifold M^n in an (n+p)-dimensional Riemannian manifold N^{n+p} of positive curvature bounded below by a constant $c > 0$ is compact with diameter $\leq \frac{\pi}{\sqrt{c}}$

if the second fundamental form satisfies (3.7).

Remark. Another proof of this theorem in the case when N^{n+p} is of constant positive curvature is given by M.Okumura [3] .

Theorem 4.3. Let M^n be an n-dimensional non-compact complete connected submanifold in an (n+p)-dimensional Riemannian manifold N^{n+p}

If at each point x of M^n, for which $\frac{1}{n-1}\sum_\lambda (h^\lambda)^2 < S$ the inequality

$k_N(\sigma) \geqq \frac{1}{2}S - \frac{1}{2(n-1)}\sum_\lambda (h^\lambda)^2$ is fulfilled or if at each point x ,

for which $S \leqq \frac{1}{n-1}\sum_\lambda (h^\lambda)^2$ the inequality $k_N(\sigma) \geqq 0$ holds , then

there exists in M^n a compact totally geodesic and totally convex submanifold Q_M without boundary such that M^n is diffeomorphic to the normal bundle of Q_M . In the case when N^{n+p} is of positive curvature which is not bounded bellow by a positive constant then

M^n is diffeomorphic to R^n if $S \leqq \frac{1}{n-1}\sum_\lambda (h^\lambda)^2$.

We prove this theorem using corollary 3.1 and the theorems of Gheeger and Gromoll [5] and Gromoll and Meyer [6].

REFERENCES

1. Spivak, M.: A Comprehensive Introduction to Differential Geometry. Vol. four. Berkeley: Publish or Perish (1979; Zbl. 439.53004).

2. Chen, B. Y.: Geometry of Submanifolds. New York: Marcel Dekker. (1973; Zbl. 262.53036).

3. Okumura, M.: A pinching problem on the second fundamental tensors and submanifolds of a sphere. Tohoku Math. J. 25, 461 - 467. (1973; Zbl. 284.53039).

4. Chern, S. S.: Differential Geometry; its past and its future. Actes, Congres intern. math., Tome 1, 41 - 53. (1970; Zbl. 232.53001).

5. Cheeger, J.; Gromoll, D.: The structure of complete manifolds of nonnegative curvature. Bull. Amer. Math. Soc. 74, 1147 - 1150. (1968; Zbl. 169.241).

6. Gromoll, D.; Meyer, W.: On complete open manifolds of positive curvature. Annals of Math., II. Ser. 90, 75 - 90. (1969; Zbl. 191.199).

EMBEDDED MINIMAL SURFACES, COMPUTER GRAPHICS AND ELLIPTIC FUNCTIONS

by

David Hoffman
University of Massachusetts

A longstanding problem in the theory of minimal surfaces is the construction of complete embedded minimal surfaces with finite topology. Many examples of complete properly embedded minimal surfaces in Euclidean three space have been constructed, but except for the plane, the catenoid and the helicoid, all the known surfaces have infinite genus. Since it is natural to try to develop a theory of the global geometry of embedded minimal surfaces of finite type, the lack of examples is a major obstacle. In fact, it has often been conjectured that these three examples are the only complete embedded minimal surfaces in R^3 of finite topological type.
We have recently proved that this conjecture is false. This is joint work with W.H. Meeks III. We have established that there exist complete embedded minimal surfaces of every genus $k > 0$ with three ends. These surfaces are described in [3],[4] and [5]. The model and inspiration for the construction of these surfaces came from looking carefully, using computer graphics, at an example that had been described in the thesis of C. Costa, [2] (see also [1], Section 6, and Appendix.) Here we will describe that surface in detail and indicate how the proof of embeddedness goes. (For complete details, see [4].)

In 1980, Jorge and Meeks [3] developed a theory that could be used to study the geometry and topology of complete embedded minimal surfaces in R^3 of finite total curvature. (By a classical theorem of R. Osserman [8] a complete minimal surface of finite total curvature is conformally a compact Riemann surface $\overline{M}$ of genus k with a finite number, $r \geq 1$, of points removed. Also the Gauss map, meromorphic on M extends to a meromorphic function on $\overline{M}$.) Among other things, they pointed out that for a complete embedded minimal surface with finite total curvature, one must have equality in the classical estimate

$$\int_M K\,da \leq -4\pi(k + r - 1)$$

of Osserman-Cohn-Vossen. Furthermore, after a rotation, the extended Gauss map must take one of the values $(0,0 \pm 1)$ at each of the ends; that is, the asymptotic tangent planes at infinity are parallel. Using this, they were able to prove that the only complete embedded minimal surface of finite total curvature with one end $(r = 1)$ is the flat plane, a known folk theorem. Furthermore they showed that embeddedness was impossible for complete minimal surfaces with finite total curvature, genus zero and three, four or five ends $(k = 0,$ $3 \leq r \leq 5)$.

Recently, R. Schoen [5] has proved that the only embedded minimal surface of any genus with finite total curvature and two ends $(r = 2)$ is the catenoid. The catenoid has genus zero, two ends and total curvature -4π.

Here is a scorecard of the aforementioned nonexistence results. P stands for the plane, C for the catenoid and X for a topology that is known not to exist for a complete embedded minimal surface of finite total curvature. The integer k is the genus and the integer r is the number of ends.

r \ k	0	1	2	3	4	5	6	..	
0	X	X	X	X	X	X	X	X	...
1	P	X	X	X	X	X	X	X	...
2	C	X	X	X	X	X	X	X	...
3	X								
4	X								
5	X								
6									

According to the results of Meeks-Jorge, an example with "(k,r) topology" would have to have total curvature $-4\pi(k+r-1)$. The plane and the catenoid have total curvatures 0 and -4π, respectively. Reading the scorecard we see readily that -8π is not a possibility and the next largest total curvature that could occur is -12π, when $(k,r) = (1,3)$.

In 1982, Costa [2] established the existence of a complete minimal surface of genus $k = 1$ with $r = 3$ parallel ends: The Riemann surface M is the torus associated with the square lattice on $\mathbb{C}$. Taking as fundamental domain the unit-square minus the points 0, $\omega_1 = 1/2$ and $\omega_2 = i/2$, let P be the Weierstrass P-function. Let $\eta = P dz$ and $g = a/P'$, where $a = 2\sqrt{2\pi}P(\omega_1)$. The surface is given by using η and g in the Weierstrass Representation Formula. (See Hoffman-Osserman [6].)

(1) For $\Phi = \eta(1-g^2,\ i(1+g^2),\ 2g)$.

$$X(z) = \mathrm{Re}\int_{z_0}^{z} \Phi \quad .$$

The Gauss map of X is (after stereographic projection) given by g, which in this case is a/P'. Since P' is of order three on the torus, the total curvature is -12π, the right amount for X to be an embedding. Also, $P'(0) = \infty$, $P'(\omega_1) = P'(\omega_2) = 0$ which implies that g is 0 or ∞ at the ends; the ends must be parallel. The pole structure of Φ at the ends makes it easy to prove that X is complete. The choice of constant a is forced by the requirement that X have no real periods.

From the formula (1) it is very hard to visualize what the surface looks like, though it is easy to establish that near ω_1 and ω_2 it looks like opposite ends of a vertical catenoid and near O it looks like a horizontal plane. In particular, it can be shown (see B. below) that the surface is embedded outside of some compact set. In order to check for embeddedness, we used a computer to calculate points on the surface and then used a computer graphics package, developed by J. Hoffman, to draw the surface from any viewing point we desired. We observed that the surface was in fact embedded, and we could see that it possessed certain symmetries. We then proved these things mathematically.

Theorem 1. The minimal surface (1) is embedded.

1. Its symmetry group is the dihedral group D_4 with eight elements, generated by

$$L = \begin{bmatrix} 0 & -1 & 0 \\ 1 & 0 & 0 \\ 0 & 0 & -1 \end{bmatrix} \quad \text{and} \quad K = \begin{bmatrix} 1 & 0 & 0 \\ 0 & -1 & 0 \\ 0 & 0 & 1 \end{bmatrix} .$$

2. The surface contains two straight lines in the (x_1,x_2)-plane, crossing at right angles. (The lines are $x_1 \pm x_2 = x_3 = 0$.)
3. The surface is made up of eight congruent pieces, each lying in a distinct octant of R^3. Each piece is a simply connected graph.

The mathematical proof of symmetry is based on properties of the Weierstrass P-function. During the course of our investigation, we were able to test ideas for proofs that we had by doing numerical computations. Also, we verified our proof that the surface decomposed into eight pieces, each in its own octant, by actually drawing the appropriate eighth part of the surface.

The special properties of the surface of Theorem 1 are consequences of special properties of the P-function for the square lattice:

1. If $\omega_3 = \omega_1 + \omega_2$, $P(\omega_3 + iz) = -P(\omega_3 + z)$ and

$$P(\omega_3 + \bar{z}) = \bar{P}(\omega_3 + z) \quad .$$

II. $P'^2 = 4P(P - e_1)(P + e_1)$, where $e_1 = P(\omega_1)$.

Property I provided a group of motions of the torus that lifted to the Gauss map as isometries of S^2. Property I, which makes all the calculations come out neatly, could be viewed as the defining relation for the torus. The function P' is used to construct the Gauss map, while P is used to create the auxiliary function $\eta = f\,dz$ in the Weierstrass representation.

We wish present an outline of the embeddedness proof somewhat different form our original one. We begin by focusing on Property II. Consider the Riemann surface M defined by the equation

(2) $$y^2 = x(x^2 - 1)$$

deleting the points p_{-1}, p_1 and p_∞ where x equals 1, -1 and infinity, respectively. This is in fact the square torus and up to a multiplicative constant:

$$x = aP \quad \text{and} \quad y = aP' \ , \ a \in R.$$

The points where x equals 1, -1 and infinity correspond to the points ω_1, ω_2, and 0 in $\mathbb{C}/L$. On this surface, let

(3) $$\eta = f\,dz = (x/y)dx \text{ and } g = b/y$$

for some real constant b. From (2) it follows that

$$\eta = (P/P')P'\,dz = P\,dz \quad \text{and} \quad g = b/P'.$$

Now put the values of f and g from (3) into the formula for Φ in the Weierstrass formula (1). We are looking at another incarnation of the example described by Costa. (Again, the constant b may be chosen so that there are no real periods. For details see [1], Section 6.) We will now sketch a proof of the embeddedness, using this model for the surface.

A. Outside of a compact set $C \subset M$, the surface is embedded.

Choosing for the base-point of integration in (1) the point p_0 where $x = 0$ on M, it is easy to see that the third component of X(p) in (1) is given by

$$(4) \quad x_3 = \mathrm{Re}\int \eta g = \mathrm{Re}\int (bx/y^2)\,dx = \mathrm{Re}\, b\int dx/(x^2-1) =$$

$$= (b/2)\ \ln|(x(p)-1)/(x(p)+1)|.$$

From (4) it is clear that

$$(5) \quad x_3(p) \to \begin{cases} \infty \\ \infty \\ 0 \end{cases} \quad \text{as} \quad p \to \begin{cases} p_{-1} \\ p_1 \\ p_0 \end{cases}.$$

Since we have established that ends are separately embedded and parallel, and (5) implies that the ends are disjoint it follows that outside of a compact set of M, X is an embedding.

It will be useful to have the following notation.

$$M_t = (p \in M \,|\, x_3(p) = t)$$

$$S_t = X(M_t)$$

$$M_{a,b} = \bigcup_{a \le t \le b} M_t$$

$$S_{a,b} = X(M_{a,b})$$

$$S = X(M)$$

It follows from A. and (5) that

B. For N large, $S_{N,\infty}$ and $S_{-\infty,-N}$ are embedded.

We will show that

C. The symmetry group of M is the dihedral group D_4, generated by the matrices L and K.

Consider the conformal diffeomorphism of M given by

$$(6) \qquad \lambda(x,y) = (-x, iy).$$

Here, we identify points $p \in M$ with points (x,y) in $\mathbb{C}^2$ satisfying equation (2).

Since $g(\lambda(x,y)) = -ig(x,y)$, and $\eta(\lambda(x,y)) = i\eta(x,y)$, a computation shows that

(7) $$\Phi^* = (-\Phi_2,\Phi_1,-\Phi_3) = L\Phi.$$

Since p_0 is fixed by λ, it follows from (1),(3) and (7) that

$$X(\lambda(p)) = L\,X(p)\ .$$

A similar computation using conjugation $K(x,y) = (\overline{x},\overline{y})$ will show that $X(K(p)) = K\,X(p)$. The matrices K and L generate the dihedral group D_4 which has eight elements.

The next step is to show that

D. X is an embedding of M_0 onto $S_0 = (x_1+x_2 = x_3 = 0)$.

From (4) it is immediate that $x_3(p) = 0$ if and only if $p(x,y) \in M$, where $x = it$ for $t \in R$. This forces $f(t)$ to be of the form $\rho r(t)$ where

$$r(t) = |t|(t^2+1), \text{ and}$$

where ρ is a square root of $+i$ or $-i$. If ρ is a root of $-i$, then the point $p = (it,\rho r(t))$ is fixed by λK, which means that $X(p) = LKX(p)$. The fixed point set of LK is the line $x_1 - x_2 = x_3 = 0$. An analogous argument shows that if ρ is a root of $+i$, $p = (it,\rho r(t))$ is mapped into $x_1 + x_2 = x_3 = 0$, the fixed point set of L^3K. An easy calculation shows that X is an embedding of S_0 onto the union of these two lines in the (x_1,x_2)-plane.

For our future use, we note at this point that if

(8) $$\alpha(t) = (it, e^{3i\pi/4}r(t))\ , \quad 0 \le t < 1\ :$$

E. X maps $\alpha(t)$ one to one and onto the ray $x_1 - x_2 = x_3 = 0$, $x_1 \le 0$.

F. For $\varepsilon > 0$ sufficiently small, X is an embedding of $M_{-\varepsilon,\varepsilon}$.

As a consequence of A., all we need to show that X is an embedding on $M_{-\varepsilon,\varepsilon} \cap C$, where C is a compact set in M. By D., X embeds the subset of C where $x_3 = 0$. But X is an immersion and so is locally one to one. Hence H embeds a closed neighborhood (in C) of the set where $x_3 = 0$. Making $\varepsilon > 0$ suitably small it follows directly that X is an embedding of $M_{-\varepsilon,\varepsilon} \cap C$.

G. For $\varepsilon > 0$ sufficiently small and $N > 0$ sufficiently large, X is an embedding of the complement of $M_{\varepsilon,N} \cup M_{-N,-\varepsilon}$.
This is immediate from F. and B.

H. The set $M_{\varepsilon,N}$ is an annulus. X is an embedding of M_ε and M_N, the two boundary curves of this region.

If we choose N large enough, Assertion B. guarantees that S_N is a single embedded Jordan curve. (In fact this curve must eventually be convex since we are approaching the end corresponding to P_{-1} as $N \to \infty$.) In particular, X is one-to-one on M_N and M_N is a Jordan curve in M.
We may conclude from formula (4) that the function x_3 has a single critical point on M at $p_0 = (0,0)$, and that $X(p_0) = 0$. Hence the region $M_{\varepsilon,N}$ contains no critical points of x_3. Therefore $M_{\varepsilon,N}$ is an annular region and the curve M_ε is a simple closed curve on M. By F., this curve is embedded by X, if we choose $\varepsilon > 0$ small enough.
Note. Since $\lambda(M_{\varepsilon,N}) = M_{-N,-\varepsilon}, S_{-N,-\varepsilon} = L(S_{\varepsilon,N})$. Therefore $M_{-N,-\varepsilon}$ is also an annulus whose boundary curves are embedded.
We now wish to prove that $S_{\varepsilon,N}$ is embedded. Toward this end, we first observe that S_ε and S_N are the intersection of S with the planes $x_3 = \varepsilon$ and $x_3 = N$ respectively. Moreover, they must be symmetric with respect to reflection through the (x_1,x_3)-plane and the (x_2,x_3)-plane, since these are planes of symmetry of S. Therefore S_ε meets each of these planes in precisely two points, symmetric with respect to the x_3-axis. The same is true for S_N, $S_{-\varepsilon}$ and S_{-N}.

Consider the curves

$$(9) \qquad \beta(t) = (t, iw(t)) \text{ and } \tilde{\beta}(t) = K\beta(t) = (t, -iw(t)) \;, \; 0 \le t < 1,$$

where $w(t)$ is the positive square root of $(t(1-t^2))$. These curves actually meet smoothly on M.

<u>K. The curves (9) are embedded by X onto a smooth curve which lies in the half plane $(x_1 = 0,\ x_3 \leq 0)$, and is symmetric with respect to the x_3-axis, meeting that axis only at $0 = X\beta(0) = X\tilde{\beta}(0)$.</u>

It is easy to verify that β and $\tilde{\beta}$ are invariant under λ^2K from which it follows that

$$(10) \qquad B = X\circ\beta \quad \text{and} \quad B = KB = X\circ\tilde{\beta}$$

are invariant under L^2K which is reflection through the plane $(x_1 = 0)$. Also K is reflection in the plane $(x_2 = 0)$. This implies that B and $\tilde{B}$ lie in the (x_2,x_3)-plane and reflect one to the other across the x_3-axis. A calculation shows directly that $x_3\circ\beta$ and $x_3\circ\tilde{\beta}$ are monotone decreasing functions of t, taking values between 0 and $-\infty$. Another calculation establishes that $x_2\beta(t) \geq 0$ and $x_2\tilde{\beta}(t) \leq 0$, with equality if and only if $t = 0$.

Let P denote the (x_2,x_3)-plane. B and $\tilde{B}$ lie in P, which is a plane of symmetry of S. We will now show that B and $\tilde{B}$ slice up $S_{-N,-\varepsilon}$ into two disks.

<u>J. The part of the curves β and $\tilde{\beta}$ in (9) lying in $M_{-N,-\varepsilon}$ divide that annulus into two simply-connected regions. The image of one of these regions lies on one side of P. The image of the other region lies on the other side of P.</u>

Let us call $M_{-N,-\varepsilon} \cap \beta$ and $M_{-N,-\varepsilon} \cap \tilde{\beta}$, β_1 and β_2 respectively, and let $B_1 = x\circ\beta_1$ and $B_2 = x\circ\beta_2$. Reflection in the plane P induces an orientation-preserving isometry of the annulus $M_{-N,-\varepsilon}$ whose fixed point set F contains β_1 and β_2.

As discussed before equation (9), there are precisely two points on $S_{-N} = X(M_{-N})$ which lie in P and these must be the points $B \cap S_{-N}$ and

$\tilde{B} \cap S_{-N}$. The same is true for $S_{-\varepsilon}$. These points are the endpoints of B_1 and B_2.

It follows from the paragraph above that β_1 and β_2 divide $M_{-N,-\varepsilon}$ into two simply-connected regions. Choose one region. It is bounded by β_1, β_2, a subarc of M_{-N} which we shall call C_{-N}, and a subarc of $M_{-\varepsilon}$ which we shall call $C_{-\varepsilon}$. The arcs B_1 and B_2 lie in the plane of symmetry P. Therefore, if we stay sufficiently close to β_1, the curves C_{-N} and $C_{-\varepsilon}$ are mapped by X onto the same side of P. But S_{-N} and $S_{-\varepsilon}$ cannot recross P until they come to the place(s) where they meet B_2. Hence all of C_{-N} and $C_{-\varepsilon}$ are mapped into the same half-space. By the convex-hull property for minimal surfaces, the region itself is mapped into the same half-space, meeting the boundary only along $B_1 \cup B_2$.

Let

(11) $\gamma(t) = (t, t(t^2-1))$, $\hat{\gamma}(t) = \lambda^2 K\gamma(t)$, $1 < t < \infty$, and

(12) $G(t) = x \circ \gamma(t)$, and $\hat{G}(t) = x \circ \hat{\gamma}(t)$.

The surves $G(t)$ and $\hat{G}(t)$ lie in the plane of symmetry $(x_2 = 0)$. From (11), (12) and C., it follows that G and $\hat{G}$ are symmetric with respect to the x_3-axis.

<u>K. The curves G and $\hat{G}$ are plane curves of symmetry lying in the plane $(x_2 = 0)$. They never meet the x_3-axis. On both curves, $x_3 < 0$. The curves $M_{-M,-\varepsilon} \cap \gamma$ and $M_{-N,-\varepsilon} \cap \hat{\gamma}$ together with β_1 and β_2 divide $M_{-N,-\varepsilon}$ into four simply-connected regions. Each region is mapped into a distinct octant in the part of R^3 below the plane $(x_3 = 0)$.</u>

The first statement is clear. The second statement follows from J. The third statement follows from (11) and (4). The last two statements follow by repeating the proof of J. for $\gamma(t)$ and $\hat{\gamma}(t)$.

Let T be one of the four simply connected regions desribed in K.

<u>L. X(T) is embedded. In fact, X(T) is a graph.</u>

$S_{-N,-\varepsilon}$ is embedded.

All of the four regions comprising $M_{-N,-\varepsilon}$ are congruent and the congruence between any two is an element of the dihedral group D_4. Therefore, if one of them is embedded by X, they all are. Their union is $M_{-N,-\varepsilon}$, and their images are all disjoint thus statement L. is proved once we show that any one of the regions is embedded. We choose the region whose boundary contains β_1 and $M_{-M,-\varepsilon} \cap \gamma$. We refer to this one as T'.

Thinking of M as a two-sheeted covering of $\mathbb{C}$, we identify T' with a subset of a quadrant of one of the sheets. If we let ε go to zero and N go to infinity, T' fills out the quadrant, gaining α as a third boundary curve. The stereographic projection of the Gauss map of M is given by $g = b/y$, where b is a real constant and y satisfies (2). A computation using this fact shows that the quadrant has Gaussian image in a sector W of S^2 with vertices at $(0,0,\pm 1)$ and a width of $3\pi/2$-radians.

Since no point in M except $p_0 = (0,0)$ has a normal equal to $(0,0,\pm 1)$ and $p_0 \notin T'$, it follows that the Gaussian image of T' lies in the interior of any hemispere determined by a plane with the following property:it contains the x_3-axis and is disjoint from W $((0,0,\pm 1))$. Since T' is simply-connected and its boundary projects in a one-to-one fashion onto the equatorial plane, the fact that its Gaussian image lies in an open hemisphere implies that it is a graph (over the equatorial plane).

Since $S_{\varepsilon,N}$ is congruent to and disjoint from $S_{-N,-\varepsilon}$, it follows from L. and G. that

M. S is embedded.

To complete the proof of Theorem 1 we note the following:

From K. it follows, by letting ε go to zero and N go to infinity, that M consists of eight congruent simply-connected regions, one in each octant of R^3. The congruences involved are the elements of D_4. This is a group of symmetries of S as proved in C., and it is transitive on the octants. It can be shown that this group is the symmetry group of S. The surface also contains two straight lines as described in D. This completes the proof of the theorem.

REFERENCES

[1] J. Lucas M. Barbosa and A. Gervasio Colares, Exemplos de Superficies Minimas no R^3, Fifth Differential Geometry Workshop, University of São Paulo, July 30-August 4, 1984.

[2] C. Costa, Imersões minimas completas em R^3 de genero um e curvatura total finita, Doctoral thesis, IMPA, Rio de Janeiro, Brasil, 1982.
(Also in "Example of a complete minimal immersion in R^3 of genus one and three embedded ends", to appear in Boletim da Sociedade Brasiliera Mathematica, 15, No.1)

[3] D. Hoffman and W. Meeks III, A complete embedded minimal surface with genus one, three ends and finite total curvature. (preprint)

[4] D. Hoffman and W. Meeks III, Complete embedded minimal surfaces of finite total curvature. Bull.A.M.S., January 1985.

[5] D. Hoffman and W. Meeks III, to appear

[6] D. Hoffman and R. Osserman, The geometry of the generalized Gauss map, Mem. Amer Math. Soc. No. 236, 1980.

[7] L. Jorge and W. Meeks III, The topology of complete minimal surfaces of finite total Gaussian curvature, Topology, 22 No. 2, 203-221, 1983.

[8] R. Osserman, Global properties of complete minimal surfaces in E^3 and E^n, Ann. of Math., (2) 80, 340-364, 1964.

[9] R. Schoen, Uniqueness, symmetry, and embeddedness of minimal surfaces, J. Diff. Geom. 18, 791-809, 1983.

THE BERNSTEIN PROBLEM FOR FOLIATIONS*

by F. W. Kamber and Ph. Tondeur

Consider a foliation F on a smooth Riemannian manifold M, given by an involutive subbundle $L \subset TM$. The normal bundle $Q = TM/L$ inherits from the metric g_M on TM a metric g_Q and a torsion-free connection ∇ extending the partial Bott connection. Thus there is a Laplacian defined on Q-valued forms. The foliation F is harmonic, if the canonical projection $\pi : TM \to Q$ is a harmonic 1-form [KT1]. An equivalent property is that all leaves of F are minimal submanifolds of (M,g_M) [KT1]. The simplest such foliations are those which are totally geodesic, i.e. all leaves are totally geodesic submanifolds of (M,g_M).

In this note we consider foliations of codimension one. The complete totally geodesic foliations of codimension $q = 1$ in $\mathbb{R}^n$ are the foliations by parallel hyperplanes. The Bernstein problem is the question if a complete minimal graph in $\mathbb{R}^n$ is a hyperplane [B]. This is known to be true if $n \leq 8$ by the work of Bernstein, De Giorgi, Almgren and Simons, and false for $n > 8$ by Bombieri, De Giorgi and Giusti [BGG].

The natural Bernstein problem for foliations is to ask if a harmonic foliation of codimension one is necessarily totally geodesic. For negatively curved space forms this is certainly not the case, since Wang and Wei [WW] have given an example of a harmonic foliation F of codimension one on hyperbolic n-space H^n, with one leaf totally geodesic, while all other leaves are minimal, but not totally geodesic. For the compact flat space forms the Bernstein conjecture for foliations does hold without any restriction on the dimension of the space form. This is a consequence of the theorem stated below.

Recall Reinhart's concept [R] of a bundle-like metric g_M on M with respect to F: the induced metric g_Q in Q satisfies the holonomy invariance condition

$$Xg_Q(s,t) = g_Q(\nabla_X s,t) + g_Q(s,\nabla_X t)$$

for all $X \in \Gamma L$; $s,t \in \Gamma Q$. Here ∇ denotes the partial Bott connection in Q along L. A foliation F is Riemannian if it admits a bundle-like metric. The following result is then a partial answer to the Bernstein conjecture for foliations.

THEOREM. Let (M,g_M) be a compact oriented Riemannian manifold with non-negative Ricci curvature, and F a transversally orientable foliation of codimension one on M.

*Work supported in part by NSF Grant DMS 8302558.

(i) _If_ F _is harmonic, then_ F _is totally geodesic and Riemannian with respect to_ g_M.
(ii) _If the Ricci operator is positive for at least one_ $x_0 \in M$, _then_ F _cannot be harmonic._

A slightly weaker form is stated in [KT2][O]. But the arguments in these papers imply also the stronger version above. We sketch the proof. Let Z be a unit normal vectorfield. Then F is defined by the 1-form $\omega(X) = g_M(X,Z)$ for $X \in \Gamma TM$ and $\pi(X) = \omega(X).Z$.

LEMMA. [KT1, 3.9 and 3.14].
(i) F _is harmonic iff_ $d^*\omega = 0$,
(ii) F _is Riemannian iff_ $d\omega = 0$.

Yano's integral formula [K, p. 154] yields

$$||d^*\omega||^2 = \int_M Tr((\nabla^M Z)^2).\eta_M + \int_M Ric^M(Z,Z).\eta_M$$

where $\nabla^M Z : TM \to TM$ is given by $(\nabla^M Z)(X) = \nabla^M_X Z$ for $X \in \Gamma TM$, and η_M is the volume form on M. Thus the harmonicity of F together with the Ricci curvature assumption implies $Tr((\nabla^M Z)^2) = 0$. Since ∇^M restricts on L to the Weingarten map $W(Z) : L \to L$ (up to sign), it follows that $Tr(W(Z)^2) = 0$. The selfadjointness of $W(Z)$ implies then $W(Z) = 0$, and thus F is necessarily totally geodesic. The same formula shows that if the Ricci operator is positive for at least one $x_0 \in M$, then $d^*\omega \neq 0$ and thus F is not harmonic. Finally Oshikiri shows by an explicit calculation in local frames that $d\omega = 0$. By the lemma stated above, F must therefore be Riemannian.

We conclude by noting that e.g. on a torus T^n this result shows that the harmonic foliations of codimension one are precisely the linear foliations given by the constant non-trivial one-forms. We refer to [KT2] for partial results in case the codimension is greater then one.

REFERENCES

[B] S. Bernstein, Sur un théorème de géométrie et ses applications anx équations aux dérivées partielles du type elliptique, Comm. Soc. Math. Karkov 15, 38-45 (1915-1917).

[BGG] E. Bombieri, E. De Giorgi, E. Giusti, Minimal cones and the Bernstein problem, Inventiones Math. 7, 243-269 (1969;. Zbl. 183,259).

[K] S. Kobayashi, Transformation groups in differential geometry, Ergebnisse der Math. 70(1972; Zbl. 246.53031).

[KT1] F. W. Kamber and Ph. Tondeur, Harmonic foliations, Proc. NSF Conference on Harmonic Maps, Tulane 1980, Lecture Notes in Math 949, 87-121 (1982; Zbl. 511.57020).

[KT2] F. W. Kamber and Ph. Tondeur, Curvature properties of harmonic foliations, Ill. J. of Math. 28, 458-471 (1984; Zbl. 529.53027).

[O] Gen-ichi Oshikiri, A remark on minimal foliations, Tôhoku Math. J. 33, 133-137 (1981; Zbl. 437.57013).

[R] B. L. Reinhart, Foliated manifolds with bundle-like metrics, Annals of Math. 69, 119-132 (1959; Zbl. 122,166).

[WW] S. P. Wang and S. Walter Wei, Bernstein Conjecture in hyperbolic geometry, Seminar on Minimal Submanifolds, Annals of Math. Studies 103, 339-358 (1983).

Department of Mathematics
1409 West Green Street
University of Illinois
Urbana-Champaign
Urbana, Illinois 61801

Examples concerning the spectrum of a closed Riemannian manifold

Michael Kozlowski

Improving a result in [2] T. Pavlista ([4]) showed:

Theorem A. *Let (M,g) be a 2-dimensional, closed, connected, oriented Riemannian manifold with strictly positive Gaussian curvature K, $K_0 = \min K$ and $K_1 = \max K$. Assume that (M,g) is δ-pinched with $\delta \geq \frac{1}{2}$. Let $\lambda^{(1)}$ and $\lambda^{(2)}$ denote the zeros of the polynomial*

$$(0.1) \qquad p(\lambda) = \frac{1}{4}(\lambda - 12K_0)(\lambda - 6K_1)(\lambda - 2K_1) + \max \|\operatorname{grad} K\|^2$$

which fulfil $6K_1 \leq \lambda^{(1)} \leq \lambda^{(2)} \leq 12K_0$.
Then there are no eigenvalues of the Laplacian in the open interval $(\lambda^{(1)}, \lambda^{(2)})$.

In this paper we calculate $\lambda^{(1)}$ and $\lambda^{(2)}$ on an ellipsoid of revolution; furthermore, we show by examples that $\max \|\operatorname{grad} K\|^2$ may be arbitrarily large even when (M,g) is δ-pinched with δ arbitrarily close to one.

1. Ellipsoid of revolution

For a real number c, $0 < c \leq 1$, let E_c denote the following ellipsoid:

$$E_c = \{ (x,y,z) \in \mathbb{R}^3 : x^2 + y^2 + \frac{z^2}{c^2} = 1 \} .$$

Let g_c be the metric induced from $\mathbb{R}^3$. Using the parametrization

$$F_c : \{ (u,v) \in \mathbb{R}^2 : u^2 + v^2 < 1 \} \rightarrow \mathbb{R}^3$$

defined by

$$F_c(u,v) = (u, v, c(1 - u^2 - v^2)^{1/2})$$

See also T.Pavlista's article in this volume.

and setting $t = u^2 + v^2$, $0 \leq t < 1$, we get for the Gaussian curvature of (E_c, g_c)

$$K_c(t) = \frac{1}{c^2 \left[\left(1 - \frac{1}{c^2} \right) t + \frac{1}{c^2} \right]^2}, \quad 0 \leq t < 1.$$

Hence we have

$$(1.1) \qquad \min_{E_c} K_c = c^2, \quad \max_{E_c} K_c = \frac{1}{c^2}, \quad \delta_c = c^4.$$

The norm of the gradient of K_c is given by

$$\|\mathrm{grad}\ K_c\|^2 (t) = \frac{16\, c^4 \,(c^2 - 1)^2\, t\, (1 - t)}{[(c^2 - 1) t + 1]^7}$$

or using (1.1)

$$\|\mathrm{grad}\ K_c\|^2 (t) = \frac{16\, \delta_c \,(\delta_c^{1/2} - 1)^2\, t\, (1 - t)}{[(\delta_c^{1/2} - 1) t + 1]^7}, \quad 0 \leq t < 1.$$

We introduce the metric $\hat{g}_c$ with

$$\hat{g}_c = \frac{1}{\delta_c^{1/2}}\, g_c .$$

Then it follows for the Gaussian curvature $\hat{K}_c$ of $(E_c, \hat{g}_c)$ that

$\max_{E_c} \hat{K}_c = 1$ and

$$(1.2) \qquad \|\widehat{\mathrm{grad}}\ \hat{K}_c\|^2 (t) = \frac{16\, \delta_c^{5/2} \,(\delta_c^{1/2} - 1)^2\, t\, (1 - t)}{[(\delta_c^{1/2} - 1) t + 1]^7}, \quad 0 \leq t < 1.$$

Moreover, $\min_{E_c} \hat{K}_c = \delta_c$ and the polynomial (0.1) has the form

$$(1.3) \qquad p(\lambda) = \frac{1}{4} \lambda^3 - (3\delta_c + 2)\, \lambda^2 + (24\delta_c + 3)\, \lambda - 36\delta + \max_{E_c} \|\widehat{\mathrm{grad}}\ \hat{K}_c\|^2 .$$

From Theorem A, (1.2), and (1.3), we get for the ellipsoid $(E_c, \hat{g}_c)$ open intervals $(\lambda^{(1)}, \lambda^{(2)})$ which are "dark bands" in the spectrum of the Laplacian.

δ_c	$\lambda^{(1)}$	$\lambda^{(2)}$	$12\ \delta_c$
0,6	6,145	7,085	7,2
0,7	6,037	8,377	8,4
0,8	6,01	9,594	9,6
0,9	6,002	10,799	10,8
1	6	12	12

2. An example of the gradient of the Gaussian curvature

It follows from Theorem A that for a fixed δ the quotient $\lambda^{(2)}/\lambda^{(1)}$ is determined by the term $\frac{1}{K_1^3} \max \|\text{grad } K\|^2$. $\lambda^{(2)}/\lambda^{(1)}$ decreases when $\frac{1}{K_1^3} \max \|\text{grad } K\|^2$ increases. Of particular interest are examples of *closed manifolds with* δ *arbitrarily close to one and* $\frac{1}{K_1^3} \max \|grad\ K\|^2$ *arbitrarily large*. For these manifolds the intervals $(\lambda^{(1)}, \lambda^{(2)})$ will be empty.

Given a strictly positive C^∞-function $\Gamma : S^2(1) \to \mathbb{R}$ on the topological unit sphere there exists a metric γ such that $\Gamma : S^2(1) \to \mathbb{R}$ is the Gaussian curvature of $(S^2(1), \gamma)$ ([1]). By Weyl's existence theorem ([3]) $(S^2(1), \gamma)$ may be realized as an ovaloid in the Euclidean 3-space.

The examples described below are not based on given functions on $S^2(1)$; our examples consist of a whole class of surfaces of revolution which are explicitly exhibited by their generating curves.

Let n be an integer and define the mappings $\varphi_n, \psi : [0,\pi] \to \mathbb{R}$ as follows

(2.1) $$\varphi_n(v) = [\ 1 + \frac{1}{n^{5/2}} \cos(nv)\ ] \sin v\ ,$$

(2.2) $$\psi(v) = \cos v \quad \text{for all } v \in [0,\pi].$$

$\varphi_n, \psi : [0,\pi] \to \mathbb{R}$ induce a surface of revolution $x_n : [0,2\ )\times[0,\pi] \to \mathbb{R}^3$ with

(2.3) $$x_n(u,v) = \begin{pmatrix} \varphi_n(v) \cos u \\ \varphi_n(v) \sin u \\ \psi(v) \end{pmatrix} \quad \text{for all } (u,v) \in [0,2\pi)\times[0,\pi].$$

The corresponding Gaussian curvature K_n and the term $\|\text{grad } K_n\|^2$ depend only on the parameter v

(2.4) $$K_n(v) = \frac{\psi'(v)\ [\ \psi''(v)\varphi_n'(v) - \psi'(v)\varphi_n''(v)\ ]}{\varphi_n(v)\ [\ \varphi_n'(v)^2 + \psi'(v)^2\ ]^2}\ ,$$

(2.5) $$\|\text{grad } K_n\|^2 (v) = \frac{K_n'(v)^2}{\varphi_n'(v)^2 + \psi'(v)^2} \quad \text{for all } v \in (0,\pi).$$

The derivatives of the mappings $\varphi_n, \psi : [0,\pi] \to \mathbb{R}$ defined in (2.1) and (2.2) are

(2.6) $$\varphi_n'(v) = [\ 1 + \frac{1}{n^{5/2}} \cos(nv)\ ] \cos v - \frac{1}{n^{3/2}} \sin(nv) \sin v\ ,$$

(2.7) $$\varphi_n''(v) = -[\ 1 + \frac{1}{n^{5/2}} \cos(nv)\ ] \sin v - \frac{2}{n^{3/2}} \sin(nv) \cos v - \frac{1}{n^{1/2}} \cos(nv) \sin v\ ,$$

(2.8) $\psi'(v) = -\sin v$,

(2.9) $\psi''(v) = -\cos v$ for all $v \in [0,\pi]$.

Hence the sequences of mappings (φ_n), (φ_n'), and (φ_n'') have the limits

(2.10) $\overline{\lim_{n \to \infty}}\; \varphi_n(v) = \sin v$,

(2.11) $\overline{\lim_{n \to \infty}}\; \varphi_n'(v) = \cos v$,

(2.12) $\overline{\lim_{n \to \infty}}\; \varphi_n''(v) = -\sin v$ for all $v \in (0,\pi)$,

$\overline{\lim_{n \to \infty}}$ stands for the uniform convergence of the sequence. From (2.4), (2.8), (2.9), (2.10), (2.11), and (2.12), it follows that

(2.13) $$\overline{\lim_{n \to \infty}}\; K_n(v) = \frac{(-\sin v)\,[\,-\cos^2 v - \sin^2 v\,]}{\sin v\,[\,\cos^2 v + \sin^2 v\,]^2} = 1$$

for all $v \in (0,\pi)$.

Thus we get for the pinching constant δ_n of the surface defined in (2.3)

(2.14) $\lim_{n \to \infty} \delta_n = 1.$

The equation (2.4) may be written as

(2.15) $$K_n(v) = \frac{Z_n(v)}{N_n(v)}$$

with

(2.16) $Z_n(v) = \psi'(v)\,[\,\psi''(v)\varphi_n'(v) - \psi'(v)\varphi_n''(v)\,]$,

(2.17) $N_n(v) = \varphi_n(v)\,[\,\varphi_n'(v)^2 + \psi'(v)^2\,]^2$ for all $v \in (0,\pi)$.

(2.15) implies

$$(2.18)\qquad K_n'(v) = \frac{N_n(v)Z_n'(v) - Z_n(v)N_n'(v)}{N_n(v)^2} \quad \text{for all } v \in (0,\pi).$$

From (2.16) and (2.17) we get

$$(2.19)\qquad \begin{aligned} N_n'(v) &= \varphi_n'(v)\ [\ \varphi_n'(v)^2 + \psi'(v)^2\ ]^2 + \\ &+ 4\ \varphi_n(v)\ [\ \varphi_n'(v)^2 + \psi'(v)^2\ ][\ \varphi_n'(v)\varphi_n''(v) + \psi'(v)\psi''(v)\ ], \end{aligned}$$

$$(2.20)\qquad \begin{aligned} Z_n'(v) &= \psi''(v)\ [\ \psi''(v)\varphi_n'(v) - \psi'(v)\varphi_n''(v)\ ] + \\ &+ \psi'(v)\ [\ \psi'''(v)\varphi_n'(v) - \psi'(v)\varphi_n'''(v)\ ] \quad \text{for all } v \in (0,\pi). \end{aligned}$$

Using (2.8), (2.9), (2.10), (2.11), (2.12), (2.16), (2.17), we obtain the following limits

$$(2.21)\qquad \lim_{n\to\infty} N_n(v)^2 = \sin^2 v\ ,$$

$$(2.22)\qquad \lim_{n\to\infty} Z_n(v)N_n'(v) = \sin v \cos v \quad \text{for all } v \in (0,\pi).$$

(2.17), (2.1), (2.6), and (2.8) imply

$$(2.23)\qquad \begin{aligned} N_n(v) &= [\ 1 + \frac{1}{n^{5/2}} \cos(nv)\ ]\ \sin v\ \cdot \\ &\cdot \{[\ 1 + \frac{1}{n^{5/2}}\cos(nv)\ ]^2 \cos^2 v\ - \\ &- \frac{2\ [\ 1 + \frac{1}{n^{5/2}}\cos(nv)\ ]}{n^{3/2}} \sin(nv)\ \sin v \cos v\ + \\ &+ \frac{1}{n^3}\sin^2(nv)\ \sin^2 v + \sin^2 v\ \}^2 \quad \text{for all } v \in (0,\pi). \end{aligned}$$

It follows from (2.7) that

$$\varphi_n'''(v) = -\cos v + n^{1/2}\sin(nv)\sin v + F_{on}(v),$$

(2.24)

$$\text{where } \lim_{n\to\infty} F_{on}(v) = 0 \quad \text{for all } v \in (0,\pi).$$

By virtue of (2.20), (2.8), (2.9), (2.11), (2.12), and (2.24), we have

$$Z_n'(v) = \cos v - n^{1/2}\sin(nv)\sin^3 v + F_{1n}(v),$$

(2.25)

$$\text{where } \lim_{n\to\infty} F_{1n}(v) = 0 \quad \text{for all } v \in (0,\pi).$$

(2.25) and (2.23) imply

$$Z_n'(v)N_n(v) = \sin v\cos v - n^{1/2}\sin(nv)\sin^4 v\, G_n(v) + F_{2n}(v),$$

(2.26)

$$\text{where } \lim_{n\to\infty} G_n(v) = 1,\ \lim_{n\to\infty} F_{2n}(v) = 0$$

$$\text{for all } v \in (0,\pi).$$

Equations (2.18), (2.21), (2.22), and (2.26) give

$$K_n'(v) = \frac{-n^{1/2}\sin(nv)\sin^4 v\, G_n(v) + F_{3n}(v)}{\sin^2 v + F_{4n}(v)},$$

(2.27)

$$\text{where } \lim_{n\to\infty} G_n(v) = 1,\ \lim_{n\to\infty} F_{3n}(v) = 0,$$

$$\lim_{n\to\infty} F_{4n}(v) = 0 \quad \text{for all } v \in (0,\pi).$$

Let us return to equation (2.5)

$$\text{(2.28)}\quad \|\operatorname{grad} K_n\|^2(v) = \frac{K_n'(v)^2}{\varphi_n'(v)^2 + \psi'(v)^2} \quad \text{for all } v \in (0,\pi).$$

Using (2.11) and (2.8), we have

$$\text{(2.29)}\quad \overline{\lim_{n\to\infty}}\ \varphi_n'(v)^2 + \psi'(v)^2 = 1 \quad \text{for all } v \in (0,\pi).$$

The limit in (2.13) was

(2.30) $$\overline{\lim_{n \to \infty}} K_n(v) = 1 \qquad \text{for all } v \in (0,\pi).$$

In (2.29) and (2.30) we have a uniform convergence. Hence, it follows from (2.27), (2.28), (2.29), and (2.30) that $\frac{1}{(\max K_n)^3} \max \|\text{grad } K_n\|^2$ becomes arbitrarily large when n is a sufficently large integer. For example, choose in (2.27) $n \equiv 1 \pmod 4$ and $v = \frac{\pi}{2}$.

R E F E R E N C E S

[1] Jerry L. Kazdan, F.W. Warner: A direct approach to the determination of Gaussian and scalar curvature functions, Inventiones math. 28, 227 - 230 (1975; Zbl. 241.53031).

[2] M. Kozlowski: Geometrische Abschätzungen für kleine Eigenwerte des Laplaceoperators auf fast-sphärischen Rotationsellipsoiden. Dissertation FB 3 TU Berlin 1983.

[3] L. Nirenberg: The Weyl and Minkowski problems in differential geometry in the large, Comm. Pure Appl. Math. 6, 337 - 394 (1953; Zbl. 51,124).

[4] T. Pavlista, unpublished.

TIGHT SMOOTHING OF SOME POLYHEDRAL SURFACES

W. Kühnel and U. Pinkall

1. INTRODUCTION AND RESULTS

The main goal of this paper is to approximate certain tight polyhedral surfaces in euclidean 3-space E^3 by tight smooth surfaces of the same topological type: we call this *tight smoothing*. This smoothing procedure gives a convenient tool for constructing examples of tight smooth surfaces. In particular we will construct in this way tight smooth surfaces of odd Euler characteristic χ. So far the existence of tight surfaces with odd Euler characteristic depended on a very complicated construction due to N.H.Kuiper ([9]). Although we have no doubt that the idea behind this construction is essentially correct, it seems to be very difficult to prove rigorously that the described surface is tight. Our examples are very explicit and have moreover a three-fold symmetry.

A smooth compact surface M immersed into E^3 is called *tight* if its total absolute curvature $\int |K| do$ equals the minimal possible value $2\pi(4 - \chi(M))$. Equivalent conditions (also applicable to nonsmooth surfaces) are

1. for every closed halfspace $H \subset E^3$ the preimage of H in M is connected (two-piece-property), or
2. every strict local supporting plane is a global supporting plane.

For other characterizations and general results about tightness see the survey articles [11], [12].

A *compact polyhedral surface* in E^3 is a finite complex consisting of vertices, straight edges and planar (but not necessarily convex) faces such that every edge is contained in exactly two faces and the faces around each vertex form locally a cone over a simply closed spherical polygon. All edges are assumed to be proper meaning that the adjacent faces are not coplanar. Such terms as "locally" or "ε-neighborhood" will always refer to the inner geometry of the complex. In particular there may occur selfintersections (like in the case of smooth immersions). A vertex is called *convex* if the corresponding cone is convex, otherwise it is called nonconvex. The number of edges meeting at a vertex v is called the *valence* of v.

We call v a *standard saddle vertex* if

(i) the valence of v is four,

(ii) all angles of the faces at v are strictly smaller than π,

(iii) there is no local supporting plane through v .

THEOREM 1 : Assume that M is a tight polyhedral surface in E^3 such that all of its nonconvex vertices have valence 3 . Then for every $\varepsilon>0$ there exists a tight smooth surface $M(\varepsilon)$ of the same topological type which coincides with M except in the ε-neighborhood of the union of all edges of M .

THEOREM 2 : Assume that M is a tight polyhedral surface in E^3 whose nonconvex vertices are either 3-valent or standard saddle vertices. Then for every $\varepsilon>0$ there is a tight polyhedral surface M' whose faces are in 1-1-correspondence with the faces of M such that corresponding faces are parallel in distance less than ε and such that all vertices are of valence 3 .

Combining the two theorems we see that every tight polyhedral surface whose nonconvex vertices are either 3-valent or standard saddle vertices can be approximated (in a certain sense) by tight smooth surfaces of the same topological type.
The proof of these theorems will be given in section 2 below.
In section 3 we use theorem 1 to give an improved version of a statement originally due to N.H.Kuiper ([9]):

THEOREM 3 : For any odd integer $\chi \leq -3$ there exists a tight smooth immersion of the surface with Euler characteristic χ into E^3. It may be chosen to have a three-fold symmetry.

In section 4 we discuss further applications and some open questions.

2. PROOF OF THEOREMS 1 AND 2 : TIGHT SMOOTHING

Proof of theorem 1: First of all we replace each edge by a piece of a cylinder over a certain convex smooth curve. For given $\delta>0$ let $f: \mathbb{R} \to \mathbb{R}$ be a function satisfying

(i) f is convex and C^∞ ,

(ii) f is strictly convex in the interval $(-\delta,\delta)$,

(iii) $f(x) = x$ for $x \geq \delta$,

(iv) $f(x) = f(-x)$ for all x .

The orthogonal cylinder over the graph of this function will be a suitable smoothing of an edge between two orthogonal planes P_1, P_2. Let $g_1 \subset P_1$ and $g_2 \subset P_2$ denote the generators of the cylinder through the points $(\delta,f(\delta))$ and $(-\delta,f(-\delta))$. The distance of g_1 from P_2

(and of g_2 from P_1) will be $\sqrt{2}\cdot\delta$. Now for every angle between two planes we take an affine image of this smoothed orthogonal angle. We can do this in such a way that the distances of g_1 from P_2 (and of g_2 from P_1) after the affine transformation will be exactly this same value $\sqrt{2}\cdot\delta$. Furthermore by compactness for an arbitrary given $\varepsilon>0$ one can find a $\delta>0$ such that all these cylindrical pieces will lie in the ε-neighborhood of the corresponding edge. The smoothing is now clear everywhere except in the ε-neighborhood of the vertices.

We first consider the nonconvex vertices of valence three. It is easy to see that exactly one of the three angles between edges is greater than π. The plane of this angle will be labeled as "horizontal", the direction of the remaining edge as "vertical" (see fig. 1). For both horizontal cylinders (coming from the smoothing of the horizontal edges) the upper boundary is a horizontal straight line lying in one of the vertical faces. Our assumption that both of these lines have distance $\sqrt{2}\cdot\delta$ from the horizontal plane implies that these two lines meet on the vertical edge.
Using an affine transformation of E^3 we can assume without loss of generality that all these planes are orthogonal and that the two horizontal cylinders are congruent. Let $t \mapsto g_t$, $0 \leq t \leq 1$ be a smooth parametrization of the family of generators of the vertical cylinder C_v, N_t the normal plane to C_v through the generator g_t. Let γ_o be the curve of intersection of the first cylinder with the plane N_o, γ_1 the intersection of the second cylinder with N_1. Then the orthogonal trajectories of the family N_t through the points of γ_o intersect each plane N_t in a congruent copy $\tilde{\gamma}_t$ of γ_o. Because of $\tilde{\gamma}_1 = \gamma_1$ the curves $\tilde{\gamma}_t$ provide a smooth interpolation between γ_o and γ_1 (see fig. 1). The surface swept out by the $\tilde{\gamma}_t$ is usually called a "molding surface" or "Gesimsfläche" (see [6]). The $\tilde{\gamma}_t$ are lines of curvature on this surface and it is easy to check that its Gaussian curvature is nonpositive everywhere. Clearly the molding surface fits smoothly the cylinders and the planes.

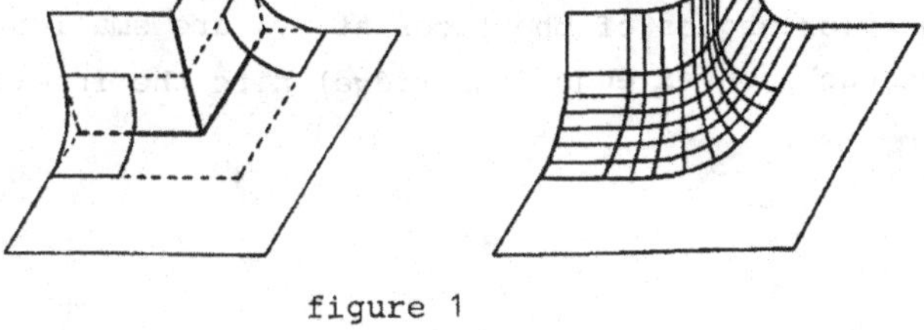

figure 1

Now for a <u>convex</u> vertex we can proceed similarly. In order to preserve a possible rotational symmetry of the vertex cone we suggest the following: truncate the vertex by a plane (preserving the symmetry, if there is any). Then the resulting vertices will have valence three. Now smooth the new edges by cylinders as described above (but with essentially smaller δ). Then we can apply the same kind of construction near the vertices. The only difference is that this time the molding surface will have nonnegative Gaussian curvature.

Altogether we get a smooth approximation of the given polyhedral surface M which coincides with M outside the ε-neighborhood of the union of the edges and such that positive Gaussian curvature occurs only in the ε-neighborhood of the convex vertices. If the polyhedral surface is tight these vertices lie on the boundary of the convex hull and consequently the smoothed surface will be tight as well.

<u>Proof of theorem 2</u>: The idea behind theorem 2 is that generically all vertices are 3-valent: if the planes spanned by the various faces lie in general position then no more than three planes can meet at a vertex. On the other hand the tightness itself is a very special situation. Therefore we have to be very careful when changing our given tight surface.

<u>Lemma:</u> *The vertex cone of a standard saddle vertex is the boundary of the union of two "roofs" where we mean by a roof the intersection of two closed halfspaces. The ridge of each roof will hit the interior of the other roof.*

Proof of the lemma: Locally there is an orientation of the surface. This enables us to talk about "convex" and "concave" edges meeting at a standard saddle vertex v . If there are two subsequent edges (in cyclic order) of the same kind then the plane spanned by those two edges will be a supporting plane. This implies that the edges of a standard saddle vertex have to be creased in an alternating manner: convex, concave, convex, concave. Let us take the two convex edges as ridges. Then the two roofs are built up by the two pairs of planes adjacent to the two convex edges. The condition that all interior angles of the faces at v are smaller than π implies that each convex edge (the ridge) hits the interior of the other roof.

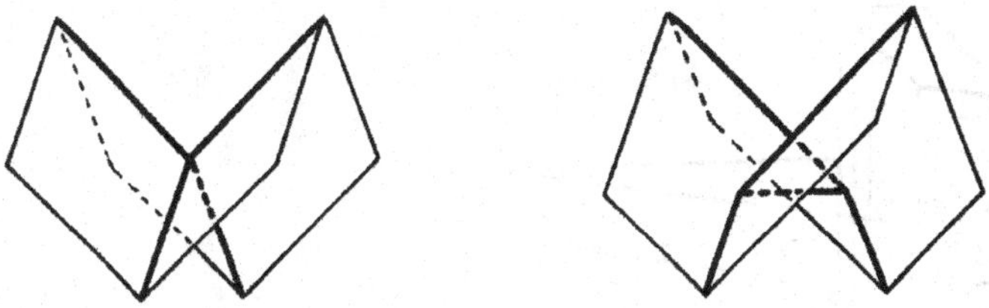

figure 2

To finish the proof of theorem 2 we observe what happens if we move one of the four planes a little bit in the normal direction. This is the same as saying that one of the roofs is moved against the other one. First of all we see that the boundary of the union of the two roofs will still be a surface. Hence the topology does not change by this process. Secondly we see that the saddle vertex splits into two vertices, each being the intersection point of a ridge with the other roof. The vertex cone of each of the new vertices will be the boundary of the union of a roof and a closed halfspace. Therefore these new vertices are nonconvex and of valence three with exactly one local supporting plane (see figure 2 for the situation before and after this process). After iterated application of this process the surface finally will have only 3-valent vertices. The convex vertices will split into convex vertices of valence 3 and the former nonconvex 3-valent vertices will keep their geometry: they will just be moved by a euclidean translation. The important point is that by this process no open sets of local supporting planes are created. Therefore if the old surface was tight the new surface will be tight as well.

3. A TIGHT POLYHEDRAL PROJECTIVE PLANE WITH TWO HANDLES

The main part of the surface to be described is a polyhedral Boy surface with a three-fold symmetry (a modified version of the one given in [1]). This has the property that all its strict local supporting planes concentrate at the vertices of three convex faces on the boundary of the convex hull. Then one has to cut out these three faces and to join the resulting surface with an exterior tetrahedral surface.

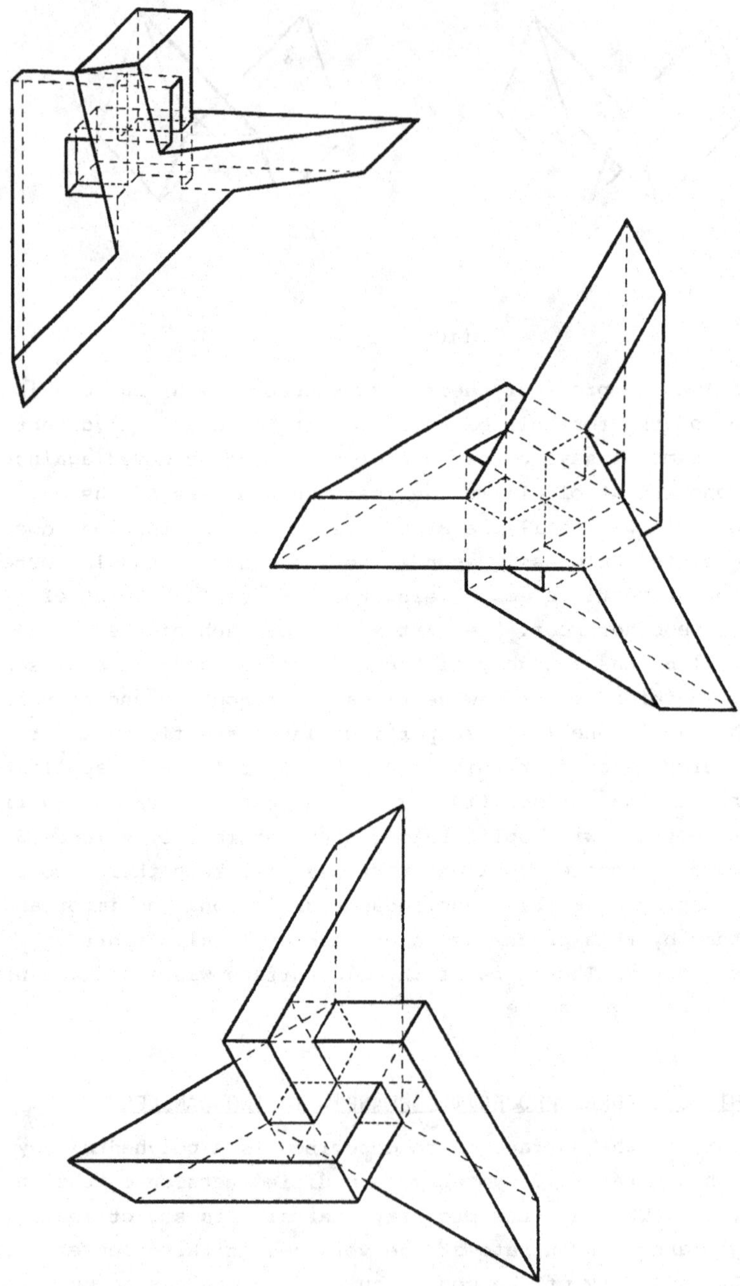

figure 3

type 1 (lying in the plane x=-1)

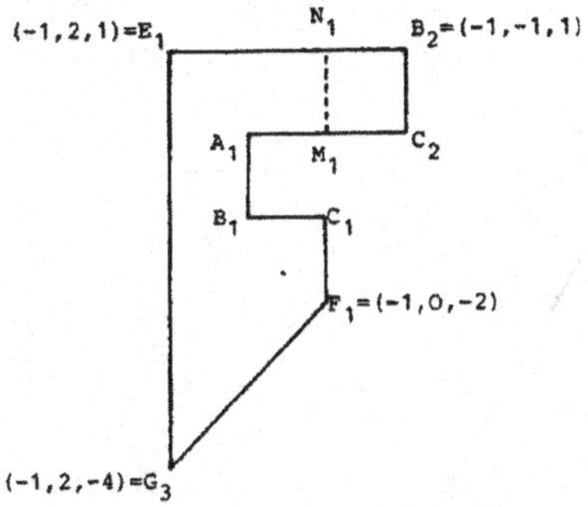

type 2 (in the plane x=0)

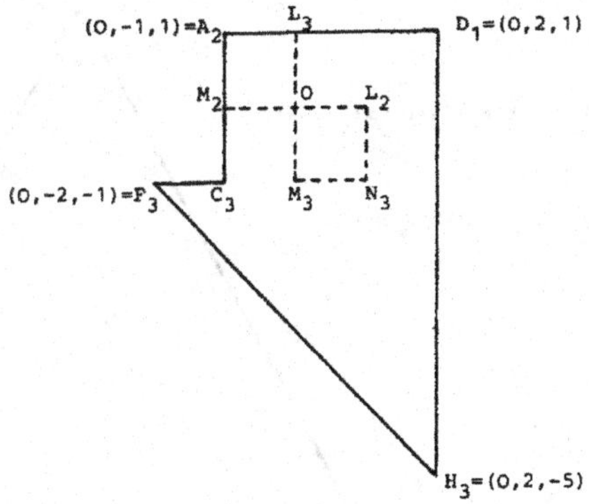

type 3 (in the plane z=1)

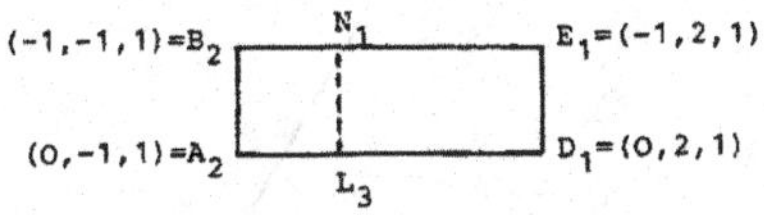

type 4 (in the plane y=2)

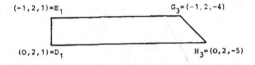

type 5 (in the plane x+y+z=-3)

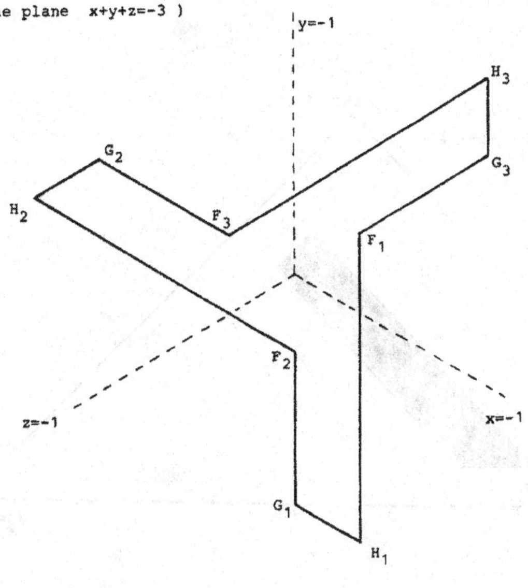

figure 4

type 6 (in the plane x+y+z=-4)

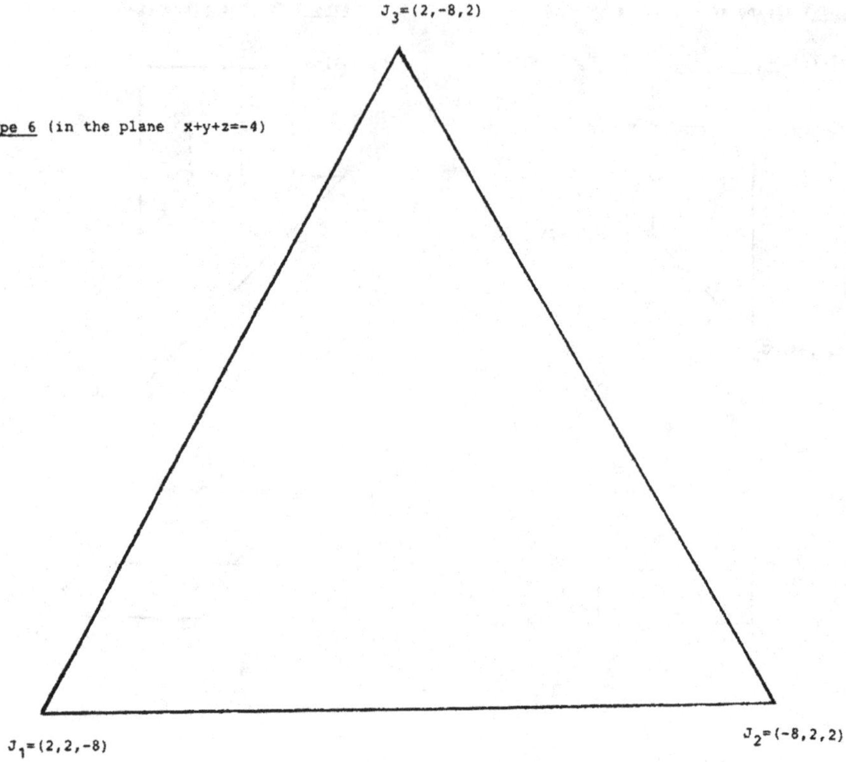

type 7 (in the plane y=2)

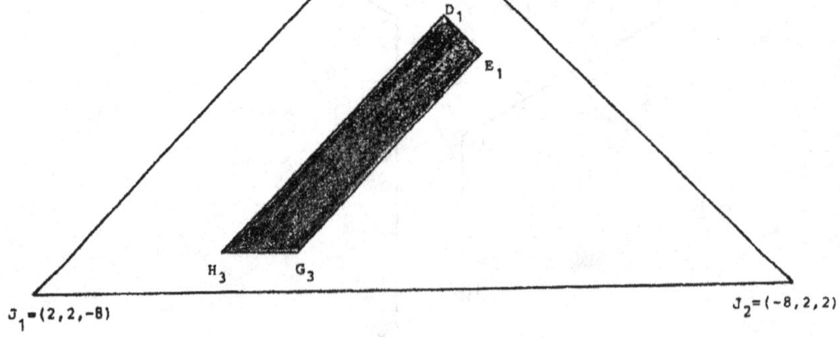

figure 5

All the strict local supporting planes of this polyhedral Boy surface with two handles will concentrate at the vertices of the tetrahedron which implies that the surface will be tight. Every vertex lies in the cubic lattice $\mathbb{Z}^3 \subset E^3$ where the origin is the triple point of the Boy surface (which is not a vertex of the surface). All of the nonconvex vertices are in fact of valence three. The figure 3 shows three different views onto the polyhedral Boy surface (the tetrahedron does not show up there). Note one can distinguish the curve of selfintersections from the edges by a different drawing.

We describe the surface explicitly by the coordinates of the vertices and the shape of the faces. The three-fold symmetry appears as cyclic shift of the three coordinates $x \to y \to z \to x$ corresponding to the cyclic shift of the indices $1 \to 2 \to 3 \to 1$ of the following vertices:

$A_1 = (-1, 1, 0)$	$A_2 = (0,-1, 1)$	$A_3 = (1, 0,-1)$
$B_1 = (-1, 1,-1)$	$B_2 = (-1,-1, 1)$	$B_3 = (1,-1,-1)$
$C_1 = (-1, 0,-1)$	$C_2 = (-1,-1, 0)$	$C_3 = (0,-1,-1)$
$D_1 = (0, 2, 1)$	$D_2 = (1, 0, 2)$	$D_3 = (2, 1, 0)$
$E_1 = (-1, 2, 1)$	$E_2 = (1,-1, 2)$	$E_3 = (2, 1,-1)$
$F_1 = (-1, 0,-2)$	$F_2 = (-2,-1, 0)$	$F_3 = (0,-2,-1)$
$G_1 = (-4,-1, 2)$	$G_2 = (2,-4,-1)$	$G_3 = (-1, 2,-4)$
$H_1 = (-5, 0, 2)$	$H_2 = (2,-5, 0)$	$H_3 = (0, 2,-5)$
$J_1 = (2, 2,-8)$	$J_2 = (-8, 2, 2)$	$J_3 = (2,-8, 2)$
	$K = (2, 2, 2)$	

The vertices J_1, J_2, J_3, K span a tetrahedron which contains a polyhedral Boy surface spanned by the 24 vertices $A_i, B_i, \ldots, H_i$, $i = 1,2,3$. The 13 faces of this Boy surface split into 5 different types under the three-fold symmetry. Figure 4 shows these 5 types where type 5 is invariant. The dashed lines indicate the curve of selfintersections. This curve is the closed space polygon

$$(O\ L_1\ N_2\ M_2\ O\ L_2\ N_3\ M_3\ O\ L_3\ N_1\ M_1\ O)$$

where the points are the following:

$L_1 = (1, 0, 0)$	$L_2 = (0, 1, 0)$	$L_3 = (0, 0, 1)$
$M_1 = (-1, 0, 0)$	$M_2 = (0,-1, 0)$	$M_3 = (0, 0,-1)$
$N_1 = (-1, 0, 1)$	$N_2 = (1,-1, 0)$	$N_3 = (0, 1,-1)$
	$O = (0, 0, 0)$	

The faces of the tetrahedron are shown in figure 5 . In order to join the Boy surface with the tetrahedron we have to cut out the three faces of type 4 from both of them. This is already indicated in the drawing of type 7 .

The Euler characteristic of the Boy surface is easily computed to be $\chi = 24 - 36 + 13 = 1$. After gluing it together with the tetrahedron we get $\chi = 28 - 42 + 11 = -3$ (Note that the three noncontractible faces of the tetrahedron don't make a contribution to the Euler characteristic). It is not hard to see that all the vertices of this surface are in fact of valence three.

In order to complete the proof of theorem 3 one has to attach an arbitrary number of handles to this tight polyhedral surface such that the threefold symmetry is preserved. This is easy to do if this number has a residue 0 or 1 modulo 3. In the case of a residue 2 modulo 3 one can use the boundary of a hexagonal truncated pyramid where three quadrilaterals have been removed which are not adjacent to each other (see figure 6).

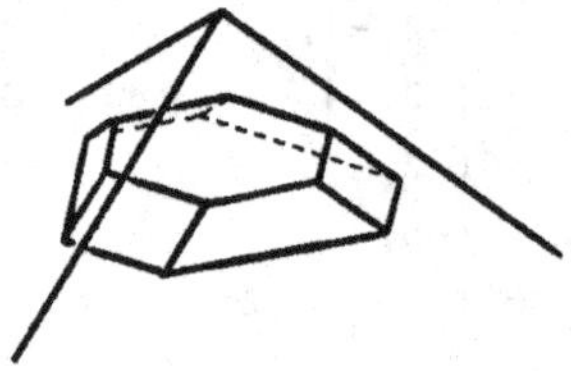

figure 6

4. CONCLUDING REMARKS

1. Many examples of tight polyhedral surfaces in the literature are of the kind to which our theorems apply. In fact such a smoothing process has already been used implicitly in Kuiper's example of a tight Klein bottle with handle (see [9]). Banchoff's tight square tori (see [4]) satisfy the assumptions of our theorem 2 if the meridian curve has a horizontal top and bottom edge. If there is a top or bottom vertex one has to truncate it horizontally a little bit.

Other examples are the boundaries of the difference set of two convex 3-polytopes. Such a construction has been used in [5] to get tight polyhedral surfaces with geometrical degree four. Assume that P and P' are 3-polytopes which are dual to each other and such that the vertices of each lie outside of the other. Then the boundary of each difference set $P \setminus P'$, $P' \setminus P$ will be a tight polyhedral surface satisfying the assumptions of our theorem 1. Quite symmetrical examples can be constructed using the Platonic or Archimedean solids and their duals (see figure 7 which shows the difference of a dodecahedron and an icosahedron). In the case of an octahedron minus a cube the resulting manifold of genus 7 happens to be a so-called "Platonic manifold" $\{5,4;7\}$ (see [14]).

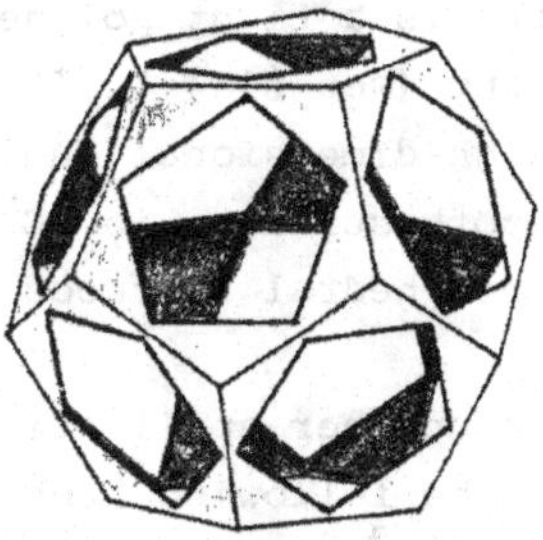

figure 7

2. However, the question is what happens for tight polyhedral surfaces in general. The 7-vertex Császár torus (cf. [8]) and Brehms's flat torus are tight and contain vertices of other types, in particular the "mixed curvature type" (see figure 8) where a nonconvex vertex admits an open set of local supporting planes.

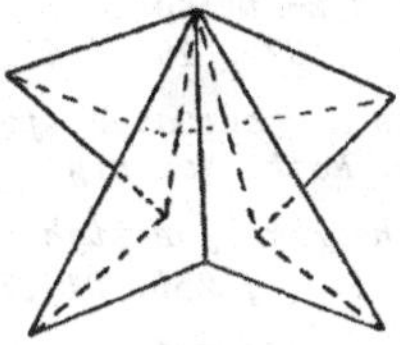

figure 8

There is the notion of a polyhedral curvature $K = K^{+} - K^{-}$ which corresponds to the Gaussian curvature, and an absolute curvature $K = K^{+} + K^{-}$ which corresponds to the absolute Gaussian curvature of a smooth surface (see [3],[7]). In terms of those a tight smoothing has to preserve the positive part K^{+} and the negative part K^{-} simultaneously. In [7] it is shown that one can approximate an arbitrary polyhedral surface in such a way that the various curvatures converge. This however does not imply that a tight polyhedral surface will be smoothed tightly because additional open sets of local supporting planes will be created.

We would like to ask the following

Question: Does an arbitrary tight polyhedral surface in E^3 admit a tight smoothing?

The corresponding question for tight surfaces with boundary is known to be false because there is a tight polyhedral Möbius band in E^3 but not tight smooth one (see [10]). It is also known to be false for surfaces in high dimensional euclidean space because there are no tight smooth surfaces in E^n for $n \geq 6$ whereas there are such tight polyhedral surfaces (cf. [2]).

3. Finally we want to mention another application of our theorem. In [13] there will be studied the following problem: which regular homotopy classes of surfaces in E^3 contain a tight smooth one? In particular it will be shown that there are tight smooth surfaces of non-standard regular homotopy types. Certain tightly smoothed tight polyhedral surfaces will be used in this construction, in particular the tight Boy surface with two handles described in section 3.

References

1 F. Apery, *La surface de BOY*, Thèse Univ. Louis Pasteur, Strasbourg 1984

2 T. F. Banchoff, *Tightly embedded 2-dimensional polyhedral manifolds*, Amer. J. of Math. 87 , 462 - 472 (1965; Zbl. 136,210).

3 - *Critical points and curvature for embedded polyhedral surfaces*, Amer. Math. Monthly 77 , 475 - 485 (1970; Zbl. 191,528).

4 - *Non-rigidity theorems for tight polyhedra*, Arch. Math. 21 , 416 - 423 (1970; Zbl. 216,188).

5 T. F. Banchoff and N. H. Kuiper, *Geometrical class and degree for surfaces in three-space*, J. Diff. Geom. 16 , 559-576 (1981; Zbl. 496.53003).

6 W. Blaschke und K. Leichtweiß, *Elementare Differentialgeometrie*, Springer 1973 (Grundlehren der mathematischen Wissenschaften Band 1) (1973; Zbl. 264.53001).

7 U. Brehm and W. Kühnel, *Smooth approximation of polyhedral surfaces regarding curvatures*, Geometriae Dedicata 12 , 435 - 461 (1982), short version in : Proc. Conf. Global Diff. Geom. and Global Analysis, Berlin 1979, 64 - 68 (1981) (Lecture Notes in Mathematics Vol. 838)(Zbl. 437.53040 and 483.53046).

8 A. Császár, *A polyhedron without diagonals*, Acta Sci. Math. Szeged 13 , 140 - 142 (1949; Zbl. 38,307).

9 N. H. Kuiper, *Convex immersions of closed surfaces in* E^3, Comment. Math. Helv. 35 , 85 - 92 (1961; Zbl. 243.53043).

10 - *Tight topological embeddings of the Moebius band*, J. Diff. Geom. 6 , 271 - 283 (1972; Zbl. 234.53048).

11 - *Tight embeddings and maps, Submanifolds of geometrical class three in* E^N, Proc. Chern Symp. Berkeley 1979, 97 - 145, Springer (1980; Zbl. 461.53033).

12 - *Geometry in total absolute curvature theory*, in: Perspectives in Mathematics, Anniversary of Oberwolfach, 377 - 392, Birkhäuser 1984

13 U. Pinkall, *Regular homotopy classes of tight surfaces* (in preparation)

14 J. M. Wills, *Reguläre polyedrische Mannigfaltigkeiten*, Mitteilungen Math. Ges. Hamburg 11 , 171 - 177 (1983).

Wolfgang Kühnel
Fachbereich Mathematik
der TU Berlin
Str. d. 17. Juni 136
D - 1000 Berlin 12

Ulrich Pinkall
Max-Planck-Institut
für Mathematik
Gottfried-Claren-Str. 26
D - 5300 Bonn 3

ON THE NUMBER OF TRITANGENCIES OF A SURFACE IN $\mathbb{R}^3$

Tetsuya OZAWA
Department of Mathematics
Faculty of Science
Nagoya University
Chikusa-ku, Nagoya, 464 JAPAN

§0. Introduction.

A tritangency of a surface in $\mathbb{R}^3$ is, by definition, a plane which is tangent to the surface at three different points. Generically a closed surface in $\mathbb{R}^3$ has a finite number of tritangencies, and has no planes which are tangent at four points or more. In this paper we are interested in the number $N_{1,1,1}(\varphi)$ of the tritangencies of a generic immersion φ of a closed surface into $\mathbb{R}^3$, relating to the number $N_3(\varphi)$ of the godron points on φ counted with indices (the definitions are in §1). We will prove the

Theorem 1.5. Generically $N_{1,1,1}(\varphi) \equiv N_3(\varphi) \pmod 2$, provided the surface has an even Euler number.

The outline of the proof is as follows. $C^\infty(M)/\sim$ is the space obtained from the function space $C^\infty(M)$ of a given surface M identifying f and $-f$ $(f \in C^\infty(M))$. For any immersion $\varphi : M \to \mathbb{R}^3$, the map $H_\varphi : S^2 \to C^\infty(M)$ is defined by the scalar product of $\mathbb{R}^3$:

$$H_\varphi(z) : M \ni x \to H_\varphi(z)(x) = \varphi(x)\cdot z \in \mathbb{R},$$

where S^2 is the unit sphere of $\mathbb{R}^3$. H_φ is reduced to the map $H_\varphi/\sim : P^2 \to C^\infty(M)/\sim$ of the projective plane $P^2 = S^2/\sim$. Two submanifolds $S(1,1,1)/\sim$ and $S(3)/\sim$ of $C^\infty(M)/\sim$ are so introduced that the intersection number of $H_\varphi/\sim$ and $S(1,1,1)/\sim$ (resp. $S(3)/\sim$) is equal to $N_{1,1,1}(\varphi)$ (resp. $N_3(\varphi)$). First we prove, in §4, the

Proposition A. Suppose the Euler number of M is even. For any two immersions φ and ψ in general position, we have

$$N_{1,1,1}(\varphi) + N_3(\varphi) \equiv N_{1,1,1}(\psi) + N_3(\psi) \pmod 2.$$

The proposition is deduced from the fact that $H_\varphi/\sim$ and $H_\psi/\sim$ are homotopic in $C^\infty(M)/\sim - \{0\}$, and from the analysis of the stratum of codimension 3 of the canonical stratification of the closure of $S(1,1,1)/\sim \cup S(3)/\sim$. Finally we show that for a surface M of even Euler number, there exists an immersion φ in general position such that $N_{1,1,1}(\varphi) + N_3(\varphi) \equiv 0 \pmod 2$.

For surfaces of odd Euler number, examples which are obtained in contrast with the calculation of $I(f)$ with $f \in \mathcal{O}_1^{(3)}(4)$ (see (IV) in §4) show that there exist two immersions φ and ψ such that $N_{1,1,1}(\varphi) \equiv N_3(\varphi)$ and $N_{1,1,1}(\psi) \not\equiv N_3(\psi) \pmod 2$.

§1. Godron point

In what follows, we denote by M a closed surface.

Let $f : M \to \mathbb{R}^3$ be an immersion of a closed surface M, and p be a point on M. Taking one of the unit normal vector of f at $f(p)$, say ν, we define the height function using the scalar product of $\mathbb{R}^3$;

$$M \ni x \to (f(x) - f(p))\cdot\nu \in \mathbb{R},$$

whose germ at p we denote by $h(p)$.

It is well know that if the Gaussian curvature of M induced by f is nonzero at p, then $h(p)$ is equivalent to the following function germ;

(a_1) $\quad h(p) \sim \pm(x^2 \pm y^2)$ $\quad$ (codim = 0)

(the codimension of a function germ will be defined in the following section). Here the first double sign $\pm$ depends on the choice of the normal vector ν, and the second depends on the sign of the curvature at p. Let M_0 denote the set of all points where the curvature is zero. If $p \in M_0$, then $h(p)$ may be equivalent to one of the following germs;

(a_2) $\quad h(p) \sim \pm x^2 + y^3$ $\quad$ (codim = 1)

(a_3) $\quad h(p) \sim \pm(x^2 \pm y^4)$ $\quad$ (codim = 2)

(a_*) $\quad h(p) \sim$ germ of codim ≥ 3.

Theorem 1.1 (Bleecker, Willson [4]). For a generic immersion f,

(1) every point p on M satisfies one of the first three: (a_1), (a_2) or (a_3).

(2) M_0 is a closed submanifold of M of dimension 1.

(3) The points which satisfy (a_3) are finitely many.

Definition 1.2. A point which satisfies (a_3) is called a godron point of f.

Remark 1.3. 1) "Godron point" is the terminology of Kergosian and Thom in [2].

2) A godron point is, in other words, a point where the Gauss map : $M \to S^2$ (or $\mathbb{R}P^2$) has a Whitney cusp (see [3]).

3) In the polynomial in (a_3), the second double sign depends on the geometric property of the godron point. If this sign is a plus (resp. a minus), then we call it elliptic (resp. hyperbolic). See the figure 1.1.

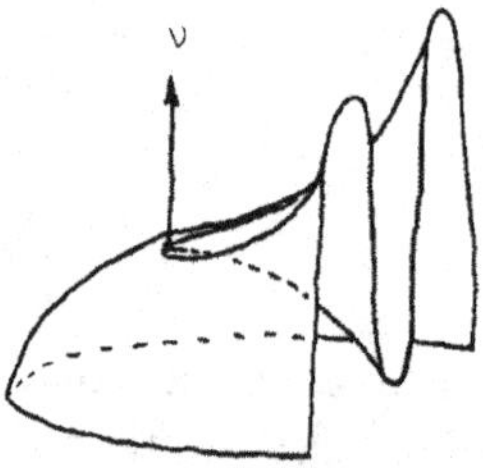

hyperbolic

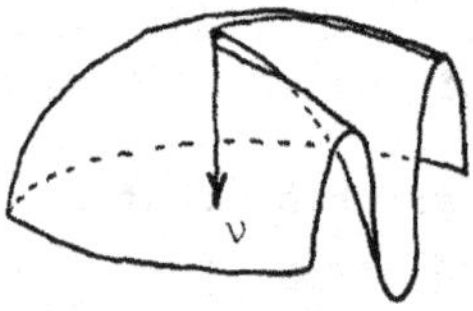

elliptic

Fig. 1.1

We had the freedom of the choice of the unit normal ν. In what follows, we choose the unit normal at each godron point in such a way that we will have $h(p) \sim \pm x^2 + y^4$. Here we respected the sign of the term y^4 to choose ν.

Definition 1.4. (index of a godron point). Fix the unit normal ν at a godron point p as above, and denote by $M(p)$ the inverse image of the open half space of the side ν separated by $T_p f(M)$;

$$M(p) = \{x \in M;\ (f(x) - f(p))\cdot\nu > 0\}.$$

Define the index $\mathrm{Ind}(p) = \mathrm{Eul}(M(p))$, where $\mathrm{Eul}(\ )$ denotes the Euler number.

Let $N_{1,1,1}(f)$ denote the number of the tritangencies of f, and $N_3(f)$ the sum of the indices $\mathrm{Ind}(p)$ of all godron points p of f.

Theorem 1.5. If the Euler number of a closed surface is even, then there exists an open and dense subset A in the set of all immersions (with the Whitney C^∞ topology) such that for any $f \in A$ we have

$$N_{1,1,1}(f) \equiv N_3(f) \pmod 2.$$

§2. Function space and the submanifolds associated to tritangencies

$C^\infty(M)$ is the space of all smooth functions on a manifold M. Let c be a critical value of $f \in C^\infty(M)$. We say that c is excellent, if f has only one critical point on the level $f^{-1}(c)$, and which is nondegenerate.

For a point $p \in M$, $C_p^\infty(M)$ is the ring of the germs at p of all functions in $C^\infty(M)$. The codimension of a germ $f \in C_p^\infty(M)$ is, by definition, the dimension over $\mathbb{R}$ of the quotient space

$$C_p^\infty(M)/\{C_p^\infty(M)(\partial f/\partial x) \oplus f^*(C_{f(p)}^\infty(\mathbb{R}))\},$$

where $C_p^\infty(M)(\partial f/\partial x)$ is the Jacobian ideal of f.

Suppose M of dimension 2. The codimension of $f \in C_p^\infty(M)$ is equal to 0, if and only if f is not critical, or f is critical and nondegenerate. The codimension of $f \in C_p^\infty(M)$ is equal to 1, 2 or 3, if and only if there exist local coordinates (x,y) of M at p such that

$$(A_2)\quad f(x,y) = \pm x^2 + y^3 + f(p) \iff \mathrm{codim} = 1$$

$$(A_3)\quad f(x,y) = \pm x^2 \pm y^4 + f(p) \iff \mathrm{codim} = 2$$

$$\left.\begin{array}{ll}(A_4) & f(x,y) = \pm x^2 + y^5 + f(p) \\ (D_4) & f(x,y) = \pm x^2 y + y^3 + f(p)\end{array}\right\} \iff \mathrm{codim} = 3.$$

Let p be a critical point of $f \in C^\infty(M)$. The codimension of the critical point p is, by definition, the codimension of the germ of f at p.

We introduce two submanifolds of codimension 2 in $C^\infty(M)$, denoted by S(1,1,1) and S(3), which will play important roles in the proof of Theorem 1.5. Define

$S(1,1,1) = \{f \in C^\infty(M);$ f has one and only one nonexcellent critical value c, and on the level $f^{-1}(c)$, f has three nondegenerate critical points$\}$,

$S(3) = \{f \in C^\infty(M);$ f has one and only one nonexcellent critical value c, and on the level $f^{-1}(c)$, f has only one critical point which is of codimension 2$\}$.

For each $f \in S(1,1,1) \cup S(3)$, $c(f)$ denotes the unique nonexcellent critical value of f. $S(1,1,1)$ and $S(3)$ are submanifolds of codimension 2 in $C^\infty(M)$, since they are locally determined by two independent equations.

Let f be an immersion of a closed surface M into $\mathbb{R}^3$ such that the map $H_f : S^2 \to C^\infty(M)$ definedby the scalar product of $\mathbb{R}^3$ (see Introduction) is transversal to $S(1,1,1)$ and $S(3)$ (this condition is generically satisfied, see Lemma 3.3). If we denote by $h(v,c)$ the plane $\{x \in \mathbb{R}^3;\ x\cdot v = c\}$, then for each $v \in H_f^{-1}(S(1,1,1))$ we have a tritangency $h(v, c(H_f(v)))$, and for each $v \in H_f^{-1}(S(3))$ we have the tangent plane $h(v, c(H_f(v))) = T_p f(M)$ at a godron point p.

We associate an index $\mathrm{Ind}(f)$ to each element f in $S(1,1,1)$ and $S(3)$. For each $f \in S(1,1,1)$, we set $\mathrm{Ind}(f) = 1$. Let $f \in S(3)$, and p be the unique degenerate critical point of f. Since p is of codimension 2, f is locally written as $f(x,y) = \pm x^2 + ay^4 + c(f)$, where $a = \pm 1$. Define

$$\mathrm{Ind}(f) = \mathrm{Eul}(f^{-1}(]c(f), a\infty[)), \quad (2.1)$$

where $\mathrm{Eul}(\ )$ is the Euler number, and $]c(f),a\infty[$ means $]c(f),\infty[$ or $]-\infty,c(f)[$, according as a=1 or -1. Since $\mathrm{Ind}(f)=\mathrm{Ind}(-f)$ for any f, the indices of elements in the quotient spaces $S(1,1,1)/\sim$ and $S(3)/\sim$ are well defined, which we denote by the same notation.

Definition 2.1. Let X be a 2-dimensional manifold, and $H: X \to C^\infty(M)/\sim$ be transversal to $S(*)/\sim$ ($(*) = (1,1,1)$ or (3)). Define the intersection number $\mathrm{int}(H, S(*)/\sim) \in \mathbb{Z}_2$ by

$$\mathrm{int}(H,S(*)/\sim) \equiv \sum_{x \in H^{-1}(S(*)/\sim)} \mathrm{Ind}(H(x)) \pmod 2.$$

The following lemma is evident;

Lemma 2.2. For $(*) = (1,1,1)$ or (3), we have

$$N_*(f) \equiv \mathrm{int}(H_f/\sim, S(*)/\sim) \pmod 2.$$

§3. Preparations

This section contains some lemmata to be used in the proof of Proposition A.

The product $\mathrm{Diff}(M) \times \mathrm{Diff}(\mathbb{R})$ of the groups of diffeomorphisms acts on $C^\infty(M)$ by the composition of maps;

$$\mathrm{Diff}(M) \times \mathrm{Diff}(\mathbb{R}) \times C^\infty(M) \ni (\rho,\sigma,f) \to \sigma\circ f\circ\rho^{-1} \in C^\infty(M).$$

Let $\mathcal{O}^{(i)}$ denote the union of all $\mathrm{Diff}(M)\times\mathrm{Diff}(\mathbb{R})$-orbits of codimension i in $C^\infty(M)$ $(i=0,1,2,\cdots)$. In particular, $\mathcal{O}^{(0)}$ is the set of all excellent Morse functions. $\mathcal{O}^{(i)}$ is not necessarily of codimension i in $C^\infty(M)$, since the orbits form, in general, moduli for large i.

Lemma 3.1. The closure $\overline{\mathcal{O}^{(4)}} = \cup_{i\geq 4}\mathcal{O}^{(i)}$ is of codimension 4 in $C^\infty(M)$, and the partition $\{\mathcal{O}^{(i)}\}_{i=0}^{3}$ gives a Whitney stratification of $C^\infty(M) - \overline{\mathcal{O}^{(4)}}$.

Proof. In the k-jet space $J^k(M,\mathbb{R})$ with k sufficiently large, the jets of simple germs (in the sense of Arnold, c.f. [1]) form a subset of codimension $\geq \dim(M) + 6$. The lemma follows from this fact using the argument with multi-jet spaces. q.e.d.

$\mathcal{O}_j^{(i)}$ denotes the subset of $\mathcal{O}^{(i)}$ consisting of functions with j nonexcellent critical values. Let c be a critical value of $f \in C^\infty(M)$, and $\{p_1,\cdots, p_k\}$ be the set of all critical points of f on the level $f^{-1}(c)$, whose codimensions be equal to $n_1-1, n_2-1,\cdots$ and n_k-1 respectively. We say that $(n_1,\cdots, n_k)$ is the type number of the critical value c. In case $j = 1$, $\mathcal{O}_1^{(i)}(n_1,\cdots, n_k)$ denotes the subset of $\mathcal{O}_1^{(i)}$ consisting of functions whose unique nonexcellent critical value has the type number $(n_1,\cdots, n_k)$. If $i \leq 3$, then $n_1 +\cdots+ n_k = i + 1$, since the germ at any critical point of $f \in \mathcal{O}^{(i)}$ $(i = 1,2,3)$ is simple (if a germ $g \in C_p^\infty(M)$ is simple, then $g^*(\mathfrak{m}_{g(p)}(\mathbb{R}))$ is contained in the Jacobi ideal $C_p^\infty(M)(\partial g/\partial x)$. c.f. [5]). Therefore we have the partition of $\mathcal{O}_1^{(i)}$ $(i = 1,2,3)$ which corresponds to the partition of the number $i + 1$ by positive integers.

Using the above notations, we can write

$$S(1,1,1) = \mathcal{O}_1^{(2)}(1,1,1) \quad \text{and} \quad S(3) = \mathcal{O}_1^{(2)}(3).$$

From Lemma 3.1, it follows that the closure $\overline{S}$ of $S = S(1,1,1) \cup S(3)$ in $C^\infty(M) - \overline{\mathcal{O}^{(4)}}$ admits a Whitney stratification, and the 3-stratum $\overline{S}^{(3)}$ has the partition

$$\overline{S}^{(3)} = (\overline{S}^{(3)} \cap \mathcal{O}_2^{(3)}) \cup \mathcal{O}_1^{(3)}(1,1,1,1) \cup \mathcal{O}_1^{(3)}(2,1,1) \cup \mathcal{O}_1^{(3)}(3,1) \cup \mathcal{O}_1^{(3)}(4) \tag{3.1}$$

About the stratification and the partition, the same holds for $\overline{S}/\sim$ in $(C^\infty(M) - \overline{\mathcal{O}^{(4)}})/\sim$.

If a map $f \in C^\infty(M, \mathbb{R}^3)$ is an immersion, then the map $H_f/\sim$: $P^2 \to C^\infty(M)/\sim$ doesn't meet the origin $0 \in C^\infty(M)/\sim$. $H_f/\sim$ is not homotopic to a constant map in $C^\infty(M)/\sim - \{0\}$. In fact the image of a generator of $\pi_1(P^2)$ by $H_f/\sim$ is not contractible in $C^\infty(M)/\sim - \{0\}$. But any two $H_f/\sim$ and $H_g/\sim$ are homotopic to each other, that is,

<u>Lemma 3.2</u>. For any two immersions f and $g \in \mathrm{Imm}(M, \mathbb{R}^3)$, $H_f/\sim$ and $H_g/\sim : P^2 \to C^\infty(M)/\sim - \{0\}$ are homotopic.

<u>Proof</u>. We remark that, for a map $f \in C^\infty(M, \mathbb{R}^3)$ and a unit vector $z \in S^2$, we have $H_f(z) = 0$ if the image $f(M)$ is contained in the plane $\{x \in \mathbb{R}^3;\ x \cdot z = 0\}$. If f and g are immersions, then we can suppose, after some translations of $\mathbb{R}^3$ and a diffeomorphism on M, that there exists a point $p \in M$ such that $T_pf = T_pg : T_pM \to \mathbb{R}^3$, and that their image $T_pf(T_p(M))$ doesn't pass through the origin of $\mathbb{R}^3$. In this case, if we set $h_t(x) = (1-t)\cdot g(x)+t\cdot f(x)$, then $t \to H_{h_t}/\sim$ is a required homotopy between $H_f/\sim$ and $H_g/\sim$. Since the given map f is an immersion, $H_f/\sim$ doesn't meet $0 \in C^\infty(M)/\sim$, and the same holds after applying translations in $\mathbb{R}^3$ and diffeomorphisms on M.

q.e.d.

$G(M,\mathbb{R}^3)$ denotes the subset of $C^\infty(M, \mathbb{R}^3)$ consisting of maps f such that $H_f : S^2 \to C^\infty(M)$ doesn't meet $\overline{\mathcal{O}^{(3)}}$, and is transverse to the stratification $\{\mathcal{O}^{(i)}\}_{i=1,2}$ of $C^\infty(M) - \overline{\mathcal{O}^{(3)}}$.

<u>Lemma 3.3</u>. $G(M, \mathbb{R}^3)$ is open and dense in $C^\infty(M, \mathbb{R}^3)$.

Proof. The map : ${}_rJ^k(M, \mathbb{R}^3) \times S^2 \to {}_rJ^k(M, \mathbb{R})$ of the multi-jet spaces is defined in the same way as H_f using the scalar product of $\mathbb{R}^3$. This map is a subversion. The stratification $\{\mathcal{O}^{(i)}\}_{i=1,2}$ and the subset $\overline{\mathcal{O}^{(3)}}$ are well characterized by the associated stratifications of ${}_rJ^k(M, \mathbb{R})$. Hence this lemma follows from Thom's transversality theorem applied to maps ${}_rj^kf : M^{(r)} \to {}_rJ^k(M, \mathbb{R}^3)$. q.e.d.

Remark 3.4. Since $\mathrm{Imm}(M, \mathbb{R}^3)$ is open in $C^\infty(M, \mathbb{R}^3)$, $A = \mathrm{Imm}(M, \mathbb{R}^3) \cap G(M, \mathbb{R}^3)$ is open and dense in $\mathrm{Imm}(M, \mathbb{R}^3)$. Of course this subset A in $\mathrm{Imm}(M, \mathbb{R}^3)$ plays the role of A in Theorem 1.5.

§4. Proof of Proposition A

$S^{(3)}$ denotes the stratum of codimension 3 of the closure $\overline{S} = \overline{S(1,1,1) \cup S(3)}$ in $C^\infty(M) - \overline{\mathcal{O}^{(4)}}$. To each $f \in S^{(3)}$, we associate a number $I(f) \in \mathbb{Z}_2$ as follows. Let $u : \mathbb{R}^3 \to C^\infty(M)$ with $u(0) = f$ be transversal to $\overline{S}$ at f. Take a small 2-sphere S^2 in $\mathbb{R}^3$ centered at the origin of $\mathbb{R}^3$, and suppose that $u|_{S^2}$ is transversal to $\overline{S}$. Define $I(f)$ by

$$I(f) = \mathrm{int}(u|_{S^2}, S(1,1,1)) - \mathrm{int}(u|_{S^2}, S(3)) \in \mathbb{Z}_2.$$

Let f and g be in $\mathrm{Imm}(M, \mathbb{R}^3) \cap G(M, \mathbb{R}^3)$. From Lemma 3.1 and 3.2, it follows that there exists a map $h : P^2 \times [0,1] \to C^\infty(M)/\sim - \{0\}$ such that

(i) $h|_{P^2\times\{0\}} = f$, and $h|_{P^2\times\{1\}} = g$,

(ii) the image of h is contained in $(C^\infty(M) - \overline{\mathcal{O}^{(4)}})/\sim$, and h is transversal to $\overline{S}/\sim$.

Therefore Proposition A is deduced from the fact that $I(f) = 0$ for any $f \in S^{(3)}$, which we are going to verify following the partition in (3.1)

Let $f_0 \in S^{(3)}$ be fixed, and $u : \mathbb{R}^3 \to C^\infty(M)$ be as above. Except in (V) below, c_0 denotes the unique nonexcellent critical value of f_0.

(I) In case that $f_0 \in \mathcal{O}_1^{(3)}(1,1,1,1)$: Let p_1, p_2, p_3 and p_4 be the critical points of f_0 with $f(p_i) = c_0$ $(i=1,2,3,4)$. There exist a small neighborhood $\mathcal{U}$ of f_0 in $C^\infty(M)$ and functions $\lambda_i : \mathcal{U} \to \mathbb{R}$ with $\lambda_i(f_0) = 0$ $(i=1,2,3,4)$ such that each $f \in \mathcal{U}$ can be written, near p_i, by

$$f(x_i, y_i) = \pm x_i^2 \pm y_i^2 + \lambda_i(f) + c_0 \quad (i = 1,2,3,4)$$

using local charts (x_i, y_i) of M at p_i which depend continuously on $f \in \mathcal{U}$. From the transversality of u to $S^{(3)}$, if follows that $\{\mu_i = \lambda_i \circ u - \lambda_4 \circ u\}_{i=1,2,3}$ is a local chart of $\mathbb{R}^3$ at the origin. The inverse image $u^{-1}(S(1,1,1))$ is equal, near the origin, to

$$\{\mu_1 = \mu_2 = 0,\ \mu_3 \neq 0\} \cup \{\mu_1 = \mu_3 = 0,\ \mu_2 \neq 0\} \cup$$

$$\{\mu_2 = \mu_3 = 0,\ \mu_1 \neq 0\} \cup \{\mu_1 = \mu_2 = \mu_3 \neq 0\}.$$

Hence $u^{-1}(S(1,1,1))$ is the union of eight curves starting from $0 \in \mathbb{R}^3$. On the other hand, it is clear that $u^{-1}(S(3)) = \phi$ near $0 \in \mathbb{R}^3$. Hence we have $I(f_0) = 0$.

(II) In case that $f_0 \in \mathcal{O}_1^{(3)}(2,1,1)$: Let p_1 and p_2 be nondegenetate and p_3 be the degenerate critical point of f_0 with $f_0(p_i) = c_0$ $(i = 1,2,3)$. Since the universal unfolding of the function: $y \to y^3$ is given by $y^3 + 3\lambda_1 y + \lambda_2$, there exist a neighborhood $\mathcal{U}$ of f_0 and functions $\lambda_i : \mathcal{U} \to \mathbb{R}$ with $\lambda_i(f_0) = 0$ $(i = 1,2,3,4)$ such that for any $f \in \mathcal{U}$, we have

$$\begin{cases} f(x_i, y_i) = \pm x_i^2 \pm y_i^2 + \lambda_i(f) + c_0 & (\text{near } p_i,\ i = 1,2) \\ f(x_3, y_3) = \pm x_3^2 + y_3^3 + 3\lambda_3(f) y_3 + \lambda_4(f) + c_0 & (\text{near } p_3) \end{cases}$$

By the transversality of u, $\{\mu_i = \lambda_i \circ u\}_{i=1,2,3}$ is a local chart of $\mathbb{R}^3$ at 0. Using this chart we have, near $0 \in \mathbb{R}^3$,

$$\begin{cases} u^{-1}(S(1,1,1)) = \{\mu_1 = \mu_2 = 4(-\mu_3)^{3/2}\ ;\ \mu_3 < 0\} \\ u^{-1}(S(3)) = \phi \end{cases}$$

Hence we have $I(f_0) = 0$.

(III) In case that $f_0 \in \mathcal{O}_1^{(3)}(3,1)$; Let p_0 and q_0 be the critical points of f_0 with $f_0(p_0) = f_0(q_0) = c_0$ which have the codimension 3 and 1 respectively. Since the universal unfolding of the function: $y \to y^4$ is given by $y^4 + 4\lambda_1 y^2 + \lambda_2 y + \lambda_3$, there exist a neighborhood $\mathcal{U}$ of f_0 and functions $\lambda_i : \mathcal{U} \to \mathbb{R}$ with $\lambda_i(f) = 0$ $(i = 1,2,3,4)$ such that for $f \in \mathcal{U}$ we have

$$\begin{cases} f(x_1, y_1) = \pm x_1^2 \pm (y_1^4 + 4\lambda_1(f) y_1^2 + \lambda_2(f) y_1 + \lambda_3(f)) + c_0 & (\text{near } p_0) \\ f(x_2, y_2) = \pm x_2^2 \pm y_2^2 + \lambda_4(f) + c_0 & (\text{near } q_0). \end{cases}$$

$\{\mu_i = \lambda_i \circ u\}_{i=1,2,3}$ is a local chart of $\mathbb{R}^3$ at 0. We find, near $0 \in \mathbb{R}^3$,

$$\begin{cases} u^{-1}(S(1,1,1)) = \{\mu_2 = 0,\ \mu_3 = -3\mu_1^2;\ \mu_1 < 0\} \\ u^{-1}(S(3)) = \{\mu_1 = \mu_2 = 0,\ \mu_3 \neq 0\}. \end{cases}$$

Since $u^{-1}(S(1,1,1))$ is a curve starting from $0 \in \mathbb{R}^3$, the intersection number $\mathrm{int}(u|_{S^2}, S(1,1,1)) = 1$ for a small sphere S^2 centered at $0 \in \mathbb{R}^3$.

If we take a small positive number $a > 0$, the functions $f_\pm = u(\mu_1,\mu_2,\mu_3)$ with $(\mu_1,\mu_2,\mu_3) = (0,0,\pm a)$ have pairs of critical points $(p_\pm,q_\pm)$ such that $p_\pm$ and $q_\pm$ are close to p_0 and q_0 respectively, and that p_+ (resp. q_-) is above q_+ (resp. p_-) with respect to f_+ (resp. f_-). Hence the difference of the Euler numbers $\mathrm{Eul}(f_\pm^{-1}(]c(f_\pm), a\infty[))$ is equal to 1 (see (2.1)). Therefore we have $\mathrm{int}(u|_{S^2}, S(3)) = 1$. These imply that $I(f_0) = 0$.

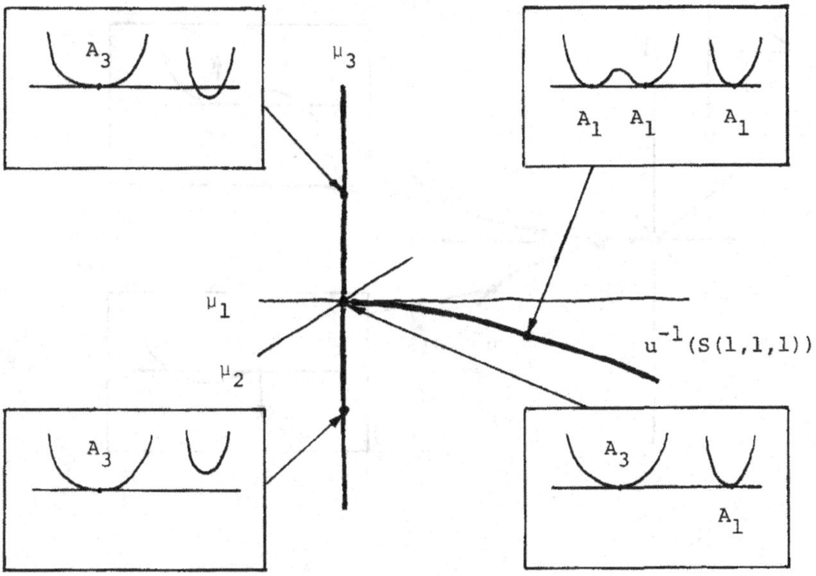

Fig. 4.1.

(IV) In case that $f_0 \in \mathcal{O}_1^{(3)}(4)$: f_0 has only one critical point p_0 on the level $f_0^{-1}(c_0)$ which is of codimension 3. We have two types of critical points of codimension 3; A_4 and D_4.

(a) Suppose p_0 is of the type A_4. Since the universal unfolding of the function: $y \to y^5$ is given by $y^5 + \lambda_3 y^3 + \lambda_2 y^2 + \lambda_1 y + \lambda_0$, there exist a neighborhood $\mathcal{U}$ of f_0 and functions $\lambda_i : \mathcal{U} \to \mathbb{R}$ with $\lambda_i(f_0) = 0$ $(i = 0,1,2,3)$ such that near p_0, $f \in \mathcal{U}$ is written by

$$f(x,y) = \pm x^2 + y^5 + \lambda_3(f)y^3 + \lambda_2(f)y^2 + \lambda_1(f)y + \lambda_0(f) + c_0$$

$\{\mu_i = \lambda_i \circ u\}_{i=1,2,3}$ is a local chart of $\mathbb{R}^3$ at 0. Hence we have

$$\begin{cases} u^{-1}(S(1,1,1)) = \phi \\ u^{-1}(S(3)) = \{(\mu_1,\mu_2,\mu_3) = (-15\alpha^4,\ 20\alpha_1^3 - 10\alpha^2);\ \alpha \neq 0\}. \end{cases}$$

For a small sphere S^2 centered at $0 \in \mathbb{R}^3$, $u(S^2)$ and $S(3)$ have two intersection points f_+ and f_-, and their indices are congruent modulo 2 provided $\mathrm{Eul}(M) \equiv 0 \pmod 2$. Therefore $I(f_0) = 0$.

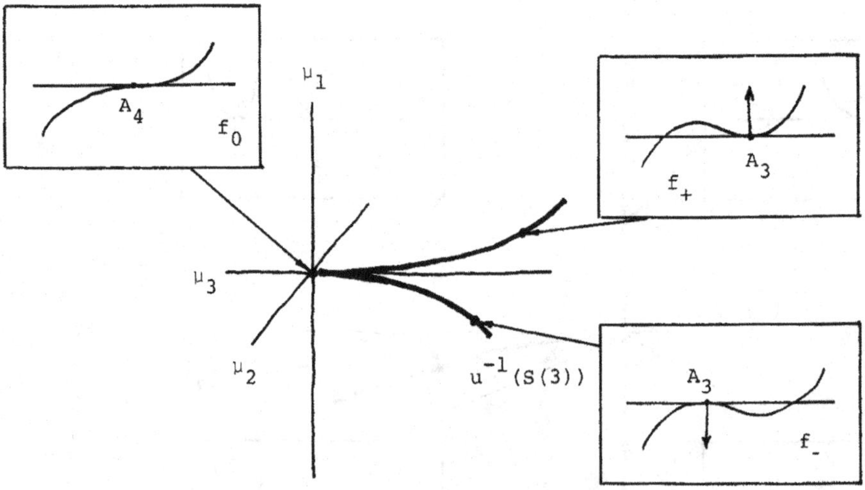

Fig. 4.2.

(b) Suppose p_0 is of the type D_4: The universal unfolding of the function: $(x,y) \to \pm x^2 y + y^3$ is given by $\pm x^2 y + y^3 + \lambda_3 x + \lambda_2 y^2 + \lambda_1 y + \lambda_0$. Hence there exist a neighborhood $\mathcal{U}$ of f_0 and functions λ_i: $\mathcal{U} \to \mathbb{R}$ with $\lambda_i(f_0) = 0$ $(i = 0,1,2,3)$ such that each $f \in \mathcal{U}$

is written, near p_0, by

$$f(x,y) = \pm x^2 y + y^3 + \lambda_3(f)x + \lambda_2(f)y^2 + \lambda_1(f)y + \lambda_0(f) + c_0.$$

Using the local chart $\{\mu_i = \lambda_i \circ u\}_{i=1,2,3}$ of $\mathbb{R}^3$ at 0, we obtain

$$\begin{cases} u^{-1}(S(1,1,1)) = \phi \\ u^{-1}(S(3)) = \{\mu_1 = \mu_3 = 0, \ \mu_2 \neq 0\}. \end{cases}$$

For a small sphere S^2 centered at $0 \in \mathbb{R}^3$, $u(S^2)$ and $S(3)$ have two intersection points f_+ and f_-, whose indices are congruent modulo 2 provided $\mathrm{Eul}(M) \equiv 0 \pmod 2$. Therefore $I(f_0) = 0$.

(V) In case that $f_0 \in S^{(3)} \cap \mathcal{O}_2^{(3)}$: In a small neighborhood $\mathcal{U}$ of f_0, $S^{(3)}$ is not an essential stratum, that is, $\overline{S} \cap \mathcal{U}$ is a smooth submanifold of codimension 2. Any functions f in $S^{(2)} \cap \mathcal{U}$ have the same type of the unique nonexcellent critial values, and also have the same index. See the fiqure 4.3. Hence we have $I(f_0) = 0$.

q.e.d.

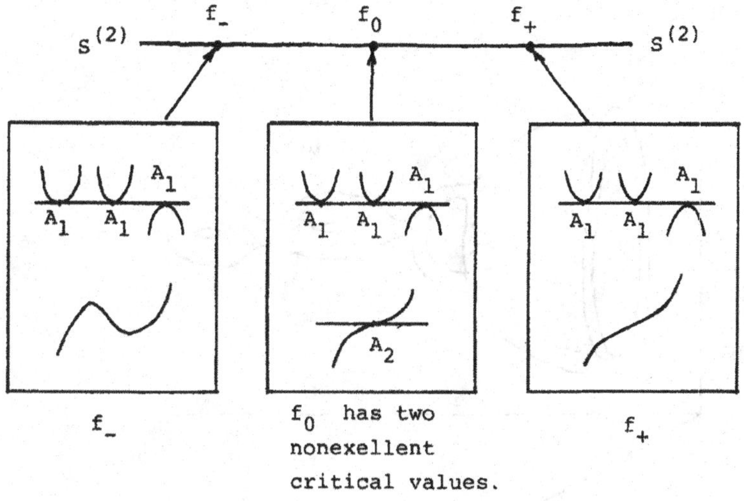

Fig. 4.3.

§5. Completion of the proof of Theorem 1.5.

Any closed surface with even Euler number is a double of a bounded surface, where double meens the quotient space of two copies of a surface obtained by identifying the corresponding boundary points. For example, the 2-sphere, the 2-torus, Klein bottle ⋯ are the doubles of the 2-disc, the cylinder $S^1 \times [0,1]$, the projective plane removed one disc, ⋯, respectively.

Let M be a surface with boundary $\partial M \neq \phi$, and take an immersion $f : M \to \mathbb{R}^3$ such that

1) $f(M)$ is contained in the half space $\{z \geq 0\}$ (where z is a coordinate of $\mathbb{R}^3$), and $f(\partial M)$ is contained in the plane $\{z = 0\}$.
2) $f(\partial M)$ is a disjoint union of convex circles $c_1, \cdots, c_k$ and c_i is contained in the interior of c_{i-1} $(i = 2, \cdots, k)$.
3) The map $f' : M' \to \mathbb{R}^3$ of the double M' of M obtained by the reflection of $\mathbb{R}^3$ with respect to $\{z = 0\}$ is an immersion.
4) The Gaussian curvature of M' induced by f' is positive on a neighborhood of $\partial M \subset M'$.
5) If $\pi : \mathbb{R}^3 \to \{z = 0\}$ is the orthogonal projection, then the singular set of $\pi \circ f|_{(M - \partial M)}$ is contained in the interior of c_k.

See the Figure 5.1.

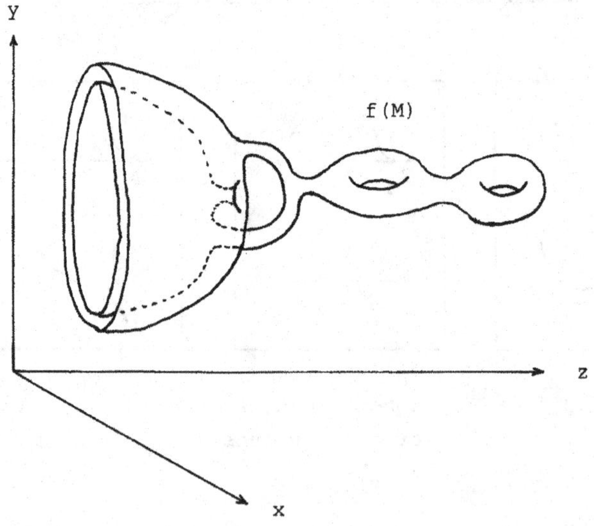

Fig. 5.1.

If we slightly deform f' to f'' preserving the symmetry with respect to $\{z = 0\}$, then f'' can be supposed to be in general position with respect to all planes in $\mathbb{R}^3$ except to planes which are perpendicular to $\{z = 0\}$. In general f'' has quadri-tangencies which are perpendicular to $\{z = 0\}$, but no tritangencies perpendicular to $\{z = 0\}$ (this follows from the hypothesis (1), (2) and (5) above). A quadri-tangencies branches into four tritangencies when we deform again f'' to f''' breaking the symmetry.

Since the invariant planes under the above reflection are the plane $\{z = 0\}$ and the planes perpendicular to $\{z = 0\}$, it follows, from the above construction, that a closed surface with even Euler number has at least one immersion $f : M \to \mathbb{R}^3$ such that $N_{1,1,1}(f) \equiv N_3(f) \equiv 0 \pmod 2$.
q.e.d.

Acknowledgment: I thank Tom Banchoff for the interesting discussions about the problem in this paper, during the conference, and for pointing out to me that Terry Gaffney had also obtained a formula for the number of tritangencies for surfaces in $\mathbb{R}^3$, in a paper which I don't have received from him.

REFERENCES

[1] V.I. Arnold; Normal Forms for Functions near Degenerate Critical points, the Weyl Groups of A_k, D_k, E_k and Lagrangian Singularities, Functional Anal. Appl. 6, 254-272 (1972; Zbl. 278.57011).

[2] Y.L. Kergosien and R. Thom; Sur les Points Paraboliques des Surfaces, C.R. Acad. Sc. Paris t. 290 Sèrie A, 705-710 (1980; Zbl.435.58005).

[3] T.F. Banchoff, T. Gaffney and C. McCrory; Cusps of Gauss Mappings, Pitman. Per. Notes Math.55 (1982;Zbl. 478.53002).

[4] D. Bleecker and L.C. Wilson; Stability of Gauss Maps, Illinois J. Math. 22, 279-289 (1978; Zbl. 382.58004).

[5] F. Sergeraert; Un théorème de fonctions implicites sur certains espaces de Frechet et quelques applications, Ann. Scient. Ec. Norm. Sup. 4^eserie t.5, 599-660 (1972;Zbl. 246.58006).

SMALL EIGENVALUES OF THE LAPLACIAN

AND EXAMPLES

by TARGO PAVLISTA

0 Introduction. Let (M,g) be a 2-dimensional, closed, connected, oriented Riemannian manifold with strictly positive Gaussian curvatue K, $\kappa_0 := \min_M K$, $\kappa_1 := \max_M K$. The differential equation $\Delta f + \lambda f = 0$ determines the eigenfunctions $f \in C^\infty(M)$ and the eigenvalues $0 = \lambda_0 < \lambda_1 < \lambda_2 < ..$ of the Laplacian Δ. The eigenvalues of the sphere $S^2(K)$ are given by $\lambda_\ell = \ell(\ell+1)$, $\ell \in \mathbb{N}$, but for manifolds which are "not too far from the sphere" like an ellipsoid of revolution one does not even know the first eigenvalue λ_1.

The following estimates are sharp in the case that M is a sphere:

$$\lambda_1 \in [2\kappa_0, 2\kappa_1] \quad [3]\ [2] \quad , \qquad \lambda \notin (2\kappa_1, 6\kappa_0) \quad [1] .$$

These results suggests investigation of eigenvalues λ with $\lambda > 6\kappa_1$. The following theorem improves a result of Kozlowski [5].

1 Theorem ([7]). Let (M,g) be a 2-dimensional, closed, oriented, connected Riemannian manifold with strictly positive Gaussian curvature K. Assume furthermore $6\kappa_1 < 12\kappa_0$ and

$$\max_M ||\mathrm{grad}\ K||^2 < \frac{4}{27} [10\kappa_1^3 + 18\kappa_0\kappa_1^2 - 216\kappa_0^2(\kappa_1 - \kappa_0) + (7\kappa_1^2 - 24\kappa_1\kappa_0 + 36\kappa_0^2)^{3/2}] .$$

Then the polynomal

$$P(\lambda) = \frac{1}{4}(\lambda - 12\kappa_0)(\lambda - 6\kappa_1)(\lambda - 2\kappa_1) + \max_M ||\mathrm{grad}\ K||^2$$

has two zeros a,b, which fulfil $6\kappa_1 \leq a < b \leq 12\kappa_0$ and no eigenvalue of the Laplacian lies in the interval (a,b).

The steps of the proof are:

1. Choise of a tensor A_{ijkl}.

For related investigations see M. Kozlowski's article in this volume.

2. Calculation of $\int ||A_{ijkl}||^2 do$.

3. Application of integral formulas and estimates.

4. Discussion of the inequality $0 \le \int ||A_{ijkl}||^2 do \le P(\lambda) \int ||f_i||^2 do$ where $P(\lambda)$ is a polynomal in λ.

The last step promises sharp results only if $||A_{ijkl}||^2$ is as small as possible.
In chapter 2 of [6] it is outlined that this condition leads to the choice of a traceless and totally symmetric tensor A_{ijkl}.
In the case of constant Gaussian curvature K the zeros of $P(\lambda)$ are just the eigenvalues λ_1, λ_2, λ_3 of the sphere $S^2(K)$.
The resulting integral formula gives also a very short proof ([7], part 5) of a result of Kozlowski and Simon [6] concerning minimal immersions:
Let (M^2, g) be a closed, connected manifold with curvature K, $\frac{1}{6} \le K \le \frac{1}{3}$. If $\tilde{x}: M^2 \to S^N(1)$ is an isometric minimal immersion, then there are only two possibilities: $K = \frac{1}{3}$ or $K = \frac{1}{6}$.

2 <u>Preliminaries</u>. Let g_{ij} resp. g^{ij} denote the components of the metric tensor g resp. g^{-1} in local coordinates. ∇ and do denote the corresponding covariant differentiation and the volume element on M. The components $R_{ijkl} = K(g_{ik}g_{jl} - g_{il}g_{jk})$ have the sign of [4]. As usual raising and lowering of indices are defined. For $f \in C^\infty(M)$ $f_{ij} := \nabla_j \nabla_i f$ denotes the components of the Hessian and $\nabla f := g^{ij} f_{ij}$ denotes the Laplacian. Finally we need the first four Ricci identities:

$$f_{ij} = f_{ij},$$

$$f_{ijk} = f_{ikj} + f_h R^h{}_{ijk},$$

$$f_{ijkl} = f_{ijlk} + f_{hj} R^h{}_{ikl} + f_{ih} R^h{}_{jkl},$$

$$f_{ijklr} = f_{ijkrl} + f_{hjk} R^h{}_{ilk} + f_{ihk} R^h{}_{jlr} + f_{ijh} R^h{}_{klr}$$

3. <u>Definition</u>. Let $f \in C^\infty(M)$. The tensor fields C_{ij}, B_{ijk}, $A^{(1)}_{ijkl}$, $A^{(2)}_{ijkl}$, $A^{(3)}_{ijkl}$ and A_{ijkl} are defined by

$$C_{ij} := f_{ij} - \frac{1}{2} f_s{}^s g_{ij},$$

$$B_{ijk} := \frac{1}{3}(f_{ijk} + f_{ikj} + f_{jki})$$

$$- \frac{1}{12}(f_{s\ i}^{\ s}\ g_{kj} + f_{s\ j}^{\ s}\ g_{ik} + f_{s\ k}^{\ s}\ g_{ij})$$

$$- \frac{1}{6}(f_{is}^{\ \ s} g_{kj} + f_{js}^{\ \ s}\ g_{ik} + f_{ks}^{\ \ s}\ g_{ij}) \ ,$$

$$A^{(1)}_{ijkl} := f_{ijkl} + f_{ijlk} + f_{ikjl} + f_{iklj} + f_{iljk} + f_{ilkj}$$

$$+ f_{jkil} + f_{jkli} + f_{jlik} + f_{jlki} + f_{klij} + f_{klji} \ ,$$

$$A^{(2)}_{ijkl} := -\frac{1}{6}\{[f_{s\ ij}^{\ s} + f_{s\ ji}^{\ s} + 2(f_{si\ j}^{\ \ s} + f_{sj\ i}^{\ \ s} + f_{sij}^{\ \ \ s} + f_{sji}^{\ \ \ s} + f_{ijs}^{\ \ \ s})]g_{kl}$$

$$+ [f_{s\ il}^{\ s} + f_{s\ li}^{\ s} + 2(f_{si\ l}^{\ \ s} + f_{sl\ i}^{\ \ s} + f_{sil}^{\ \ \ s} + f_{sli}^{\ \ \ s} + f_{ils}^{\ \ \ s})]g_{jk}$$

$$+ [f_{s\ ik}^{\ s} + f_{s\ ki}^{\ s} + 2(f_{si\ k}^{\ \ s} + f_{sk\ i}^{\ \ s} + f_{sik}^{\ \ \ s} + f_{ski}^{\ \ \ s} + f_{iks}^{\ \ \ s})]g_{jl}$$

$$+ [f_{s\ jl}^{\ s} + f_{s\ lj}^{\ s} + 2(f_{sj\ l}^{\ \ s} + f_{sl\ j}^{\ \ s} + f_{sjl}^{\ \ \ s} + f_{slj}^{\ \ \ s} + f_{jls}^{\ \ \ s})]g_{ik}$$

$$+ [f_{s\ jk}^{\ s} + f_{s\ kj}^{\ s} + 2(f_{sj\ k}^{\ \ s} + f_{sk\ j}^{\ \ s} + f_{sjk}^{\ \ \ s} + f_{skj}^{\ \ \ s} + f_{jks}^{\ \ \ s})]g_{il}$$

$$+ [f_{s\ lk}^{\ s} + f_{s\ kl}^{\ s} + 2(f_{sl\ k}^{\ \ s} + f_{sk\ l}^{\ \ s} + f_{slk}^{\ \ \ s} + f_{skl}^{\ \ \ s} + f_{lks}^{\ \ \ s})]g_{ij}\},$$

$$A^{(3)}_{ijkl} := \frac{1}{24}\{(4f_{s\ t}^{\ s\ t} + 8f_{st}^{\ \ st})(g_{ij}g_{lk} + g_{il}g_{jk} + g_{ik}\ g_{jl})\} \ ,$$

$$A_{ijkl} := \frac{1}{12}(A^{(1)}_{ijkl} + A^{(2)}_{ijkl} + A^{(3)}_{ijkl}) \ .$$

4. Remark. These tensors are totally symmetric and C_{ij}, B_{ijk} and A_{ijkl} are traceless.

The Ricci identities imply the following formulas.

5. Lemma. Let $f \in C^\infty(M)$ be an eigenfunction. Then

$$f_{ik}^{\ \ k} = (K - \lambda)f_i \ ,$$

$$f_{ijk}^{\ \ \ k} = (4K - \lambda)f_{ij} + 2\lambda K f g_{ij} + K_j f_i + f_j K_k - f_k K^k g_{ij} \ ,$$

$$f_{ijk}^{\ \ \ j} = (3K - \lambda)f_{ik} + K_k f_i + \lambda K f g_{ik} \ ,$$

$$f_{i1jk}{}^{1} = -\lambda f_{ijk} + Kf_{kij} + \nabla_k(Kf_{ij})$$
$$+ f_{1hj} R^{h}{}_{ik}{}^{1} + \nabla_k(f_{1h} R^{h}{}_{ij}{}^{1})$$
$$+ f_{1ih} R^{h}{}_{jk}{}^{1} + \nabla_k \nabla_j(K f_i) \quad .$$

6. Lemma. Let $f \in C^\infty(M)$ be an eigenfunction. Then

$$||C_{ij}||^2 = ||f_{ij}||^2 - \frac{\lambda^2}{2} f^2 \quad ,$$

$$||B_{ijk}||^2 = ||f_{ijk}||^2 - \{\frac{3}{4}\lambda^2 - \lambda K + K^2\}||f_i||^2 \qquad [1] \; .$$

7 Proof. $||C_{ij}||^2 := C_{ij} C^{ij} = (f_{ij} + \frac{\lambda}{2} fg_{ij})(f^{ij} + \frac{\lambda}{2} fg^{ij})$

$$= ||f_{ij}||^2 - \frac{2\lambda^2}{2} f^2 + \frac{\lambda^2}{2} f^2 \quad .$$

The Green's formula and the Ricci identities imply the following equations.

8 Lemma. Let $f \in C^\infty(M)$ be an eigenfunction. Then

$$\int ||C_{ij}||^2 do = \frac{1}{2} \int (\lambda - 2K)||f_i||^2 do$$

$$\int ||f_i||^2 do = \lambda \int f^2 do \; ,$$

$$\int fK^i f_i do = \int \lambda K f^2 do - \int K||f_i||^2 do \; ,$$

$$\int KfK^i f_i do = \frac{1}{2}\int \lambda K^2 f^2 do - \frac{1}{2}\int K^2||f_i||^2 do \; ,$$

$$\int f^{ij} K_i f_j do = \int (\lambda K - K^2)||f_i||^2 do - \int K||f_{ij}||^2 do \; ,$$

$$\int Kf^{ij} K_i f_j do = \frac{1}{2}\int (\lambda K^2 - K^3)||f_i||^2 do - \frac{1}{2}\int K^2||f_{ij}||^2 do \; ,$$

$$\int ||f_{ijk}||^2 do = \int (\lambda - 2K)||C_{ij}||^2 do + \int (\frac{\lambda^2}{2} - \lambda K + 2K^2)||f_i||^2 do$$

$$\int ||B_{ijk}||^2 do = \frac{1}{3}\int (\lambda - 6K)||C_{ij}||^2 do$$
$$+ \frac{1}{12} \int (\lambda - 6K)(\lambda - 2K)||f_i||^2 do$$

Proof. $\int ||C_{ij}||^2 do = \int ||f_{ij}||^2 do - \frac{\lambda^2}{2}\int f^2 do$

$$= -\int f_i f^{ij}{}_j do - \frac{\lambda}{2}\int ||f_i||^2 do$$

$$= \int (\lambda - K)||f_i||^2 do - \frac{\lambda}{2}\int ||f_i||^2 do ,$$

$$\int ||f_i||^2 do = -\int f \; f_i{}^i \; do = \lambda \int f^2 do ,$$

$$\int f \, K^i \, f_i \; do = -\int K||f_i||^2 do - \int K \, f \, f^i{}_i do ,$$

$$\int K \, f \, K^i \, f_i do = -\int K \, \nabla^i (K \, f \, f_i) do$$

$$= -\int K \, K^i \, f \, f_i \; do - \int K^2 \, f \, f^i do - \int K^2 ||f_i||^2 do ,$$

$$\int f^{ij} \, K_i \, f_j do = -\int K||f_{ij}||^2 do - \int K \, f_k \, f^{ij}{}_i do$$

$$= -\int K||f_{ij}||^2 do + \int (\lambda K - K^2)||f_i||^2 do ,$$

$$\int K \, f^{ij} \, K_i \, f_j do = -\int K_i \, f^{ij} \, K \, f_j \; do - \int K^2 ||f_{ij}||^2 do$$

$$-\int K^2 \, f_j f^{ij}{}_i do$$

$$= -\int K_i \, f^{ij} \, K \, f_j do - \int K^2 ||f_{ij}||^2 do$$

$$+\int (\lambda K^2 - K^3)||f_i||^2 do ,$$

$$\int ||f_{ijk}||^2 do = -\int f_{ij} f^{ijK} \; K \; do$$

$$= -\int (4K - \lambda)||f_{ij}||^2 do + \int 2\lambda^2 K f^2 do - \int 2K_j f_i f^{ij} do$$

$$-\int \lambda f \; f_k K^k do$$

$$= \int (\lambda - 2K)||f_{ij}||^2 do + \int \lambda^2 K \; f^2 do$$

$$+ \int 2K(K - \lambda)||f_i||^2 do + \int \lambda K||f_i||^2 do$$

$$= \int (\lambda - 2K)(||f_{ij}||^2 - \frac{\lambda^2}{2} f^2) do$$

$$+ \frac{\lambda^3}{2}\int f^2 do + \int (2K^2 - \lambda K)||f_i||^2 do$$

$$= \int (\lambda - 2K)||C_{ij}||^2 do + \int (\frac{\lambda^2}{2} - \lambda K + 2K^2)||f_i||^2 ,$$

$$\int ||B_{ijk}||^2 do = \int ||f_{ijk}||^2 do + \int (- \frac{3}{4}\lambda^2 + \lambda K - K^2)||f_i||^2$$

$$= \frac{1}{3}\int (3\lambda - 6K)||C_{ij}||^2 do - \frac{2}{3}\int \lambda ||C_{ij}||^2 do$$

$$+ \int (-\tfrac{1}{4}\lambda^2 + K^2)||f_i||^2 do + \tfrac{2}{3}\lambda \int (||f_{ij}||^2 - \tfrac{\lambda^2}{2} f^2) do$$

$$= \tfrac{1}{3}\int (\lambda - 6K)||C_{ij}||^2 do + \tfrac{1}{12}\int (\lambda - 6K)(\lambda - 2K)||f_i||^2 do \ .$$

10 Proof of the theorem. Since A_{ijkl} is traceless, we have

$$||A_{ijkl}||^2 = \left(\tfrac{1}{12}\right)^2 A^{(1)}{}_{ijkl} \cdot A^{ijkl}$$

$$= \tfrac{1}{144} A^{(1)ijkl}(A^{(1)}{}_{ijkl} + A^{(2)}{}_{ijkl} + A^{(3)}{}_{ijkl}) \ .$$

Using the symmetry of $A^{(1)}{}_{ijkl}$, $A^{(2)}{}_{ijkl}$ and $A^{(3)}{}_{ijkl}$ we get

$$\begin{aligned}||A^{(1)}{}_{ijkl}||^2 &= A^{(1)}{}_{ijkl}(f^{ijkl} + f^{ijlk} + f^{ikjl} + f^{iklj} \\ &\quad + f^{iljk} + f^{ilkj} + f^{jkil} + f^{jkli} \\ &\quad + f^{jlik} + f^{jlki} + f^{klij} + f^{klji} \\ &= 12A^{(1)}{}_{ijkl}f^{ijkl} \ ,\end{aligned}$$

$$\begin{aligned}A^{(1)}{}_{ijkl}A^{(2)ijkl} &= -A^{(1)}{}_{ijkl}(A^{(1)stkl}\, g_{st})g^{ij} \\ &= -(A^{(1)}{}_{ijkl}\, g^{ij})(A^{(1)stkl}\, g_{st}) \\ &= -||A^{(1)}{}_{ijkl}\, g^{ij}||^2 \\ &= -||f_{i}{}^{i}{}_{kl} + f_{i}{}^{i}{}_{lk} + (2f_{ik}{}^{i}{}_{l} \\ &\quad + f_{il}{}^{i}{}_{k} + f_{ikl}{}^{i} + f_{ilk}{}^{i} + f_{kli}{}^{i})||^2 \ ,\end{aligned}$$

$$\begin{aligned}A^{(1)}{}_{ijkl}\, A^{(3)ijkl} &= \tfrac{1}{8} A^{(1)}{}_{ijkl}g^{ij}\, g^{kl}(4f_{s}{}^{s}{}_{t}{}^{t} + 8f_{st}{}^{st}) \\ &= \tfrac{1}{8}(4f_{s}{}^{s}{}_{t}{}^{t} + 8f_{st}{}^{st})g^{ij}\, g^{kl} \\ &\quad (f^{ijkl} + f^{ijlk} + f^{ikjl} + f^{iklj} \\ &\quad + f^{iljk} + f^{ilkj} + f^{jkil} + f^{jkli} \\ &\quad + f^{jlik} + f^{jlki} + f^{klij} + f^{klji} \\ &= \tfrac{1}{8}(4f_{s}{}^{s}{}_{t}{}^{t} + 8f_{st}{}^{st})^2 \ ,\end{aligned}$$

so we obtain

$$\|A_{ijkl}\|^2 = \frac{1}{12} f^{ijkl} A^{(1)}_{ijkl} - \frac{1}{144}\|f_i{}^i{}_{kl} + f_i{}^i{}_{lk} + 2(f_{ik}{}^i{}_l + f_{il}{}^i{}_k + f_{ikl}{}^i + f_{ilk}{}^i + f_{kli}{}^i\|^2 + \frac{1}{1152}(4f_s{}^s{}_t{}^t + 8f_{st}{}^{st})^2 . \tag{11}$$

The tools for the following calculations are the Ricci identities, Green's formula, Lemma 5 and Lemma 7.

$$\begin{aligned} &\|f_i{}^i{}_{kl} + f_i{}^i{}_{lk} + 2(f_{ik}{}^i{}_l + f_{il}{}^i{}_k + f_{ikl}{}^i + f_{ilk}{}^i + f_{kli}{}^i)\|^2 \\ &= \|-12\lambda f_{kl} + 6\nabla_l(f_h R^h{}_{ik}{}^i) + 2f_{hl} R^h{}_{ik}{}^i \\ &\quad + 4\nabla_k(f_h R^h{}_{il}{}^i) + 4f_{hk} R^k{}_{il}{}^i + 2\nabla^i(f_h R^h{}_{kli}) \\ &\quad + 4f_{ih} R^h{}_{kl}{}^i + 2f_{ih} R^h{}_{lk}{}^i\|^2 \\ &= \|(24K - 12\lambda)f_{kl} + 6f_k K_l + 6f_l K_k + (8\lambda K f - 2f_i K^i)g_{kl}\|^2 \\ &= (24K - 12\lambda)^2\|f_{ij}\|^2 + 72\|f_i\|^2\|K_j\|^2 \\ &\quad + 128\lambda^2 K^2 f^2 - 64\lambda K f f_i K^i \\ &\quad + 8(f_i K^i)^2 + 24(24K - 12\lambda)f_{ij} f^i K^j \\ &\quad + 4\lambda(24K - 12\lambda)f_i K^i f - 16\lambda^2 K(24K - 12\lambda)f^2 \\ &\quad + 72(f_i K^i)^2 + 192\lambda K(f_i K^i)f - 48(f_i K^i)^2 . \end{aligned} \tag{12}$$

By similar calculations one finds

$$(4f_s{}^s{}_t{}^t + 8f_{st}{}^{st})^2 = (12\lambda^2 - 8\lambda K)^2 f^2 + 64(f_i K^i)^2 + 16(12\lambda^2 - 8\lambda K)f f_i K^i \tag{13}$$

and

$$\int f^{ijkl} A^{(1)}_{ijkl}\, do = \int [(4K - 12\lambda)\|f_{ijk}\|^2 + (42\lambda K - 118K^2)\|f_{ij}\|^2 \tag{14}$$

$$+ (54\lambda K^2 - 42\lambda^2 K - 34K^3)||f_i||^2$$
$$+ 62\lambda^2 K^2 f^2 - 2(K_i f^i)^2$$
$$- (28\lambda K + 12\lambda^2) f(f_i K^i) - 12||K_i||^2||f_i||^2$$
$$+ (18\lambda - 116K)\ K_i f_j f^{ij}]do \quad .$$

From 11, 12, 13 and 14 we obtain

15
$$\int ||A_{ijkl}||^2 do = \int (6K^2 - \lambda^2)||f_{ij}||^2 do + \int (\frac{3}{24}\lambda^4 - 3\lambda^2 K^2) f^2 do + \int (\lambda - 3K)||f_{ijk}||^2 do + \int \frac{5}{2}\lambda K(\lambda - K)\cdot ||f_i||^2 do + \int \frac{1}{2}||K_i||^2||f_j||^2 do \ .$$

Lemma 6 gives

16
$$\int ||A_{ijkl}||^2 do = 2\int (3K^2 - \frac{1}{2}\lambda^2)||C_{ij}||^2 do - \int \frac{9}{24}\lambda^4 f^2 do + \int (\lambda - 3K)||B_{ijk}||^2 do + \int (-\frac{3}{4}\lambda^2 K^2 + \frac{3}{2}\lambda K^2 + \frac{3}{4}\lambda^3 - 3K^3)||f_i||^2 do + \int \frac{1}{2}||K_i||^2||f_j||^2 do \ .$$

By using Lemma 6 we get the needed linear factors which vanish in the case that M is a sphere.

17
$$\int ||A_{ijkl}||^2 do = \frac{1}{4}\int (\lambda - 12K)||B_{ijk}||^2 do + \frac{1}{12}\int (\lambda - 12K)(\lambda - 6K)||C_{ij}||^2 do + \frac{1}{48}\int (\lambda - 12K)(\lambda - 6K)(\lambda - 2K)||f_i||^2 do + \frac{1}{2}\int ||K_i||^2||f_j||^2 do \ .$$

Since $\lambda \leq 12\kappa_0$ we have $(\lambda - 12K) \leq 0$ on M and get with Lemma 8

18 $$\begin{aligned}\int ||A_{ijkl}||^2 do &\leq \tfrac{1}{4}(\lambda - 12\kappa_0)\int ||B_{ijk}||^2 do \\ &+ \tfrac{1}{12}(\lambda - 12\kappa_0)(\lambda - 6\kappa_1)\int ||C_{ij}||^2 do \\ &+ \tfrac{1}{48}(\lambda - 12\kappa_0)(\lambda - 6\kappa_1)(\lambda - 2\kappa_1)\int ||f_i||^2 do \\ &+ \tfrac{1}{2}\int ||K_i||^2 ||f_j||^2 do \\ &= \tfrac{1}{12}(\lambda - 12\kappa_0)\int (\lambda - 6K)||C_{ij}||^2 do \\ &+ \tfrac{1}{48}(\lambda - 12\kappa_0)\int (\lambda - 6K)(\lambda - 2K)||f_i||^2 do \\ &+ \tfrac{1}{24}(\lambda - 12\kappa_0)(\lambda - 6\kappa_1)\int (\lambda - 2K)||f_i||^2 do \\ &+ \tfrac{1}{48}(\lambda - 12\kappa_0)(\lambda - 6\kappa_1)(\lambda - 2\kappa_1)\int ||f_i||^2 do \\ &+ \tfrac{1}{2}\int ||K_i||^2 ||f_j||^2 do \quad , \\ &\leq \tfrac{1}{12}(\lambda - 12\kappa_0)(\lambda - 6\kappa_1)\int ||C_{ij}||^2 do \\ &+ \tfrac{2}{24}(\lambda - 12\kappa_0)(\lambda - 6\kappa_1)(\lambda - 2\kappa_1)\int ||f_i||^2 do \\ &+ \tfrac{1}{2}\int ||K_i||^2 ||f_j||^2 do \\ &= \tfrac{1}{24}(\lambda - 12\kappa_0)(\lambda - 6\kappa_1)\int (\lambda - 2K)||f_i||^2 do \\ &+ \tfrac{2}{24}(\lambda - 12\kappa_0)(\lambda - 6\kappa_1)(\lambda - 2\kappa_1)\int ||f_i||^2 do \\ &+ \tfrac{1}{2}\int ||K_i||^2 ||f_j||^2 do \\ &\leq \{\tfrac{1}{8}(\lambda - 12\kappa_0)(\lambda - 6\kappa_1)(\lambda - 2\kappa_1) + \tfrac{1}{2}\max_M ||K_i||^2\}\int ||f_i||^2 do\end{aligned}$$

so we have

19 $$0 \leq 2\int ||A_{ijkl}||^2 do \leq \{\tfrac{1}{4}(\lambda - 12\kappa_0)(\lambda - 6\kappa_1)(\lambda - 2\kappa_1) + \max_M ||K_i||^2\} \int ||f_i||^2 do \quad .$$

Since $6\kappa_1 < 12\kappa_0$ the polynomial $P_0(\lambda) := (\lambda - 12\kappa_0)(\lambda - 6\kappa_1)(\lambda - 2\kappa_1)$ has three different real zeros. For $\lambda \in [6\kappa_1, 12\kappa_0]$ we get $P_0(\lambda) \leq 0$. This implies the existence of a local minimum $\lambda_m \in [6\kappa_1, 12\kappa_0]$. A simple calculation gives

$$P_0(\lambda_{min}) = \frac{16}{27}[-10\kappa_1^3 - 18\kappa_0\kappa_1^2 + 216\kappa_0^2(\kappa_1 - \kappa_0) - (7\kappa_1^2 - 24\kappa_1\kappa_0 + 36\kappa_0^2)^{3/2}] .$$

If

$$\max_M ||K_i||^2 < \frac{4}{27}[10\kappa_1^3 + 18\kappa_0\kappa_1^2 - 216\kappa_0^2(\kappa_1 - \kappa_0) + (7\kappa_1^2 - 24\kappa_1\kappa_0 + 36\kappa_0^2)^{3/2}] ,$$

then

$$P(\lambda) := \frac{1}{4}(\lambda - 12\kappa_0)(\lambda - 6\kappa_1)(\lambda - 2\kappa_1) + \max_M ||K_i||^2$$

has two zeros a,b with $6\kappa_1 \leq a < b \leq 12\kappa_0$ such that $\lambda \in (a,b)$ implies $P(\lambda) < 0$.

REFERENCES

[1] K. Benko, M. Kothe, K. D. Semmler, U. Simon: Eigenvalues of the Laplacian and curvature. Colloquium math. 42, 19-31 (1979 ; Zbl. 437.53032).

[2] M. Berger, P. Gauduchon, E. Mazet: Le spectre d'une variété Riemannienne. Lecture Notes in Math., Vol. 194, Berlin-Heidelberg-New-York: Springer-Verlag 1971.

[3] J. Hersch: Quatre propiêtés isopérimétriques de membranes sphériques homogenes, C. R. Acad, Sci. Paris, Ser. A 270, 1645-1648 (1970 ; Zbl. 224.73083).

[4] S. Kobayashi, K. Nomizu: Foundations of differential geometry Vol. I, Interscience Publishers New York, London 1963.

[5] M. Kozlowski: Geometrische Abschätzungen für kleine Eigenwerte des Laplaceoperators auf fast-sphärischen Rotationsellipsoiden. Dissertation FB 3 TU Berlin 1983.

[6] M. Kozlowski, U. Simon: Minimal immersions of 2-manifolds into spheres. Math. Z. 186, 377-382 (1984)

[7] T. Pavlista: Geometrische Abschätzungen kleiner Eigenwerte des Laplaceoperators. Dissertation FB 3 TU Berlin 1984.

Horizontal lifts of isometric immersions into the bundle space of a pseudo-Riemannian submersion

Helmut Reckziegel

In differential geometry there are important examples of manifolds which occur as base space of a submersion and whose geometric structure is obtained by projecting the structure of the bundle space. For instance, the geometric structure of the complex projective space $\mathbb{C}P^n$ arises in this way from the Sasakian structure of the sphere S^{2n+1} .

It is the purpose of this article to point to a close relation between immersed submanifolds of the bundle space N and the base space B of a pseudo-Riemannian submersion $\pi : N \to B$. We show (Theorem 1): If $g : M \to N$ is a horizontal isometric immersion, then its second fundamental form is closely related to the second fundamental form of its projection $f = \pi \circ g$. For instance, f is totally geodesic, minimal or totally umbilical, if and only if g has the corresponding property.

In order to make this correspondence more profitable we prove a criterion (Theorem 2) which characterizes those submanifolds of the base space B which (at least locally) occur as projection of horizontal submanifolds. This criterion is formulated in terms of the so-called *horizontal integrability structure* $\mathfrak{J}$ of π , a family of special subsets $\mathfrak{J}_b \subset \mathrm{End}(T_bB)$, $b \in B$. While in general the $\mathfrak{J}_b$'s may be rather bizarre, $\mathfrak{J}$ is shown to be a (weakly) smooth family of linear spaces, if the fibres of π are locally homogeneous, cf. Theorem 3. In many concrete examples (cf. sect.5 and 6) $\mathfrak{J}$ turns out to be a well-known subbundle of the tensor bundle of B , for instance, a complex or quaternion structure. In the last situation exactly the totally real (= anti-invariant) submanifolds are locally representable by horizontal submanifolds. If we apply this result to the canonical fibration of a Sasakian manifold (cf. sect. 5), we obtain a correspondence between the anti-invariant submanifolds of Kählerian and Sasakian manifolds. These types of submanifolds were treated detailed in Chap. III and V of the book [YK]. But now we see that many of the results of Chap. V of [YK] can immediately be deduced from Chap. III, and conversely. A forthcoming paper will be devoted to this matter.

1. Preliminaries

Let N and B be connected pseudo-Riemannian manifolds[1] and $\pi : N \to B$ a submersion. The *vertical subbundle* of the tangent bundle TN (with resp. to π) is

$$\mathcal{V} = (\mathcal{V}_p := \operatorname{Kern} \pi_{*|p})_{p \in N} .$$

We suppose that for every $p \in N$ the linear subspace $\mathcal{V}_p$ is non-degenerate with resp. to the inner product of T_pN. Hence we have the splitting

$$T_pN = \mathcal{V}_p \oplus \mathcal{H}_p \quad \text{with} \quad \mathcal{H}_p := \mathcal{V}_p^{\perp} .$$

$\mathcal{H} = (\mathcal{H}_p)_{p \in N}$ is called the *horizontal subbundle* of π. The symbols $\mathcal{V}$ and $\mathcal{H}$ also denote the corresponding projections, i.e.,

$$\forall v \in T_pN : v = \mathcal{V}v + \mathcal{H}v \quad \text{with} \quad \mathcal{V}v \in \mathcal{V}_p \quad \text{and} \quad \mathcal{H}v \in \mathcal{H}_p .$$

The modules of the vertical and horizontal vector fields are denoted by $\Gamma(\mathcal{V})$ and $\Gamma(\mathcal{H})$, respectively. The skew-symmetric bilinear bundle map Ω on $\mathcal{H}$ with values in $\mathcal{V}$ characterized by

$$\Omega(X,Y) = \mathcal{V}[X,Y] \quad \text{for} \quad X,Y \in \Gamma(\mathcal{H})$$

is called the *curvature form* of $\mathcal{H}$. Obviously, $\mathcal{H}$ is integrable if and only if $\Omega = 0$. In the case of pseudo-Riemannian submersions we have the following fundamental

1) In this article all manifolds, maps, ... are supposed to be of class C^∞.

Lemma *Suppose* $\pi : N \to B$ *to be a pseudo-Riemannian submersion, i.e., for every* $p \in N$

$$\pi_* : \mathcal{H}_p \to T_{\pi(p)}B \quad \text{is a linear isometry}$$

(of pseudo-euclidean vector spaces); let $g : M \to N$ *be a differentiable map,* X *a vector field of* M *such that* g_*X *is horizontal, and* η *a further horizontal vector field along* g *(i.e.,* $\forall p \in M : \eta_p \in \mathcal{H}_{g(p)}$*). Then the covariant derivative* $\overset{N}{\nabla}_X \eta$ *is characterized by*

$$2 \cdot \mathcal{V}\, \overset{N}{\nabla}_X \eta = \Omega(g_*X, \eta) \quad \text{and} \tag{1}$$

$$\pi_* \overset{N}{\nabla}_X \eta = \overset{B}{\nabla}_X \pi_* \eta \quad . \tag{2}$$

($\overset{N}{\nabla}$ and $\overset{B}{\nabla}$ denote the Levi-Civita connection of N and B , respectively.)

In case $M = N$ and $g = id_M$ formula (1) is exactly Lemma 2 of [O1], and formula (2) follows easily from [O1] Lemma 1.3. To obtain the general version one represents η locally as $\sum \phi_i \cdot Y_i \circ g$ with $Y_i \in \Gamma(\mathcal{H})$ and $\phi_i \in C^\infty(M)$ and uses the Koszul axioms for ∇ .

2. Horizontal immersions

In the situation of sect. 1 a differentiable map $g : M \to N$ is said to be horizontal, iff $g_*T_pM \subset \mathcal{H}_{g(p)}$ for all $p \in M$.

Theorem 1 *Let* $\pi : N \to B$ *be a pseudo-Riemannian submersion,* M *a further pseudo-Riemannian manifold and* $g : M \to N$ *a horizontal isometric immersion. Then:*

a) $\Omega(g_*T_pM, g_*T_pM) = 0$ *for every* $p \in M$.

b) $f := \pi \circ g$ *is an isometric immersion* $M \to B$.

c) *The second fundamental form* h^g *of* g *takes its values in* $\mathcal{H}$ *, and* $\pi_* h^g$ *equals the second fundamental form* h^f *of* f .

d) *Let* $\perp(g)$ *and* $\perp(f)$ *denote the normal bundles of* g *and* f *, and* $\nabla^\perp$ *the normal connection of* $\perp(g)$ *and* $\perp(f)$ *. For every* $\eta \in \Gamma(\perp(g))$ *we have*

$$\pi_* \eta \in \Gamma(\perp(f)) \quad , \tag{3}$$

<u>and in case that</u> η <u>is horizontal</u>

$$2 \cdot \mathcal{V}\, \nabla^{\perp}_{X}\eta \;=\; \Omega(g_* X, \eta) \quad \underline{\text{and}} \tag{4}$$

$$\pi_* \nabla^{\perp}_{X}\eta \;=\; \nabla^{\perp}_{X}\pi_*\eta \tag{5}$$

<u>for all vector fields</u> X <u>of</u> M .

<u>Proof</u> Let X and Y be vector fields of M . Combining Gauß' formula $\overset{N}{\nabla}_X g_* Y = g_* \overset{M}{\nabla}_X Y + h^g(X,Y)$ (cf. [C]) and formula (1) we obtain $\Omega(g_* X, g_* Y) = 2 \cdot \mathcal{V} h^g(X,Y)$. As the term on the right (resp. left) is symmetric (resp. skew-symmetric), both terms vanish. Therefore a) and the first half of c) are proved. b) and formula (3) are trivial. For the second half of c) we substitute Gauß' formula (applied to f and g) in (2). Formula (4) is immediate from (1), because $\overset{N}{\nabla}_X \eta - \nabla^{\perp}_X \eta$ is tangent to g (according to Weingarten's formula, cf. [C]), hence horizontal. For the proof of (5) we use the shape operators A^f and A^g of f and g . From c) we obtain $A^f_{\pi_*\eta} = A^g_\eta$ (cf. [C] p.41), hence

$$\pi_*(\overset{N}{\nabla}_X \eta - \nabla^{\perp}_X \eta) = \overset{B}{\nabla}_X \pi_* \eta - \nabla^{\perp}_X \pi_* \eta$$

by means of Weingarten's formula. Formula (5) is now clear from (2) . □

Let H^f and H^g denote the mean curvature vector fields of f and g , cf. [C] p.43. Theorem 1 c) implies: H^g is horizontal and $\pi_* H^g = H^f$. Therefore Theorem 1 has the following corollary:

<u>Corollary 1</u> <u>In the situation of Theorem 1 the immersion</u> f <u>is totally geodesic, minimal, totally umbilical or pseudoumbilical</u> (cf. [C]) , <u>if and only if</u> g <u>has the corresponding property</u>.

Let $\sigma_p(g)$ denote the *first osculating space* of g at $p \in M$, i.e.,

$$\sigma_p(g) := g_* T_p M \oplus \operatorname{span}\{ h^g(u,v) \mid u,v \in T_p M \} \subset T_{g(p)} N \quad ,$$

and ∇h^g the covariant derivative of h^g , i.e.,

$$(\nabla_X h^g)(Y,Z) = \nabla^{\perp}_X (h^g(Y,Z)) - h^g(\overset{M}{\nabla}_X Y, Z) - h^g(Y, \overset{M}{\nabla}_X Z)$$

for all vector fields X, Y, Z of M. If $\nabla h^g = 0$, then h^g is said to be parallel. Isometric immersions with parallel second fundamental form have been investigated in many articles during the last years, see, e.g., [F], [N], [BR]. Therefore the following corollary of Theorem 1c) and d) is of interest.

Corollary 2 In the situation of Theorem 1

a) $\sigma_p(g) \subset \mathcal{H}_{g(p)}$ and $\pi_* \sigma_p(g) = \sigma_p(f)$

b) $2 \cdot \mathcal{V}(\nabla_X h^g)(Y,Z) = \Omega(g_* X, h^g(Y,Z))$ and
$\pi_*(\nabla_X h^g)(Y,Z) = (\nabla_X h^f)(Y,Z)$

c) If $\Omega(g_* T_p M, \sigma_p(g)) = 0$ for all $p \in M$, then:

$$\nabla h^f = 0 \iff \nabla h^g = 0 .$$

3. On the existence of horizontal lifts

Let $\pi : N \longrightarrow B$ be a surjective pseudo-Riemannian submersion and Ω the curvature form of its horizontal subbundle, cf. sect. 1. Then for every $p \in N$ and $v \in \mathcal{V}_p$ we define a skew-symmetric linear map

$$\phi^v : T_{\pi(p)}B \longrightarrow T_{\pi(p)}B$$

by

$$\langle \phi^v(\pi_* u), \pi_* w \rangle = \frac{1}{2} \cdot \langle \Omega(u,w), v \rangle \quad \text{for} \quad u,w \in \mathcal{H}_p . \tag{6}$$

If Y is a vector field of B, $\tilde{Y} \in \Gamma(\mathcal{H})$ its horizontal lift (i.e., $\pi_* \tilde{Y} = Y \circ \pi$) and $\xi \in \Gamma(\mathcal{V})$, then formula (1) implies

$$\phi^{\xi(p)} Y_{\pi(p)} = - \pi_* \overset{N}{\nabla}_{\tilde{Y}(p)} \xi \tag{7}$$

for every $p \in N$. (The family of all ϕ^v's, therefore, reflects the second fundamental tensor of $\mathcal{H}$.) For each $b \in B$ we define

$$\mathcal{J}_b := \{ \phi^v \mid v \in \mathcal{V}_p , p \in \pi^{-1}(b) \} .$$

Because of the following theorem and remark I will call $\mathcal{J} = (\mathcal{J}_b)_{b \in B}$ the *horizontal integrability structure* of π.

Theorem 2 If $\pi : N \to B$ is a surjective pseudo-Riemannian submersion, M a further pseudo-Riemannian manifold and $f : M \to B$ an isometric immersion, then the following assertions are equivalent:

a) f satisfies

$$\forall p \in M, \ \phi \in \mathcal{J}_{f(p)} \ : \ f_* T_p M \perp \phi(f_* T_p M) \ .$$

b) For every initial data $(p,q) \in M \times N$ with $f(p) = \pi(q)$ there exists a neighbourhood $U = U(p) \subset M$ and a horizontal isometric immersion $g : U \to N$ with $g(p) = q$ such that $f|U = \pi \circ g$.

With regard to applications of Cor.2c) we add: If a) is replaced by the stronger condition

$$\forall p \in M, \ \phi \in \mathcal{J}_{f(p)} \ : \ \sigma_p(f) \perp \phi(f_* T_p M) \ ,$$

then every horizontal lift $g : U \to N$ of assertion b) satisfies

$$\forall \tilde{p} \in U \ : \ \Omega(g_* T_{\tilde{p}} M, \sigma_{\tilde{p}}(g)) = 0 \ .$$

The proof is immediate because of (6) and Cor.2a).

Proof of Theorem 2 To prove b) $\Rightarrow$ a) let $p \in M$, $u,w \in T_p M$, $q \in \pi^{-1}(f(p))$ and $v \in \mathcal{V}_q$ be given. If $g : U \to N$ is a horizontal lift of $f|U$ as described in b), then Theorem 1a) implies

$$2 \cdot \langle \phi^v(f_* u), f_* w \rangle = \langle \Omega(g_* u, g_* w), v \rangle = 0 \ .$$

For the proof of a) $\Rightarrow$ b) let $\tilde{M}$ be the fibre product

$$\tilde{M} := \{ (p,q) \in M \times N \mid f(p) = \pi(q) \} = \bigcup \left(\{p\} \times \pi^{-1}(f(p)) \right) ;$$

it is a regular submanifold of $M \times N$ (cf. [B] 5.11.2), and the canonical projection $\tilde{\pi} : \tilde{M} \to M$ resp. $\tilde{f} : \tilde{M} \to N$ is a surjective submersion resp. an immersion. Since every fibre $\tilde{\pi}^{-1}(p)$ is mapped diffeomorphically onto the fibre $\pi^{-1}(f(p))$ by $\tilde{f}$, for every $x \in \tilde{M}$ the vertical subspace $\tilde{\mathcal{V}}_x$ (with resp. to $\tilde{\pi}$) is mapped isomorphically onto $\mathcal{V}_{\tilde{f}(x)}$ by $\tilde{f}_*$. We, therefore, obtain

$$T_x \tilde{M} = \tilde{\mathcal{V}}_x \oplus \tilde{\mathcal{H}}_x \quad \text{with} \quad \tilde{\mathcal{H}}_x := \{ v \in T_x \tilde{M} \mid \tilde{f}_* v \in \mathcal{H}_{f(x)} \} \ .$$

Now it is easy to show that $\tilde{f}$ becomes an isometric immersion with resp. to a suitable pseudo-Riemannian metric and that (with resp. to it) $\tilde{\pi} : \tilde{M} \longrightarrow M$ becomes a surjective pseudo-Riemannian submersion. Of course, $\tilde{\mathcal{H}}$ is its horizontal subbundle. In order to calculate its curvature form $\tilde{\Omega}$ let $\tilde{\nabla}$ be the Levi-Civita connection of $\tilde{M}$ and $\tilde{h}$ the second fundamental form of $\tilde{f}$. From (1) and Gauß' formula we derive for all $X,Y \in \Gamma(\tilde{\mathcal{H}})$

$$\begin{aligned} \Omega(\tilde{f}_* X, \tilde{f}_* Y) - \tilde{f}_* \tilde{\Omega}(X,Y) &= 2 \cdot (\mathcal{V} \overset{N}{\nabla}_X \tilde{f}_* Y - \tilde{f}_* \tilde{\mathcal{V}} \tilde{\nabla}_X Y) \\ &= 2 \cdot \mathcal{V} (\overset{N}{\nabla}_X \tilde{f}_* Y - \tilde{f}_* \tilde{\nabla}_X Y) = 2 \cdot \mathcal{V} \tilde{h}(X,Y) , \end{aligned}$$

where the first term is skew-symmetric and the last one symmetric; hence $\tilde{f}_* \tilde{\Omega}(X,Y) = \Omega(\tilde{f}_* X, \tilde{f}_* Y)$. Therefore we obtain for every $x = (p,q) \in \tilde{M}$, $u,w \in \tilde{\mathcal{H}}_x$ and $v \in \mathcal{V}_q$

$$\langle \tilde{f}_* \tilde{\Omega}(u,w), v \rangle = 2 \cdot \langle \phi^V(\pi_* \tilde{f}_* u), \pi_* \tilde{f}_* w \rangle = 2 \cdot \langle \phi^V(f_* \tilde{\pi}_* u), f_* \tilde{\pi}_* w \rangle ,$$

cf. formula (6). Hence condition a) is equivalent to the integrability of $\tilde{\mathcal{H}}$. - If in this case some initial data $x = (p,q) \in \tilde{M}$ are given, then we can choose a (local) integral manifold $\hat{M}$ of $\tilde{\mathcal{H}}$ with $x \in \hat{M}$ such that $\tilde{\pi}$ maps $\hat{M}$ isometrically onto a neighbourhood $U = U(p) \subset M$. The composition $g := \tilde{f} \circ (\tilde{\pi}|\hat{M})^{-1}$ is then a horizontal lift of $f|U$ with $g(p) = q$. □

<u>Remark 1</u> The diagram

$$\begin{array}{ccc} \tilde{M} & \xrightarrow{\tilde{f}} & N \\ {\scriptstyle \tilde{\pi}} \downarrow & & \downarrow {\scriptstyle \pi} \\ M & \xrightarrow{f} & B \end{array}$$

is an essential tool in several investigations of submanifolds of fibrations (see,e.g., [L]). But the correspondence between f and $\tilde{f}$ is geometrically not so close as the correspondence between f and a horizontal lift of f (see Theorem 1 and its corollaries).- As we saw in the preceding proof, condition a) of Theorem 2 is equivalent to the integrability of the distribution $\tilde{\mathcal{H}}$. In this case the integral manifolds of $\tilde{\mathcal{H}}$ are totally geodesic submanifolds of $\tilde{M}$ (a consequence of Gauß' formula and (1) applied to $\tilde{\pi}$). If in addition the fibres of π are totally geodesic (as in the case of the Hopf fibrations and the canonical fibrations of Sasakian manifolds, cf. sect.6 and 5), then $\tilde{M}$ is at least locally a pseudo-Riemannian product (related to the splitting $T\tilde{M} = \tilde{\mathcal{V}} \oplus \tilde{\mathcal{H}}$; in the pseudo-Riemannian case apply Wu's version of the de Rham decomposition theorem, cf. [W]).

4. Submersions with locally homogeneous fibres

Let again $\pi : N \to B$ be a pseudo-Riemannian submersion. Then, for every open subset $U \subset N$, let us denote the Lie algebra of vertical Killing fields on U by $\mathfrak{g}(U)$. We say that π has *locally homogeneous fibres*, iff every point $p \in N$ has a neighbourhood U such that

$$\mathcal{V}_p = \{ X_p \mid X \in \mathfrak{g}(U) \} . \tag{8}$$

While in general the horizontal integrability structure $\mathfrak{J} = (\mathfrak{J}_b)$ of π may be bizarre, it is rather nice in the preceding situation.

Theorem 3 If $\pi : N \to B$ is a surjective pseudo-Riemannian submersion with connected and locally homogeneous fibres, then

$$\forall p \in N \ : \ \mathfrak{J}_{\pi(p)} = \{ \phi^v \mid v \in \mathcal{V}_p \} \ ;$$

in particular, $(\mathfrak{J}_b)$ is a family of vector spaces.

Proof It is sufficient to show that the map

$$p \mapsto L_p := \{ \phi^v \mid v \in \mathcal{V}_p \} \subset \mathrm{End}(T_{\pi(p)}B)$$

is locally constant on each fibre $\pi^{-1}(b)$. Therefore, fix a point $p \in N$, put $b := \pi(p)$ and $N_b := \pi^{-1}(b)$, and choose a neighbourhood $U = U(p) \subset N$ such that (8) is satisfied.

Now some preparations: For every $X \in \mathfrak{g}(U)$ let (Φ_t^X) denote the local 1-parameter group of *isometries* generated by X and D_t^X the domain where Φ_t^X is defined. As X is vertical we obtain

$$\pi \circ \Phi_t^X = \pi | D_t^X , \tag{9}$$

hence

$$\Phi_{t*}^X \mathcal{V}_p = \mathcal{V}_q \quad \text{and} \quad \Phi_{t*}^X \mathcal{H}_p = \mathcal{H}_q \quad \text{with} \quad q := \Phi_t^X(p) . \tag{10}$$

By analogy with the exponential map of Lie groups let $E : G \to N_b$

be the differentiable map defined on the open subset

$$G := \{ X \in \mathfrak{g}(U) \mid p \in D_1^X \}$$

by

$$E(X) := \Phi_1^X(p) \quad .$$

Then $\alpha^X : t \mapsto E(tX)$ is the maximal integral curve of X with $\alpha^X(0) = p$. Condition (8), therefore, implies that $E_* : T_0\mathfrak{g}(U) \to T_pN_b$ is surjective. Thus:

$$E(G) \text{ is a neighbourhood of } p \text{ in } N_b \; . \tag{11}$$

For a moment let us fix $X \in \mathfrak{g}(U)$ and abbreviate $\Phi_t := \Phi_t^X$ and $\alpha := \alpha^X$. If $\tilde{Y} \in \Gamma(\mathcal{H})$ is the horizontal lift of a vector field Y of B , the formulas (9) and (10) imply

$$\Phi_{t*}\tilde{Y}_p = \tilde{Y}_{\alpha(t)} \quad . \tag{12}$$

Using (7) and (9) we, therefore, obtain for every $\xi \in \Gamma(\mathcal{V})$:

$$\begin{aligned} \phi^{\xi(\alpha(t))}Y_b &= -\pi_*\overset{N}{\nabla}_{\tilde{Y}_{\alpha(t)}}\xi = -\pi_*\overset{N}{\nabla}_{\Phi_{t*}\tilde{Y}_p}\xi \\ &= -\pi_*\overset{N}{\nabla}_{\tilde{Y}_p}(\Phi_{-t*}\xi \circ \Phi_t) = \phi^v Y_b \quad \text{with} \quad v = \Phi_{-t*}\xi_{\alpha(t)} \; . \end{aligned}$$

As Φ_t maps $\mathcal{V}_p$ isomorphically onto $\mathcal{V}_{\alpha(t)}$, we have proved that for every $v \in \mathcal{V}_p$ the map $t \mapsto \phi^{v(t)} \in L_{\alpha(t)}$ with $v(t) := \Phi_{t*}v$ is constant, and that therefore $t \mapsto L_{\alpha(t)}$ is constant. Varying X in $\mathfrak{g}(U)$ we obtain $L_{E(X)} = L_p$ for all $X \in G$. Assertion (11), therefore, completes the proof. □

<u>Remark 2</u> Every point $b_o \in B$ has a neighbourhood U on which a C^∞ section $s : U \to N$ of π exists. If we choose U small enough, then there exist vector fields $\xi_1,\ldots,\xi_m \in \Gamma(\mathcal{V})$ such that for every $b \in U$ the vectors $v_i(b) := \xi_i(s(b))$ provide a basis of $\mathcal{V}_{s(b)}$. In the situation of Theorem 3 the local tensor fields $b \mapsto \phi^{v_i(b)}$, $i=1,\ldots,m$, therefore, generate the horizontal integrability structure $\mathcal{J}$ of π over U . In spite of that, $\mathcal{J}$ has not to be a subbundle of the tensor bundle of B , because the function $b \mapsto \dim \mathcal{J}_b$ may vary. But:

Proposition In the situation of Theorem 3 suppose that for every pair b_1, b_2 of points of B there exists a fibre preserving isometry $f : N \longrightarrow N$ with $f(\pi^{-1}(b_1)) = \pi^{-1}(b_2)$. Then B is a pseudo-Riemannian homogeneous space and $\mathcal{J}$ a subbundle of the tensor bundle of B.

Proof If $f : N \longrightarrow N$ is a fibre preserving isometry, then there exists an isometry F of B such that $F \circ \pi = \pi \circ f$. For every $p \in N$ and $v \in \mathcal{V}_p$ we obtain $F_* \circ \phi^v = \phi^{f_* v} \circ F_{*|\pi(p)}$. Therefore $\mathcal{J}_{F(b)} = \{ F_* \circ \phi \circ F_{*|b}^{-1} \mid \phi \in \mathcal{J}_b \}$; in particular, $\dim \mathcal{J}_{F(b)} = \dim \mathcal{J}_b$ for all $b \in B$. As $\mathcal{J}$ is locally generated by tensor fields (see Remark 2), the proposition is proved. □

Note that the hypotheses of the Proposition are satisfied by the Hopf fibrations of sect.6.

Remark 3 Sometimes a surjective submersion $\pi : N \longrightarrow B$ is given, where only N is a pseudo-Riemannian manifold; and one is interested in the question whether B can be equipped with a pseudo-Riemannian metric such that π becomes a pseudo-Riemannian submersion (cf. sect.5 and 6). We can give a positive answer, if the vertical subspaces $\mathcal{V}_p$'s are non-degenerate (cf. sect.1) and π has connected and locally homogeneous fibres. Indeed, it is sufficient to prove: If Y, Z are vector fields of B and $\tilde{Y}, \tilde{Z} \in \Gamma(\mathcal{H})$ their horizontal lifts, then the function $p \mapsto \langle \tilde{Y}_p, \tilde{Z}_p \rangle$ is locally constant along the fibres $\pi^{-1}(b)$. For that one uses the foregoing method, in particular formula (12) and assertion (11).

5. Canonical fibrations of Sasakian manifolds

Let N be a Sasakian manifold with structure vector field ξ, i.e., N is a Riemannian manifold and ξ a unit Killing field such that the curvature tensor R of N satisfies

$$R(X,Y)\xi = \langle Y,\xi \rangle \cdot X - \langle X,\xi \rangle \cdot Y ,$$

cf. [Bl] p.75. Furthermore, we suppose that there is given a manifold B and a surjective submersion $\pi : N \to B$ with connected fibres such that

$$\mathcal{V} = \mathbb{R}\xi \quad . \tag{13}$$

(Note that for a given Sasakian manifold this situation always can be realized at least locally.) As is well known (see, e.g., [M] p.91) B can be equipped with the structure of a Kählerian manifold such that π becomes a Riemannian submersion and the complex structure J of B and the structure tensor field $\phi := \overset{N}{\nabla}\xi$ of N are related by

$$\pi_* \circ \phi = J \circ \pi_* \quad . \tag{14}$$

Comparing (7) and (14) we obtain $J_{\pi(p)} = \phi^{-\xi(p)}$ for all $p \in N$. The horizontal integrability structure $\mathfrak{J} = (\mathfrak{J}_b)$ of π, therefore, is the complex structure of B, i.e.,

$$\mathfrak{J}_b = \mathbb{R}\cdot J_b \quad \text{for all } b \in B \ . \tag{15}$$

(It should be mentioned that we can arrive at the same configuration when we start from a Kählerian manifold B ; cf. [Bl] p.99.) Theorem 2 now can be formulated in well-known terms, cf. [YK], [CO]:

<u>Theorem 4</u> <u>Let N be a Sasakian manifold with structure vector field ξ and structure tensor field $\phi = \overset{N}{\nabla}\xi$, B a Kählerian manifold with complex structure J and $\pi : N \to B$ a surjective Riemannian submersion satisfying</u> (13) <u>and</u> (14). <u>Then</u>:

a) <u>An isometric immersion $g : M \to N$ is horizontal, if and only if it is anti-invariant</u>, i.e., $\forall p \in M : g_*T_pM \perp \phi(g_*T_pM)$, <u>and normal to</u> ξ .

b) <u>An isometric immersion $f : M \to B$ has locally horizontal lifts (in the sense of Theorem 2b)), if and only if f is anti-invariant</u> (= <u>totally real</u>), <u>i.e.</u>,

$$\forall p \in M \ : \ f_*T_pM \perp J(f_*T_pM) \quad .$$

<u>Proof</u> a) To say that g is horizontal exactly means that it is normal to ξ. Therefore it remains to show: g is horizontal $\Rightarrow$ g is anti-invariant. For that, use Theorem 1a), (1) and $\phi = \overset{N}{\nabla}\xi$;

see also [YK] p.40.
b) Apply Theorem 2 and formula (15).□

Remark 4 Starting from a Riemannian manifold N with a Sasakian 3-structure Ishihara constructed a Riemannian submersion π onto a quaternion Kählerian manifold B ; see [I1], [I2]. In this case the horizontal integrability structure $\mathfrak{J} = (\mathfrak{J}_b)$ of π is exactly the quaternion structure of B . The isometric immersions into B satisfying condition a) of Theorem 2 again are called totally real immersions; cf. [CH]. For instance, in order to investigate totally real immersions into the quaternion projective space $\mathbb{H}P^n$ one, therefore, can study the horizontal isometric immersions of the Hopf fibration $S^{4n+3} \to \mathbb{H}P^n$; cf. sect.6.

6. Global horizontal lifts

In this section we will try to stick together local horizontal lifts (see Theorem 2b)) to obtain global ones. But first we will give an example (a Riemannian submersion with *complete* fibres) which allows only local horizontal lifts.

Example On $N = \mathbb{R}^2$ let $\mathcal{H}$ be the subbundle of TN generated by the differential equation $y' = 1 + y^2$, i.e., $\mathcal{H}_{(x,y)}$ is the 1-dim. linear subspace of $T_{(x,y)}N$ parallel to the vector $(1,1+y^2)$. If B is the real line $\mathbb{R}$ with the canonical metric, we can equip N uniquely with a Riemannian metric such that $\pi: N \to B$, $(t,s) \mapsto t$ becomes a Riemannian submersion with the horizontal bundle $\mathcal{H}$ and the fibres $\{x\} \times \mathbb{R}$ carry the canonical metric structure. Then $f = id_B$ has only local horizontal lifts, namely $x \mapsto (x,\tan(x-x_o))$, $x_o \in \mathbb{R}$.

Now a positive result:

Theorem 5 Let $\pi : N \to B$ be a surjective pseudo-Riemannian submersion.

a) Suppose the following *curve lifting property*:

For every C^∞ curve $\alpha : [0,1] \to B$ and every initial point $q \in \pi^{-1}(\alpha(0))$ there exists a horizontal curve $\tilde{\alpha} : [0,1] \to N$ with $\tilde{\alpha}(0) = q$ and $\pi \circ \tilde{\alpha} = \alpha$. (16)

Then for every isometric immersion $f : M \to B$ (M connected) satisfying condition a) of Theorem 2 and every initial data $(p,q) \in M \times N$ with $f(p) = \pi(q)$ there exists a pseudo-Riemannian manifold $\hat{M}$, an isometric covering map $\tau : \hat{M} \to M$, a horizontal isometric immersion $g : \hat{M} \to N$ and a point $\hat{p} \in \hat{M}$ such that $\pi \circ g = f \circ \tau$, $\tau(\hat{p}) = p$ and $g(\hat{p}) = q$.

b) Property (16) is satisfied in each of the following cases:

(C1) π is a proper map.

(C2) The fibres of π are connected and compact.

(C3) There exists a connected Lie group G acting on N by isometries such that the orbits of the action are exactly the fibres of π . (Principal bundles provide examples.)

Proof a) We follow the proof of Theorem 2 a) $\Rightarrow$ b) replacing the last two sentences by the following argument : Let $\hat{M}$ be the maximal integral manifold of $\tilde{\mathcal{H}}$ passing through the initial point $\hat{p} := (p,q)$, and put $\tau := \tilde{\pi}|\hat{M}$ and $g := \tilde{f}|\hat{M}$. Then we have to show that the local isometry $\tau : \hat{M} \to M$ is in fact a covering map. For that it is sufficient to prove that for every C^∞ curve $\alpha : [0,1] \to M$ and every $x \in \tau^{-1}(\alpha(0))$ there exists a curve $\hat{\alpha} : [0,1] \to \hat{M}$ with $\tau \circ \hat{\alpha} = \alpha$ and $\hat{\alpha}(0) = x$. This last condition follows from (16); indeed: If $\tilde{\alpha} : [0,1] \to N$ is the horizontal lift of $f \circ \alpha$ with $\tilde{\alpha}(0) = g(x)$, then $\hat{\alpha} := (\alpha, \tilde{\alpha})$ is a horizontal curve in $\tilde{M}$, which in fact runs in $\hat{M}$, because $\hat{\alpha}(0) = x$.

b) We start with a general observation:

For every curve $\alpha : I \to B$ (I an open interval) and every initial data $(s,q) \in I \times N$ with $\alpha(s) = \pi(q)$ there exists a *maximal* horizontal lift $\tilde{\alpha} : \tilde{I} \to N$ with $s \in \tilde{I}$ and $\tilde{\alpha}(s) = q$; if $\tilde{\alpha}(\tilde{I})$ is relatively compact, then $\tilde{I} = I$. (17)

To verify (17) we follow again the proof of Theorem 2 a) $\Rightarrow$ b), which gives the commutative diagram

$$\begin{array}{ccc} \tilde{M} & \xrightarrow{\tilde{f}} & N \\ {\scriptstyle\tilde{\pi}}\downarrow & & \downarrow{\scriptstyle\pi} \\ I & \xrightarrow{f=\alpha} & B \end{array}$$

where $\tilde{M}$ now is only equipped with a differentiable structure; $\tilde{\pi}$ again is a surjective submersion and $\tilde{\mathcal{H}}$ a distribution complementary to $\tilde{\mathcal{V}}$. Let $X \in \Gamma(\tilde{\mathcal{H}})$ be the horizontal lift of the canonical vector field $\frac{d}{dt}$ of I and $\beta : \tilde{I} \to \tilde{M}$ the maximal integral curve of X with $\beta(s) = (s,q)$. Then $\tilde{\alpha} := \tilde{f} \circ \beta$ is the maximal horizontal lift of α with $\tilde{\alpha}(s) = q$. The "boundary behaviour" of integral curves (cf. [D] p.8) completes the proof of (17).

If now π is proper, then (17) implies (16). Secondly, if π has connected and compact fibres, then $\pi : N \to B$ is a fibre bundle (according to [E] p.31); hence π is proper. Finally, in the case (C3) suppose that G acts on the right. Then the isometry $R_g : N \to N$, $p \mapsto pg$ maps horizontal curves into horizontal curves, because $R_{g*}\mathcal{V}_p = \mathcal{V}_{pg}$ and $R_{g*}\mathcal{H}_p = \mathcal{H}_{pg}$ for all $g \in G$ and $p \in N$. Combining this fact with (17) we easily obtain (16).□

Remark 5 $\pi : N \to B$ is a fibre bundle, if (16) is satisfied (cf. [H] proof of Theorem 1). - In the case (C3), π obviously has connected, locally homogeneous fibres (cf. sect.4). Indeed, if for every element A of the Lie algebra of G the associated Killing field of N is denoted by A^* (cf. [KN] p.42), then $\{A^*\} \subset \mathfrak{g}(N)$ and $\mathcal{V}_p = \{A^*_p\}$ for all $p \in N$.

The Hopf fibrations : Examples of pseudo-Riemannian submersions with totally geodesic fibres satisfying condition (C3) of Theorem 5

Let n be a positive integer, $s \in \{0,\dots,n+1\}$, $\varepsilon \in \{1,-1\}$, $\mathbb{K}$ the field $\mathbb{C}$ of complex numbers or the field $\mathbb{H}$ of quaternions, $d := \dim_{\mathbb{R}} \mathbb{K}$, $\langle\,,\rangle$ the inner product

$$\langle p,q\rangle := \mathrm{Re}\Big(- \sum_{k=1}^{s} \overline{p}_k \cdot q_k + \sum_{k=s+1}^{n+1} \overline{p}_k \cdot q_k \Big)$$

on $\mathbb{K}^{n+1}$, and N the *hyperquadric*

$$N := \{ p \in \mathbb{K}^{n+1} \mid \langle p,p \rangle = \varepsilon \} \quad , \text{ see } [O2] .$$

(The pairs $(s,\varepsilon) = (0,-1)$ and $= (n+1,1)$ must be excluded.) With resp. to the structure induced by $\langle\, ,\, \rangle$ N is a pseudo-Riemannian manifold, the restriction of the canonical projection $\mathbb{K}^{n+1} \setminus \{0\} \to \mathbb{K}P^n$ to N is a submersion $\pi : N \to B$ onto an open part $B \subset \mathbb{K}P^n$. Moreover, the action of the group S^{d-1} on N by the canonical multiplication $(p_1,\ldots,p_{n+1})g = (p_1 g,\ldots,p_{n+1}g)$ satisfies condition (C3) of Theorem 5. Therefore π has connected and locally homogeneous fibres (cf. Remark 5), and B can uniquely be equipped with a pseudo-Riemannian metric of index $(2s-1+\varepsilon)d/2$ such that π becomes a pseudo-Riemannian submersion (cf. Remark 3). Note that B is the projective space $\mathbb{K}P^n$ in case $(s,\varepsilon) = (0,1)$ with Fubini-Study metric, and the hyperbolic space $\mathbb{K}H^n$ in case $(s,\varepsilon) = (1,-1)$.

Now let $\mathfrak{g}$ be the Lie algebra of S^{d-1} , i.e., $\mathfrak{g} = \{ A \in \mathbb{K} \mid \text{Re } A = 0 \}$, and for every $A \in \mathfrak{g}$ let A^* (resp. J^A) denote the vector field (resp. tensor field of type (1,1)) on $\mathbb{K}^{n+1}$ which up to canonical identification is given by $p \mapsto pA$ (resp. $v \mapsto vA$). Then for every $p \in N$ we derive $\mathcal{V}_p = \{ A^*_p \mid A \in \mathfrak{g} \}$, $J^A \mathcal{H}_p \subset \mathcal{H}_p$ and

$$\forall\, v \in \mathcal{H}_p \quad : \quad \overset{N}{\nabla}_v A^* = J^A v \quad .$$

By means of (7) we, therefore, can easily prove

Theorem 6 The horizontal integrability structure $(\mathfrak{J}_b)$ of the general Hopf fibration $\pi : N \to B$ coincides at every point $b \in B$ with the horizontal integrability structure of the Hopf fibration $S^{d(n+1)-1} \to \mathbb{K}P^n$, i.e., with the complex structure of $\mathbb{C}P^n$ (in the sense of (15)) resp. the quaternion structure of $\mathbb{H}P^n$ (cf. Remark 4 and [I1]).

References

[B] Bourbaki,N.: Variétés différentielles et analytiques, Fascicule de resultats. Paris: Hermann 1971

[Bl] Blair,D.E.: Contact manifolds in Riemannian geometry. Lecture Notes in Mathematics, Vol. 509; Berlin, Heidelberg, New York: Springer(1976; Zbl. 319.53026).

[BR] Backes,E., Reckziegel,H.: On symmetric submanifolds of spaces of constant curvature. Math.Ann.263,419-433(1983; Zbl. 499.53045).

[C] Chen,B.Y.: Geometry of submanifolds. New York: Marcel Decker (1973; Zbl. 262.53036).

[CH] Chen,B.Y., Houh,C.S.: Totally real submanifolds of a quaternion projective space. Ann.di Mat.(IV)120,185-199(1979; Zbl. 413.53031).

[CO] Chen,B.Y., Ogiue,K.: On totally real submanifolds. Trans.of Amer.Math.Soc.193,257-266(1974; Zbl. 286.53019).

[D] Dieudonné,J.: Treatise on analysis, Vol. IV. New York,London: Academic Press (1974; Zbl. 292.58001).

[E] Ehresmann,C.: Les connexions infinitésimales dans un espace fibré différentiable. Colloque de Topologie, Bruxelles 1950, 29-55

[F] Ferus,D.: Symmetric submanifolds of euclidean space. Math. Ann.247,81-93(1980; Zbl. 446.53041).

[H] Hermann,R.: A sufficient condition that a mapping of Riemannian manifolds be a fibre bundle. Proc.Amer.Math.Soc.11,236-242(1960; Zbl. 112,137).

[I1] Ishihara,S.: Quaternion Kählerian manifolds and fibred Riemannian spaces with Sasakian 3-structure. Kōdai Math.Sem. Rep.25,321-329(1973; Zbl. 267.53023).

[I2] Ishihara,S.: Quaternion Kählerian manifolds. J.Diff.Geometry 9,483-500(1974; Zbl. 297.53014).

[KN] Kobayashi,S., Nomizu,K.: Foundations of differential geometry, Vol. I. New York: Interscience Publishers 1963

[L] Lawson,H.B.: Rigidity theorems in rank-1 symmetric spaces. J.Diff.Geometry 4,349-357(1970; Zbl. 199,564).

[M] Morimoto,A.: On normal almost contact structures with a regularity. Tôhoku Math.J.16,90-104(1964; Zbl. 135,221).

[N] Naitoh,H.: Parallel submanifolds of complex space forms. Nagoya Math.J.90,85-117(1983) and 91,119-149(1983; Zbl. 502.53044/5).

[O1] O'Neill,B.: The fundamental equations of a submersion. Michigan Math.J.13,459-469(1966; Zbl. 145,186).

[O2] O'Neill,B.: Semi-Riemannian geometry. New York: Academic Press (1983; Zbl. 531.53051).

[W] Wu,H.: On the de Rham decomposition theorem. Illinois J.Math. 8,291-311(1964; Zbl. 122,400).

[YK] Yano,K. Kon,M.: Anti-invariant submanifolds. New York: Marcel Decker (1976; Zbl. 349.53055).

Mathematisches Institut der Universität
Weyertal 86-90
D-5000 Köln 41
Fed. Rep. of Germany

POSITIVELY CURVED MINIMAL SUBMANIFOLDS

Antonio Ros, Paul Verheyen and Leopold Verstraelen

Dedicated to Professor T.J. Willmore at the occasion of his 65th birthday.

INTRODUCTION

In this article we discuss the recent solutions of some of K. Ogiue's conjectures concerning positively curved Kaehler submanifolds of complex projective spaces. In Sections 1 and 2 we sketch the context in which these conjectures may be situated and list the results which were obtained previously in this direction together with an indication of their methods of proof. In Section 3 we settle some of these conjectures completely and in the affirmative by a new approach to this type of problems which was introduced by the first named author. In Section 4 we treat related results for totally real minimal submanifolds of complex space forms. Finally, in Section 4 we state some new theorems for minimal submanifolds of Sasakian space forms analogeous to the results mentioned in Section 3 and 4.

We want to thank D.E. Blair and B.Y. Chen for the correspondence on the topics discussed in this article.

1. The Frankel conjecture

Let M be a Riemannian manifold with metric g and Riemann-Christoffel curvature tensor R. Let π be a plane section of T_xM spanned by orthonormal vectors U and V

$(x \in M)$. Then the sectional curvature $K(\pi)$ is defined by

$$K(\pi) = K(U,V) = R(U,V;V,U) \ .$$

If $K(\pi)$ is the same for all plane sections π in the tangent space T_xM of M at a point x and for all points x of M, then M is called a *real space form* or a space of constant curvature. The curvature tensor of a space of constant curvature c is given by

$$R(X,Y)Z = c[g(Y,Z)X - g(X,Z)Y] \ .$$

Now, let M be a Kaehler manifold with complex structure J, metric g and Riemann-Christoffel curvature tensor R. Let γ be a plane section of T_xM which is invariant under J, and let X be a unit vector in γ. Then X, JX is an orthonormal basis for γ and the sectional curvature

$$H(X) = K(\gamma) = R(X,JX;JX,X)$$

is called the *holomorphic sectional curvature* of M for γ. If $K(\gamma)$ is constant for all invariant plane sections γ of T_xM and for all points x of M, then M is called a *complex space form* or a space of constant holomorphic sectional curvature. The curvature tensor of a space of constant holomorphic sectional curvature c is given by

$$R(X,Y)Z = \frac{c}{4}\,[g(Y,Z)X - g(X,Z)Y + g(Z,JY)JX - g(Z,JX)JY + 2g(X,JY)JZ] \ .$$

Let γ and γ' be two J-invariant 2-planes in T_xM. Then the *holomorphic bisectional curvature* $B(\gamma,\gamma')$ is defined by

$$B(\gamma,\gamma') = R(X,JX;JX',X') \ ,$$

where X and X' are unit vectors in γ and γ', respectively [11]. One has

$$B(\gamma,\gamma') = K(\alpha) + K(\beta)$$

where α and β are the plane sections of T_xM spanned by the vectors X, X' and X, JX', respectively.

In 1961, Andreotti and Frankel [10] proved that a Kaehler surface $M^{2(*)}$ of positive sectional curvature is biholomorphic to a complex projective space $\mathbb{C}P^2$ and Frankel conjectured that this holds for all dimensions n. In 1969, Goldberg and Kobayashi [11] proved this result for n=2 under the condition of positive holomorphic bisectional curvature. Using results of Kobayashi and Ochiai [16], Mabuchi [19] showed in 1978 that this last result also holds for n=3. The validity for arbitrary n follows from a result in algebraic geometry by Mori [20] in 1979, combined with a result of Griffiths [12] and Kobayashi and Ochiai [15] (see [23]). An alternative proof of this result, using only methods in Kaehler geometry, was given in 1980 by Siu and Yau [30]. Thus one has the following theorem.

Theorem A. *Every compact Kaehler manifold M^n of positive holomorphic bisectional curvature is biholomorphic to $\mathbb{C}P^n$.*

If the Kaehler metric on the manifold M^n in Theorem A is moreover assumed to be Einstein, or still with the weaker additional assumption of constant scalar curvature, Berger [2] and Goldberg and Kobayashi [11] proved that M^n is even biholomorphically isometric to $\mathbb{C}P^n$ endowed with a Study-Fubini metric of constant holomorphic sectional curvature. In general however, Theorem A does not yield such information on the metric of M^n; i.e. Theorem A is a statement only on complex analytical equivalence and certainly does not in general imply a uniqueness theorem of the kind stated above.

2. Some of Ogiue's conjectures

In this respect, for a *Kaehler submanifold* of $\mathbb{C}P^{n+p}$, Ogiue conjectured in 1974

(*) All dimensions for complex spaces are complex dimensions except in section 5, and all manifolds under consideration are assumed to be connected.

that under suitable conditions of positivity of certain curvatures, M^n is biholomorphically isometric with $\mathbb{C}P^n$. We will formulate some of Ogiue's conjectures of this nature in more detail later in this section.

Let M^n be a complex submanifold of a complex projective space $\mathbb{C}P^m(\tilde{c})$ of constant holomorphic sectional curvature $\tilde{c} \in \mathbb{R}_0^+$. Then M^n is *totally geodesic* in $\mathbb{C}P^{(m)}(\tilde{c})$, i.e. the *second fundamental form* σ vanishes identically, if and only if

$$H = \tilde{c}\ ,$$

(hence M^n is locally isometric to $\mathbb{C}P^n(\tilde{c})$), where H is the holomorphic sectional curvature of M^n.

Concerning isometric imbeddings of complex space forms in complex space forms, Calabi [6] proved that $\mathbb{C}P^n(c)$ can be imbedded isometrically in $\mathbb{C}P^m(\tilde{c})$ if and only if there exists a natural number $\nu \in \mathbb{N}_0$ such that $\tilde{c} = \nu c$ and $m \geq \binom{n+\nu}{\nu} - 1$. The corresponding local result was obtained by Nakagawa and Ogiue [22]. In particular, if $\mathbb{C}P^n(c)$ is a Kaehler submanifold of $\mathbb{C}P^{n+p}(c)$, then $c = \tilde{c}$ (i.e. the immersion is totally geodesic) or $c \leq \frac{1}{2}\tilde{c}$ and $p \geq \frac{1}{2}n(n+1)$ (see also O'Neill [26]).

Without loss of generality, in the following we assume that the ambient space has constant holomorphic sectional curvature 1. The results given in Theorem B were obtained by Abe [1], Ogiue [24,25], Yau [37] and two of the authors [34]. In the statements of Theorem B we mention completeness, although in the proofs actually compactness is needed; the curvature conditions which are imposed on the submanifolds however, together with the completeness, do imply the necessary compactness (by essentially using a result of Myers [21]).

<u>Theorem B.</u> *For a complete Kaehler submanifold* M^n *of* $\mathbb{C}P^{n+p}(1)$, *each of the following conditions on the sectional curvature* K *or the holomorphic sectional curvature* H *of* M^n *implies that* M^n *is a totally geodesic submanifold of* $\mathbb{C}P^{n+p}(1)$:

$$\text{(i)}\ H > \delta = \begin{cases} \dfrac{3n-1}{3n+1} & (n \leq 5) \\[2ex] \dfrac{2n-3}{2n-2} & (n > 5) \end{cases}\ ;$$

(ii) $H > 1 - \frac{n+2}{2(n+2p)}$;

(iii) $p = 1$, $H > \frac{1}{2}$;

(iv) ρ *constant,* $H > \frac{1}{2}$;

(v) $2\nu \geqslant n$, $H > \frac{1}{2}$;

(vi) $H > \frac{1}{2}$, $K > \frac{1}{8}$;

(vii) $2\nu \geqslant n$, $K > 0$;

(viii) $p = 1$, $n \geqslant 2$, $K > 0$;

(ix) $p < \frac{1}{2}n(n+1)$, ρ *constant* , $K > 0$;

(x) $n = 1$, $K > \frac{1}{2}$;

(xi) $n \geqslant 2$, $K > \delta = \begin{cases} \frac{5}{23} & (n = 5) \\ \frac{5n - 2 - \sqrt{9n^2 + 60n - 4}}{n - 5} & (n \neq 5); \end{cases}$

(xii) $K > \frac{n(2p-1) + 8p - 3}{4n(4p-1)}$;

(xiii) $K > \frac{n+3}{8n}$;

(xiv) $K > \frac{p(n+4) + 1}{4n(2p+1)}$;

(xv) M *is Bochner flat and* $K > \frac{(n+1)(n^2 + 6n + 12) + 2p(n+4)}{8n[(n+1)(n+4) + 2p]}$,

where $\nu = \min\{\nu(x) \mid x \in M^n\}$, $\nu(x) = \dim_{\mathbb{C}}\{X \in T_xM^n \mid \sigma(X,Y) = 0$ *for all* $Y \in T_xM^n\}$ *and* ρ *is the scalar curvature of* M^n.

Sketch of proofs. In his partial solution of the Bernstein problem, in 1968 Simons obtained a formula for the Laplacian of the square of the norm of the second fundamental form for a minimal submanifold in a real space form [29]. Basically, the results of Theorem B where proved by using a formula of Simons' type such as computed by Chern, do Carmo and Kobayashi [9] for minimal submanifolds of locally symmetric spaces, combined with the lemma of Hopf. We briefly indicate such a proof for statement (xiv). The second fundamental form σ of an n-dimensional Kaehler submanifold M^n of $\mathbb{C}P^{n+p}(1)$ satisfies the following differential equation :

$$\Delta\|\sigma\|^2 = 2\|\nabla'\sigma\|^2 - 4\mathrm{tr}A^{*2} - 2\mathrm{tr}\tilde{A}^2 + (n+2)\|\sigma\|^2 ,$$

where ∇' is the covariant derivative of van der Waerden-Bortolotti, Δ is the Laplace operator of M and A^*, $\tilde{A}$ are operators of Kon [17] and Simons [29], respectively, which both are defined in terms of the second fundamental tensors of the immersion. By using some inequalities involving the curvature tensor and the second fundamental tensors of M^n and also making use of what in this context is sometimes called the trick of Yau [37], we obtain that

$$(*) \qquad \Delta\|\sigma\|^2 \geq \frac{1}{p+1}\left\{4[(n-1)\delta_K + \delta_H](2p+1) - [p(n+4)+1]\right\}\|\sigma\|^2 ,$$

where δ_K and δ_H are numbers such that $K \geq \delta_K$ and $H \geq \delta_H$.
So if

$$(**) \qquad (n-1)\delta_K + \delta_H > \frac{p(n+4)+1}{4(2p+1)} ,$$

it follows that $\Delta\|\sigma\|^2 \geq 0$. Thus, when M^n is compact, the well-known lemma of Hopf implies that $\|\sigma\|^2$ is constant, and therefore we actually have $\Delta\|\sigma\|^2 = 0$. From (*) and (**), this yields that $\|\sigma\|^2 = 0$, i.e. $\sigma = 0$. Since of course $H \geq K$, we may replace (**) by the condition (xiv), in which case M^n is compact.

Statement (iv) may be proved as follows. A Kaehler submanifold M^n of $\mathbb{C}P^{n+p}(1)$ satisfying $\frac{1}{2} < H$ has positive holomorphic bisectional curvature. From the result of Berger, Goldberg and Kobayashi mentioned after Theorem A, we see that M^n is a complex space form because of the fact that ρ is constant. By the result of Calabi, Nakagawa and Ogiue mentioned before, it follows that M^n must have constant holomorphic sectional curvature equal to 1, i.e. M^n is totally geodesic. Hereby we observe that the assumption on the scalar curvature being constant is essential in making it possible to use the uniqueness theorem. ■

Remark. The following quantization phenomenon was proved by Lawson [18] : for a complete complex curve M^1 in $\mathbb{C}P^n(1)$ satisfying $\frac{1}{k} < K \leq \frac{1}{k-1}$ (respectively $\frac{1}{k} \leq K < \frac{1}{k-1}$) for some integer k, $1 < k \leq n$, we have $K = \frac{1}{k-1}$ (respectively $K = \frac{1}{k}$) and the immersion is rigid (i.e. equivalent to an immersion $\mathbb{C}P^1(c) \hookrightarrow \mathbb{C}P^n(1)$ cited above, where $c = \frac{1}{k-1}$ (respectively $\frac{1}{k}$)).

These results give motivations for the following conjectures of Ogiue concerning Kaehler submanifolds M^n of $\mathbb{C}P^{n+p}(1)$, in particular when M^n is complete :

(I) $H > \frac{1}{2} \Rightarrow \sigma = 0$;

(II) for $n \geq 2$: $K > \frac{1}{8} \Rightarrow \sigma = 0$;

(III) for $p < \frac{1}{2} n(n+1)$: $K > 0 \Rightarrow \sigma = 0$.

With respect to (II), we recall that for a complex projective space of constant holomorphic sectional curvature $H = \frac{1}{2}$, the sectional curvature K ranges between $\frac{1}{8}$ and $\frac{1}{2}$: K attains all values in $[\frac{1}{8}, \frac{1}{2}]$. Moreover $\mathbb{C}P^n(\frac{1}{2})$ can be imbedded in $\mathbb{C}P^{\frac{n(n+3)}{2}}(1)$ in a not-totally geodesic way. Thus the statements (I) and (II) are best possible, in the sense that the curvature conditions can not be weakened to $H \geq \frac{1}{2}$ or $K \geq \frac{1}{8}$.

3. Solutions of (I) and (II)

Recently, the first named author introduced a method to solve problems of the nature of Ogiue's conjectures in a way which is more direct than using computations for $\Delta\|\sigma\|^2$ and which is also more successfull [27]. As a matter of fact it completely solves conjectures (I) [27] and (II) [28]. Till now, as far as we know, conjecture (III) remains open. Next we will describe the basic steps occurring in this new approach.

Theorem 1. *Every complete Kaehler submanifold* M^n *of* $\mathbb{C}P^m(1)$ *for which* $H > \frac{1}{2}$ *is totally geodesic.*

Sketch of proof. For any unit tangent vector U to M^n, the holomorphic sectional curvature H(U) is given by

$$H(U) = 1 - 2\|\sigma(U,U)\|^2 .$$

Let $\pi : UM \longrightarrow M$ and U_xM be the unit tangent bundle of M^n and its fiber over $x \in M^n$, respectively. Consider the function $f : UM \longrightarrow \mathbb{R}$ defined by

$$f(U) = \|\sigma(U,U)\|^2 .$$

The hypothesis $H > \frac{1}{2}$ is then equivalent to $f < \frac{1}{4}$.

Since UM is compact under the assumptions of the theorem, f attains a maximum at some vector $V \in U_xM$ for some point x in M. For any U in U_xM let γ_U be the geodesic in M determined by the initial conditions $\gamma_U(0) = x$ and $\gamma_U'(0) = U$. By parallel translation of V along γ_U, we obtain a vector field V_U along γ_U. Put $f_U = f \circ V_U$. Taking the first and second derivatives of such functions, we find in particular that

$$\text{(1)} \qquad \frac{d^2 f_V}{dt^2}(0) + \frac{d^2 f_{JV}}{dt^2}(0) = 3\|\sigma(V,V)\|^2 - 12 g(\sigma(V,V), \sigma(A_{\sigma(V,V)}V,V)) + 4\|(\nabla'\sigma)(V,V,V)\|^2 ,$$

where A_ξ is the second fundamental tensor corresponding to a normal vector ξ and J is the complex structure of M. From the fact that f attains its maximum at V, we have

$$\text{(2)} \qquad \frac{d^2 f_V}{dt^2}(0) + \frac{d^2 f_{JV}}{dt^2}(0) \leq 0 .$$

For any $U \in U_xM$ such that $g(V,U) = 0$, let α be a curve in the sphere U_xM such that $\alpha(0) = V$ and $\alpha'(0) = U$. As V is a critical point of f we have

$$\frac{d(f \circ \alpha)}{ds}(0) = 4g(\sigma(V,V),\sigma(V,U)) = 0 ,$$

or equivalently

$$g(A_{\sigma(V,V)}V,U) = 0 ,$$

i.e.

(3) $$A_{\sigma(V,V)}V = \|\sigma(V,V)\|^2 V .$$

From (1) and (3) we obtain

(4) $$\frac{d^2 f_V}{dt^2}(0) + \frac{d^2 f_{JV}}{dt^2}(0) = 3f(V)[1 - 4f(V)] + 4\|(\nabla'\sigma)(V,V,V)\|^2 .$$

From (2) and (4) we conclude that

(5) $$f(V)[1 - 4f(V)] \leqslant 0 .$$

Since $f(V) < \frac{1}{4}$ it then follows that $f(V) \leqslant 0$, and by the definition of f, we find that $f(V) = 0$. Finally, f attaining its maximum at V, we see that $f = 0$, or equivalently $\sigma = 0$. ∎

Theorem 2. *Every complete Kaehler submanifold* M^n *of* $\mathbb{C}P^m(1)$ *with* $n \geqslant 2$ *and* $K > \frac{1}{8}$ *is totally geodesic.*

Sketch of proof. We will prove that under the assumptions of Theorem 2, the hypothesis that M^n is not totally geodesic leads to a contradiction. By the hypothesis $\sigma \neq 0$, it follows from (5) that $f(V) \geqslant \frac{1}{4}$ for V chosen as in the proof of Theorem 1. Moreover, for each $U \in U_xM$, the function f_U attains a maximum for $t = 0$, such that

$$\frac{d^2 f_U}{dt^2}(0) + \frac{d^2 f_{JU}}{dt^2}(0) \leqslant 0 .$$

This implies that

$$(6) \qquad 2\|\sigma(V,V)\|^2 R(U,JV;JV,V) - \frac{1}{2}\|\sigma(V,V)\|^2 - 2\|A_{\sigma(V,V)}U\|^2 \leq 0$$

for all $U \in U_xM$. Now, since $n \geq 2$, we can always choose a unit eigenvector U of $A_{\sigma(V,V)}$ such that $g(U,V) = g(U,JV) = 0$. For such a vector U, (6) becomes

$$(7) \qquad aK(U,V) + bK(U,JV) - \frac{1}{2}\|\sigma(V,V)\|^2 \leq 0 ,$$

where

$$(8) \qquad a = 2\|\sigma(V,V)\|^2 - \frac{1}{2}K(U,V) + \frac{1}{2}K(U,JV) > 0$$

and

$$(9) \qquad b = 2\|\sigma(V,V)\|^2 + \frac{1}{2}K(U,V) - \frac{1}{2}K(U,JV) > 0 .$$

Since $K > \frac{1}{8}$, from (7) we obtain the strict inequality

$$(10) \qquad \frac{1}{8}(a+b) - \frac{1}{2}\|\sigma(V,V)\|^2 < 0 .$$

But from (8) and (9) we know that

$$a + b = 4\|\sigma(V,V)\|^2 ,$$

which inserted in (10) yields the desired contradiction. ∎

4. Totally real submanifolds of $\mathbb{C}P^m$

Let M^n be a real n-dimensional *totally real* submanifold of $\mathbb{C}P^m(1)$. If M^n is minimal, then M^n is totally geodesic in $\mathbb{C}P^n(1)$ if and only if

$$K = \frac{1}{4}$$

(and hence M^n is locally $\mathbb{R}P^n(\frac{1}{4}) \hookrightarrow \mathbb{C}P^n(1)$) [8].

The following results were obtained by Ogiue, Chen and Houh [7], Yau [37] and two of the authors [32] using the method described in Section 2.

Theorem C. *Let* M^n *be a complete minimal totally real submanifold of* $\mathbb{C}P^n(1)$. *Then :*

(i) *if* $K > \frac{n-2}{4(2n-1)}$, M^n *is totally geodesic;*

(ii) *if* $K \geq \frac{n-2}{4(2n-1)}$, M^n *is totally geodesic or* $n=2$ *and* M^2 *is flat;*

(iii) *if* M^n *is conformally flat,* $n \geq 4$ *and* $K > \frac{(n-1)^2}{4n(n^2+n-4)}$, M^n *is totally geodesic.*

Recently, by using the method discussed in the previous section and working with the function $f : UM \longrightarrow \mathbb{R} : U \longmapsto f(U) = g(\sigma(U,U),JU)$, Urbano [31] obtained the following result.

Theorem 3. *Every compact minimal totally real submanifold* M^n *of* $\mathbb{C}P^n(1)$ *with* $K > 0$ *is totally geodesic.*

Also this result is best possible, in the sense that there exist compact minimal totally real submanifolds M^n of $\mathbb{C}P^n(1)$ with $K \geq 0$ and which are not totally geodesic.

5. Related results for minimal submanifolds of Sasakian space forms

A Sasakian manifold is an odd dimensional analogue of a Kaehler manifold. To be more precise, a *Sasakian manifold* consists of a real (2m+1)-dimensional Riemannian manifold $(\widetilde{M},g)$ $(m \in \mathbb{N}_o)$, a (1,1)-tensor φ on $\widetilde{M}$, a unit vector field ξ on $\widetilde{M}$ and the 1-form η dual to ξ satisfying

$$\varphi^2 = -I + \eta \otimes \xi ,$$

$$g(\varphi X,\varphi Y) = g(X,Y) - \eta(X)\eta(Y)$$

and

$$(\widetilde{\nabla}_X\varphi)Y = -g(X,Y)\xi + \eta(Y)X$$

for vector fields X and Y tangent to $\widetilde{M}$ and where $\widetilde{\nabla}$ is the Riemannian connection of $\widetilde{M}$. The following are always satisfied on a Sasakian manifold :

$$\varphi\xi = 0 ,$$

$$\widetilde{\nabla}_X\xi = \varphi X$$

and

$$\widetilde{K}(Z,\xi) = 1$$

for each unit vector Z normal to ξ, where K denotes the sectional curvature of $\widetilde{M}$ [3,35,36].

The sectional curvature of a plane section containing X and φX where X is a unit vector normal to ξ, is called the *φ-sectional curvature* determined by X, and is denoted by $\widetilde{K}_\varphi(X)$, i.e. $\widetilde{K}_\varphi(X) = \widetilde{K}(X,\varphi X)$. If the φ-sectional curvature is constant on $\widetilde{M}$, $\widetilde{M}$ is said to be a *Sasakian space form* and its curvature tensor is given by

$$\widetilde{R}(X,Y)Z = \frac{c+3}{4}[g(Y,Z)X - g(X,Z)Y]$$

$$+ \frac{c-1}{4}[\eta(X)\eta(Z)Y - \eta(Y)\eta(Z)X + g(X,Z)\eta(Y)\xi - g(Y,Z)\eta(X)\xi$$

$$+ g(\varphi Y,Z)\varphi X - g(\varphi X,Z)\varphi Y - 2g(\varphi X,Y)Z]$$

where c is the constant φ-sectional curvature of $\widetilde{M}$. As an example of a Sasakian space form we mention the unit sphere $S^{2m+1} = \{z \in \mathbb{C}^{m+1} \,\big|\, |z| = 1\}$ with standard metric tensor field, where $\xi_z = Jz$ for each $z \in S^{2m+1}$ and J is the complex structure of $\mathbb{C}^{m+1}$ and where $\varphi = \pi \circ J$ if $\pi : T_z\mathbb{C}^{m+1} \longrightarrow T_zS^{2m+1}$ denotes the orthogonal projection onto T_zS^{2m+1}. In this way S^{2m+1} has constant φ-sectional curvature 1.

A Sasakian manifold $\widetilde{M}$ such that for each point x in $\widetilde{M}$ there exists a cubical neighborhood U satisfying the condition that the integral curves of ξ passing through U pass through it only once is called a *regular Sasakian manifold*. In this case one can consider $\widetilde{M}/\xi =: \widetilde{N}$, the set of orbits of ξ, which is a Kaehler manifold, and the fibering $\widetilde{M} \longrightarrow \widetilde{N}$. In particular, each Sasakian space form M of constant φ-sectional curvature c is regular and $\widetilde{M}/\xi$ has constant holomorphic sectional curvature c+3.

A submanifold M^{n+1} of a regular Sasakian manifold $\tilde{M}^{2m+1}$ which is tangent to the structure vector field ξ of $\tilde{M}$ also satisfies this regularity condition and we can also consider $N^n := M^{n+1}/\xi$ such that we get the following commutative diagram

$$\begin{array}{ccc} M^{n+1} & \longrightarrow & \tilde{M}^{2m+1} \\ \downarrow & \odot & \downarrow \\ N^n & \longrightarrow & \tilde{N}^{2m} \end{array}$$

where M is minimal in $\tilde{M}$ if and only if N is minimal in $\tilde{N}$. Moreover M is anti-invariant in $\tilde{M}$ (i.e. for each $x \in M$ we have $\varphi T_xM \subset T_x^{\perp}M$) if and only if N is totally real in $\tilde{N}$.

For submanifolds tangent to ξ in a Sasakian space form $\tilde{M}(1)$ of constant φ-sectional curvature 1, we can "transport" the above mentioned properties of the Kaehlerian case to the Sasakian case. In particular, we can apply this for Sasakian submanifolds M^{2n+1} (for each $x \in M$ we have $\varphi T_xM \subset T_xM$) and for anti-invariant minimal submanifolds M^{m+1} in $\tilde{M}^{2m+1}(1)$ tangent to ξ. This leads to the following results obtained by Van Lindt and the two last authors. For earlier results see [3,4,5,13,14,33].

<u>Theorem 4.</u> *Let M^{2n+1} be a compact Sasakian submanifold of a Sasakian space form $\tilde{M}^{2m+1}$ of constant φ-sectional curvature 1. If the φ-sectional curvature K_φ of M satisfies $K_\varphi > -1$, then M is totally geodesic in $\tilde{M}$.*

<u>Theorem 5.</u> *Let M^{m+1} $(m \geq 2)$ be a $(m+1)$-dimensional compact minimal anti-invariant submanifold of a Sasakian space form $\tilde{M}^{2m+1}$ of constant φ-sectional curvature 1 such that M is tangent to the structure vector field ξ of $\tilde{M}$. If the sectional curvature K of M satisfies $K(X,Y) > 0$ for all $X,Y \in \mathfrak{X}M$ orthogonal to ξ, then M is locally a Riemannian direct product $\sum^m \times \sum^1$ where $\sum^m$ is a totally geodesic anti-invariant submanifold of $\tilde{M}$ and $\sum^1$ is generated by ξ.*

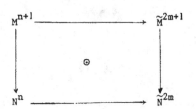

A submanifold M^n of a Sasakian manifold $\tilde{M}^{2m+1}$ such that M is normal to the structure vector field ξ of $\tilde{M}$ is automatically anti-invariant. Moreover, in this case M is an integral submanifold of the contact distribution $\mathfrak{D}$ determined by $\eta = 0$ and $n \leqslant m$. Conversely, each integral submanifold of $\mathfrak{D}$ is normal to ξ [4,5]. For integral submanifolds of $\mathfrak{D}$ of maximal dimension, one has the following result.

Theorem 6. *Let M^m be an m-dimensional compact minimal submanifold of an (2m+1)-dimensional Sasakian space form $\tilde{M}^{2m+1}$ of constant φ-sectional curvature 1 such that M is an integral submanifold of the contact distribution $\mathfrak{D}$. If M is positively curved, then M is totally geodesic in $\tilde{M}$.*

REFERENCES

[1] K. Abe, *A characterization of totally geodesic submanifolds in S^N and $\mathbb{C}P^N$ by an inequality,* Tôhoku Math. J. 23, 219-244 (1971; Zbl. 245.53053).

[2] M. Berger, *Sur les variétés d'Einstein compactes,* C.R. IIIe Réunion Math. Expression latine, Namur (1965), 35-55.

[3] D.E. Blair, *Contact manifolds in Riemannian geometry,* Lecture Notes in Mathematics 509, Springer-Verlag, Berlin, New York (1976; Zbl. 319.53026).

[4] D.E. Blair & K. Ogiue, *Geometry of integral submanifolds of a contact distribution,* Illinois J. Math. 19, 269-276 (1975; Zbl. 335.53043).

[5] —— , *Positively curved integral submanifolds of a contact distribution,* Illinois J. Math. 19, 628-631 (1975; Zbl. 317.53049).

[6] E. Calabi, *Isometric imbedding of complex manifolds,* Ann. of Math. 58, 1-23 (1953; Zbl. 51,131).

[7] B.Y. Chen & C.S. Houh, *Totally real submanifolds of a quaternion projective space,* Ann. Math. Pura Appl. CXX, 185-199 (1979; Zbl. 413.53031).

[8] B.Y. Chen & K. Ogiue, *On totally real submanifolds,* Trans. Amer. Math. Soc. 193, 257-266 (1974; Zbl. 286.53019).

[9] S.S. Chern, M. do Carmo & S. Kobayashi, *Minimal submanifolds of a sphere with second fundamental form of constant length,* Functional analysis and related

fields (Proc. Conf. for M. Stone, Univ. Chicago, Chicago, Ill., 1968), Springer-Verlag, Berlin-New York, 59-75 (1970; Zbl. 216,440).

[10] T. Frankel, *Manifolds with positive curvature,* Pacific J. Math. 11, 165-174 (1961; Zbl. 107,390).

[11] S.I. Goldberg & S. Kobayashi, *On holomorphic bisectional curvature,* J. Differential Geometry 1, 225-233 (1967; Zbl. 169,532).

[12] P.A. Griffiths, *Hermitian differential geometry, Chern classes and positive vector bundles,* Global Analysis (in honor of Kodaira), Princeton Univ. Press, Princeton, N.J.,(1970; Zbl. 201,240).

[13] M. Harada, *Sasakian space forms immersed in Sasakian space forms,* Bull. Tokyo Gakugei Univ. 24, 7-11 (1972; Zbl. 345.530.33).

[14] —— , *On Sasakian submanifolds II,* Bull. Tokyo Gakugei Univ. 25, 19-23 (1973; Zbl. 359.53010).

[15] S. Kobayashi & T. Ochiai, *On complex manifolds with positive tangent bundle,* J. Math. Soc. Japan 22, 499-525 (1970; Zbl. 197,360).

[16] —— , *Three-dimensional compact Kaehler manifolds with positive holomorphic bisectional curvature,* J. Math. Soc. Japan 24, 465-480 (1972; Zbl. 234.53051).

[17] M. Kon, *On some complex submanifolds in Kaehler manifolds,* Canad. J. Math. 26, 1442-1449 (1974; Zbl. 297.53013).

[18] H.B. Lawson, *The Riemannian geometry of holomorphic curves,* Carolina Conference Proc.(1970; Zbl. 214,470).

[19] T. Mabuchi, *ℂ-actions and algebraic threefolds with ample tangent bundle,* Nagoya Math. J.69 , 33-64 (1978; Zbl. 352.32022).

[20] S. Mori, *Projective manifolds with ample tangent bundles,* Ann. of Math. 110, 593-606 (1979; Zbl. 423.14006).

[21] S. Myers, *Riemannian manifolds with positive mean curvature,* Duke Math. J. 8, 401-404 (1941; Zbl. 25,227).

[22] H. Nakagawa & K. Ogiue, *Complex space forms immersed in complex space forms,* Trans. Amer. Math. Soc. 219, 289-297 (1976; Zbl. 273.53049).

[23] T. Ochiai, *On compact Kaehler manifolds with positive holomorphic bisectional curvature,* Proc. of Symposia in Pure Math. 27/2, 113-123 (1975; Zbl. 321.53053)

[24] K. Ogiue, *Differential geometry of Kaehler submanifolds*, Advances in Math. 13, 73-114 (1974; Zbl. 275.53035).

[25] ——, *Positively curved complex submanifolds immersed in a complex projective space* III, J. Differential Geometry 11, 613-615 (1976; Zbl. 354.53045).

[26] B. O'Neill, *Isotropic and Kaehler immersions*, Canad. J. Math. 17, 907-915 (1965; Zbl.171,205).

[27] A. Ros, *Positively curved Kaehler submanifolds*, to appear in Proceedings AMS.

[28] A. Ros & L. Verstraelen, *On a conjecture of K. Ogiue*, to appear in J. Differential Geometry, june 1984.

[29] J. Simons, *Minimal varieties in Riemannian manifolds*, Ann. of Math. 88, 62-105 (1968; Zbl. 181,497).

[30] Y.T. Siu & S.T. Yau, *Compact Kaehler manifolds of positive bisectional curvature*, Invent. Math. 59, 189-204 (1980; Zbl. 442.53056).

[31] F. Urbano, *Totally real minimal submanifolds of a complex projective space*, to appear in Proceedings AMS.

[32] P. Verheyen & L. Verstraelen, *Conformally flat totally real submanifolds of complex projective spaces*, Soochow J. Math. 6, 137-143 (1980; Zbl. 456.53035).

[33] ——, *Conformally flat C-totally real submanifolds of Sasakian space forms*, Geometriae Dedicata 12, 163-169 (1982; Zbl. 476.53035).

[34] ——, *Positively curved complex submanifolds of a complex projective space*, to appear in Publ. Math. Dep. Univ. Szczecin.

[35] K. Yano & M. Kon, *Anti-invariant submanifolds*, Marcel Dekker, New York,(1978; Zbl. 349.53055).

[36] ——, *CR-submanifolds of Kaehlerian and Sasakian manifolds*, Progress in Mathematics, Birkhäuser, Boston-Basel-Stuttgart,(1983; Zbl. 496.53037).

[37] S.T. Yau, *Submanifolds with constant mean curvature* I, Amer. J. Math. 96, 346-366 (1974; Zbl. 304.53041);—— II, Amer. J. Math. 97,76-100 (1975; Zbl. 304.53042).

Universidad de Granada
Facultad de Ciencias
Departamento de Geometria
Granada
Spain

Katholieke Universiteit Leuven
Faculteit der Wetenschappen
Departement Wiskunde
Celestijnenlaan 200B
B-3030 Leuven, Belgium.

AFFINSPHÄREN MIT EBENEN SCHATTENGRENZEN

Angela Schwenk
Technische Universität Berlin

1. Einleitung

Ziel dieses Artikels ist es, im Rahmen der affinen Differentialgeometrie kompakte, berandete Quadriken durch Eigenschaften ihrer Ränder zu charakterisieren. Die ersten lokalen Ergebnisse dazu hat Blaschke in ([1], p. 119) geliefert: die einzigen elliptisch bzw. hyperbolisch gekrümmten zweidimensionalen Flächen, für die jede Schattengrenze bei Parallelbeleuchtung eben ist, sind Quadriken. Schneider zeigte dazu, daß für n-dimensionale Affinsphären des (n+1)-dimensionalen affinen Raumes bereits (n+1) ebene Schattengrenzen bei Parallelbeleuchtung zur Charakterisierung von Quadriken ausreichen [4]. In dieser Arbeit wird gezeigt, daß schon die Existenz einer ebenen Schattengrenze ausreicht. Dies wird als Ausgangspunkt genommen, analoge Probleme zu untersuchen. Es werden Resultate angegeben, die Quadriken unter kompakten, berandeten Affinsphären durch Randbedingungen folgender Art kennzeichnen:

(i) der Rand ist eben und Schattengrenze bei Parallelbeleuchtung,
(ii) der Rand ist eben und Schattengrenze bei Zentralbeleuchtung,
(iii) der Rand ist eine totalgeodätische Untermannigfaltigkeit und Schattengrenze bei Parallelbeleuchtung,
(iv) der Rand ist eine totalgeodätische Untermannigfaltigkeit und eben.

2. Zusammenstellung der Grundbegriffe der affinen Differentialgeometrie

In der euklidischen Differentialgeometrie sind die Größen der äußeren Geometrie Invarianten bezüglich orthogonaler Abbildungen und Translationen des euklidischen Raumes. In der äquiaffinen Differentialgeometrie werden Größen betrachtet, die bezüglich volumentreuer Abbildungen des affinen Raumes invariant sind.
Die Zusammenstellung der benötigten Grundbegriffe orientiert sich an der Arbeit von Schneider [4].

A_{n+1} sei ein (n+1)-dimensionaler reeller affiner Raum, $n \geqq 2$.

V_{n+1} sei der zugehörige (n+1)-dimensionale, reelle Vektorraum mit Determinantenform det.

V^*_{n+1} sei der Dualraum von V_{n+1} mit dualer Form $\det^*$.

$x,y,z,a,\ldots$ kleine Buchstaben bezeichnen Elemente von V_{n+1}.

$X,L,\ldots$ große Buchstaben bezeichnen Elemente von V^*_{n+1}.

$< , >: V^*_{n+1} \times V_{n+1} \to \mathbb{R}$ ist das kanonische skalare Produkt $<X,y>:=X(y)$.

M sei eine n-dimensionale, orientierbare, zusammenhängende Mannigfaltigkeit und

∂M sei, falls vorhanden, der Rand von M.

$x:M \to A_{n+1}$ sei eine C^∞-Immersion; nach Wahl eines Ursprungs des A_{n+1} bezeichne x gleichzeitig die zugehörige Ortsvektorabbildung $x:M \to V_{n+1}$.

$x_i := \partial_i x$ bezeichne die erste partielle Ableitung der Abbildung x nach der i-ten Koordinate eines lokalen Koordinatensystems von M,

$\partial_j \partial_i x$ bezeichne die entsprechenden zweiten partiellen Ableitungen, $i,j = 1,\ldots,n$.

Es werden nur solche Hyperflächen $x:M \to A_{n+1}$ betrachtet, für die durch $G_{ij} := (\det \Lambda_{ij})^{-1/n+2} \cdot \Lambda_{ij}$ mit $\Lambda_{ij} := \det(\partial_j \partial_i x, x_1, \ldots, x_n)$ auf M eine Riemannsche Metrik definiert wird, d.h. $x:M \to A_{n+1}$ ist elliptisch gekrümmt.

$y := \frac{1}{n} \Delta x$ heißt äquiaffine Normalisierung von $x:M \to A_{n+1}$, dabei ist Δ der zu der Riemannschen Metrik G_{ij} gehörende Laplace-Operator. Es gilt: $\{y, x_1, \ldots, x_n\}$ ist linear unabhängig; $y_1, \ldots, y_n \in \operatorname{span}\{x_1, \ldots, x_n\}$

$X:M \to V^*_{n+1}$ heißt Konormalenvektor der Tangentialebene, wenn gilt: $<X,x_i> = 0$ für $i = 1,\ldots,n$ $\quad <X,y> = 1$.

Für alle $v \in V_{n+1}$ gilt

$$(1) \qquad <X,v> = \frac{1}{\sqrt{\det G_{ij}}} \det(v, x_1, \ldots, x_n) .$$

Eine Fläche heißt

konvex, wenn $x:M \to A_{n+1}$ eine Einbettung ist, und $x(M)$ auf dem Rand eines konvexen Gebietes des A_{n+1} liegt. Eine Fläche heißt schwach gekrümmt, wenn es einen konstanten Vektor $e \in V_{n+1}$ gibt, der zu keiner Tangentialebene von $x:M \to A_{n+1}$ parallel ist, d.h. wenn $<X,e> \neq 0$ auf M gilt.

Für Funktionen $f:M \to \mathbb{R}$ und Abbildungen $z:M \to V_{n+1}$ bzw. $Z:M \to V^*_{n+1}$ seien durch Anfügen von Indizes f_{ij}, z_{ij}, Z_{ij} die Komponenten der kovarianten Ableitungen bzgl. eines lokalen Koordinatensystems bezeichnet. Für die Metrik gilt auch $G_{ij} = \langle X, x_{ij} \rangle = -\langle X_i, x_j \rangle$.
$\{X, X_1, \ldots, X_n\}$ ist in jedem Punkt von M eine Basis des V^*_{n+1} [4]; insbesondere ist $X:M \to A^*_{n+1}$ eine Immersion, wobei A^*_{n+1} ein zu V^*_{n+1} gehörender affiner Raum ist.
Durch $A_{ijk} := \langle X, x_{ijk} \rangle = -\langle X_k, x_{ij} \rangle$ wird der kubische oder auch Darbouxsche Tensor A definiert, der symmetrisch ist und dessen Spur $A_{ij}{}^j = 0$ verschwindet; $n(n-1)J := A_{ijk}A^{ijk} = |A|^2$ bezeichnet dessen Normquadrat, J heißt Picksche Invariante.
$\alpha_{ij} := A_i{}^{rs} A_{rsj}$ definiert einen symmetrischen Tensor, für dessen Spur $\alpha_i{}^i = n(n-1)J$ gilt.
$A_{ijk\|l}$ bezeichnet die Komponenten der kovarianten Ableitung ∇A von A.
Durch $B_{ij} := -\langle X_i, y_j \rangle$ wird ein symmetrischer Tensor definiert, dessen Spur $H := -\frac{1}{n} B_i{}^i$ mittlere Krümmung heißt.
R_{ijkl} seien die Komponenten des Riemannschen Krümmungstensors mit

$$R_{ijk}{}^l \partial_l = \nabla_{\partial_i} \nabla_{\partial_j} \partial_k - \nabla_{\partial_j} \nabla_{\partial_i} \partial_k - \nabla_{[\partial_i, \partial_j]} .$$

Es gilt [4]

$$R_{ijkl} = A_{ik}{}^r A_{rjl} - A_{il}{}^r A_{rjk} + \tfrac{1}{2}(G_{jl}B_{ik} - G_{il}B_{jk} + G_{ik}B_{jl} - G_{jk}B_{il}).$$

$R_{ij} := R_{kij}{}^k = \alpha_{ij} + (n-1)HG_{ij} + \frac{2-n}{2}(B_{ij} + HG_{ij})$ seien die Komponenten des Ricci-Tensors, und $R := \frac{1}{n(n-1)} R_i{}^i = J + H$ sei die Skalarkrümmung.

Die kovarianten Ableitungen des Ortsvektors x, der Normalisierung y und der Konormalen X stellen in den Ableitungsgleichungen eine Beziehung zwischen den eben definierten Größen her. Im klassischen Tensorkalkül ist es üblich, die Gleichungen in folgender Form anzugeben:

(2a) $x_{ij} = A_{ij}{}^r x_r + G_{ij} y$

(2b) $y_i = B_i{}^r x_r$

(2c) $X_{ij} = -A_{ij}{}^r X_r + B_{ij} X$ [4].

$x:M \to A_{n+1}$ heißt eigentliche Affinsphäre mit Zentrum $z \in A_{n+1}$, wenn $y = \frac{1}{n} \Delta x = -c(x-z)$ für $c \in \mathbb{R}$ gilt. Für die mittlere Affinkrümmung H ergibt sich $H = c = \text{konst.} \neq 0$. Man unterscheidet zwischen eigentlichen Affinsphären erster Art mit $H > 0$ und eigentlichen Affinsphären zweiter Art mit $H < 0$.
$x:M \to A_{n+1}$ heißt uneigentliche Affinsphäre, wenn für die Normalisierung

$y = \frac{1}{n}\,\Delta x = \text{konst.}$ ist. Es folgt $H \equiv 0$.

(3) $x{:}M \to A_{n+1}$ ist genau dann eine Affinsphäre, wenn ∇A symmetrisch ist oder $B_{ij} + HG_{ij} = 0$ gilt.

Wie Schneider in [5] gezeigt hat, erfüllt für Affinsphären die Picksche Invariante die Differentialgleichung

(4) $$\frac{1}{2}\,n\,(n-1)\Delta J = R_{ijkl}R^{ijkl} + R_{ij}R^{ij} + A_{ijk\|l}A^{ijk\|l} - n(n-1)(n+1)H(J + H).$$

Schätzt man das Normquadrat des Krümmungstensors und des Ricci-Tensors gegenüber der Skalarkrümmung R ab [2], erhält man für ΔJ die Ungleichung

(5) $$\Delta J \geqq 2(n+1)\,J\,(J + H) + \frac{2}{n(n-1)}\,A_{ijk\|l}A^{ijk\|l}\,.$$

In [5] fehlt der Faktor $\frac{2}{n(n-1)}$ von $A_{ijk\|l}A^{ijk\|l}$.

Die Standardflächen der affinen Hyperflächentheorie sind die Quadriken, algebraische Flächen zweiter Ordnung. Dabei sind nur das Ellipsoid, eine Schale des zweischaligen elliptischen Hyperboloids und das elliptische Paraboloid elliptisch gekrümmt, die hier im folgenden mit der Bezeichnung Quadrik gemeint sind.

(6) Quadriken sind die einzigen äquiaffin normalisierten Flächen mit verschwindender Pickscher Invariante J ([1], p. 228). Es läßt sich folgendes nachrechnen:

Ist E eine Hyperebene des affinen Raumes A_{n+1}, die mit der Quadrik Q einen nicht leeren Schnitt hat, dann gilt für

a) Ellipsoide und Hyperboloide mit Zentrum $z \in A_{n+1}$:

   $z \in E \Rightarrow E \cap Q$ ist Schattengrenze bei Parallelbeleuchtung

   $z \notin E \Rightarrow E \cap Q$ ist Schattengrenze bei Zentralbeleuchtung,

b) Paraboloide mit äquiaffiner konstanter Normalisierung y:

   y ist parallel zu $E \Rightarrow E \cap Q$ ist Schattengrenze bei Parallelbeleuchtung

   y ist nicht parallel zu $E \Rightarrow E \cap Q$ ist Schattengrenze bei Zentralbeleuchtung.

Umgekehrt sind alle Schattengrenzen einer Quadrik eben.

Bei Ellipsoiden führt ein nicht leerer Schnitt mit einer Ebene stets zu geschlossenen Schattengrenzen, die sich je nach Lage der Ebene E bei Parallelbeleuchtung bzw. bei Zentralbeleuchtung ergeben. Bei Hyperboloiden und Paraboloiden ergeben sich geschlossene Schattengrenzen nur bei Zentralbeleuchtung.

Eine zweite Fragestellung behandelt die Geodätischen auf den Quadriken.

Die totalgeodätischen Untermannigfaltigkeiten des Ellipsoids und Hyperboloids erhält man durch Schnitte mit Ebenen, die durch das Zentrum gehen. Die totalgeodätischen Untermannigfaltigkeiten des Paraboloids erhält man durch Schnitte mit Ebenen, die parallel zum Affinnormalenvektor y sind.
Geschlossene totalgeodätische Untermannigfaltigkeiten besitzt nur das Ellipsoid; diese sind dann eben und Schattengrenze bei Parallelbeleuchtung. In Abschnitt 3 wird gezeigt, das Ellipsoide die einzigen eigentlichen Affinsphären sind, die geschlossene, ebene, totalgeodätische Untermannigfaltigkeiten besitzen.

3. Kennzeichnungen von Quadriken unter den berandeten Affinsphären durch Randbedingungen

Als erstes werden Schattengrenzen bei Parallelbeleuchtung betrachtet.

Satz 1 (Ebene Schattengrenzen bei Parallelbeleuchtung)

Sei M eine n-dimensionale, kompakte Mannigfaltigkeit mit geschlossenem Rand. $x:M \to A_{n+1}$ sei eine konvexe Affinsphäre, der Rand $x(\partial M)$ sei eben und Schattengrenze bei Parallelbeleuchtung. Dann ist x(M) Teil eines Ellipsoids.

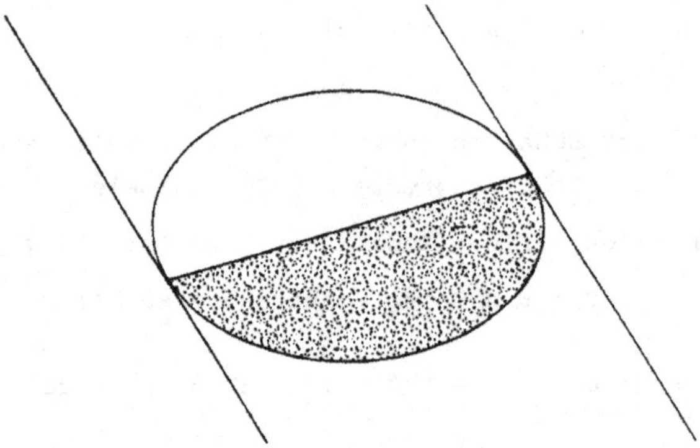

Beweis:

1. Umformulierung der Randbedingungen:

$x(\partial M)$ ist Schattengrenze bei Beleuchtung mit parallelem Licht der Richtung $a \in V_{n+1}$, $a \neq 0$. Für die Funktion $f:M \to \mathbb{R}$, $f := \langle X, a\rangle$ gilt also $f_{|\partial M} \equiv 0$.

Für die Ableitung der Funktion f ergeben sich mit den Ableitungsgleichungen (2) und mit (3)

$$f_i = \langle X_i, a\rangle$$

$$f_{ij} = -A_{ij}{}^r f_r - HG_{ij} f, \quad \Delta f + nHf = 0.$$

f ist also eine Eigenfunktion des Laplace-Operators zum Eigenwert nH, die auf dem Rand verschwindet. Damit ist $H > 0$, und $x:M \to A_{n+1}$ ist eine eigentliche Affinsphäre 1. Art.

O.B.d.A. sei der Ursprung des A_{n+1} in das Zentrum der Affinsphäre gelegt.

$x(\partial M)$ ist eben. $L \in V^*_{n+1}$ charakterisiere die Ebene, in der der Rand $x(\partial M)$ liegt, d.h., es gilt $\langle L,x\rangle$ = konst. auf ∂M. Dabei sei L so gewählt, daß in einem Punkt p von M $0 \neq f(p) = \langle L,x(p)\rangle$ gilt. Für die Funktion

$$h:M \to \mathbb{R}, \quad h := \langle L,x\rangle \text{ gilt also } h\big|_{\partial M} \equiv \text{konst.},$$

und für die Ableitungen ergibt sich analog zu oben

$$h_i = \langle L, x_i\rangle$$

$$h_{ij} = A_{ij}{}^r h_r + G_{ij}\langle L,y\rangle = A_{ij}{}^r h_r - HG_{ij} h, \quad \Delta h + nHh = 0.$$

2. Die Randebene geht durch das Zentrum der Affinsphäre:

Nach Voraussetzung sind $h\big|_{\partial M} \equiv c = \text{konst.}$ und $f\big|_{\partial M} \equiv 0$.

Beide Funktionen $f,h:M \to \mathbb{R}$ sind Eigenfunktionen des Laplace-Operators zum Eigenwert nH, d.h., es gilt $\Delta f + nHf = 0$ und $\Delta h + nHh = 0$.

Damit gilt einerseits

$$\int_M f\Delta h \, do - \int_M \Delta f \, h \, do = \int_M (-nHfh + nHfh) do = 0$$

und andererseits unter Benutzung der Integralsätze

$$\int_M f\Delta h \, do - \int_M \Delta f \, h \, do = -\int_M \langle \text{grad } f, \text{grad } h\rangle do + \int_{\partial M} f\langle \text{grad } h, \nu\rangle d\tilde{o} +$$

$$+\int_M \langle \text{grad } f, \text{grad } h\rangle do - \int_{\partial M} h\langle \text{grad } f, \nu\rangle d\tilde{o} = -c \int_{\partial M} \langle \text{grad } f, \nu\rangle d\tilde{o} .$$

Da $f\big|_{\partial M} \equiv 0$ gilt, ist ∂M eine Niveaufläche von f und grad f ist in Randpunkten parallel zum Normalenvektor ν von ∂M bzgl. M. grad f ist in ∂M nirgends Null, denn sonst würde mit

$$f = 0 = \langle X,a\rangle \text{ und } \bigwedge_{i=1}^{n} 0 = f_i = \langle X,a\rangle$$

die Lichtrichtung a verschwinden. $\langle \text{grad } f,\nu\rangle$ wechselt also auf ∂M nicht das Vorzeichen, $\int_M \langle \text{grad } f,\nu\rangle d\tilde{o} \neq 0$, es ergibt sich $c = 0$. Damit verschwindet h auf ∂M, und die Ebene, in der $x(\partial M)$ liegt, geht durch den Ursprung des A_{n+1}.

3. Die Funktionen $f, h: M \to \mathbb{R}$ fallen zusammen:

Für $h: M \to \mathbb{R}$ gilt also $h|_{\partial M} \equiv 0$ und $\Delta h + nHh = 0$.

Da $x(\partial M)$ Schattengrenze bei Parallelbeleuchtung ist, sind die Voraussetzungen des Satzes 3.2 in [4] erfüllt und es gilt wegen der geeigneten Wahl von L $f \equiv h$ auf M.
Vergleicht man die zweiten kovarianten Ableitungen von f und h

$$f_{ij} = -A_{ij}{}^r f_r - HG_{ij} f \text{ und } h_{ij} = A_{ij}{}^r h_r - HG_{ij} h \text{ ,}$$

so ergeben sich

(7) $A_{ij}{}^r f_r = A_{ij}{}^r h_r \equiv 0$ auf M und

(8) $f_{ij} = -H\, f\, G_{ij}$.

4. Die Picksche Invariante verschwindet:

(9) Auf M gilt $\langle \text{grad } J, \text{grad } f \rangle = 2H\, J\, f$:

$$J_k f^k = \frac{1}{n(n-1)} \nabla_k (A^{irs} A_{irs}) f^k = \frac{2}{n(n-1)} A^{irs}{}_{\|k} A_{irs} f^k \overset{(3)}{=} \frac{2}{n(n-1)} A^{ir}{}_{k\|}{}^{s} A_{irs} f^k$$

$$= \frac{2}{n(n-1)} A_{irs} [\nabla^s \underbrace{(A^{ir}{}_k f^k)}_{=0 \quad (7)} - A^{ir}{}_k f^{ks}] \overset{(8)}{=} \frac{2}{n(n-1)} A_{irs} A^{ir}{}_k G^{ks} Hf = 2JHf \text{ .}$$

Da f auf dem Rand ∂M verschwindet und grad f auf ∂M nirgends verschwindet, ist nach (9) grad J auf ∂M tangential an ∂M, d.h.

(10) $\langle \text{grad} J, \nu \rangle = 0$ auf M und es gilt

(11) $\int_M \Delta J \, do = \int_{\partial M} \langle \text{grad} J, \nu \rangle d\tilde{o} = 0$.

Da $H > 0$ ist, erhält man aus der Differentialgleichung (5)

$$\Delta J \geqq 2(n+1) J (J+H) + \frac{2}{n(n-1)} |\nabla A|^2 \geqq 2(n+1) J^2 \geqq 0.$$

(11) ergibt dann $\Delta J \equiv 0$ und damit $J \equiv 0$ auf M. Damit ist $x: M \to A_{n+1}$ eine der "Hälften", in die ein Ellipsoid durch Schnitt mit einer Ebene durch ihr Zentrum zerfällt.

Bemerkung 1

Für dim M = 2 läßt sich dieser Beweis wesentlich verkürzen: Für zweidimensionale Mannigfaltigkeiten gilt für den Ricci-Tensor

$$RG_{ij} = R_{ij} = \alpha_{ij} + HG_{ij} \text{ .}$$

Damit ergibt sich $\alpha_{ij} = JG_{ij}$ [6]. Mit (7) $A_{ij}{}^r f_r = 0$ auf M erhält man

$$0 = A_{ijr} f^r A^{ij}{}_s f^s = \alpha_{rs} f^r f^s = J|\text{grad } f|^2 \text{ .}$$

Es bleibt zu zeigen, daß $J \equiv 0$ gilt: Angenommen, es ist $J \not\equiv 0$, dann ist die Menge $N := \{p \in M | J(p) > 0\}$ offen und gleichzeitig Nullstellenmenge von $|\text{grad } f|^2$. Damit ist f lokal konstant auf N und mit $\Delta f + nHf = 0$ ergibt sich, daß $f = \langle X,a\rangle$ auf N identisch verschwindet. Dem widerspricht, daß $x:M \to A_{n+1}$ elliptisch gekrümmt ist.

Es soll jetzt gezeigt werden, daß die einzigen Affinsphären, deren Ränder eben und Schattengrenzen bei Zentralbeleuchtung sind, Quadriken sind. Im Gegensatz zur Parallelbeleuchtung sind hier tatsächlich die Fälle $H > 0$, $H = 0$ und $H < 0$ möglich, d.h., es ergeben sich jeweils Ellipsoide, Paraboloide und Hyperboloide.
Zum Beweis werden, wie im Fall der Parallelbeleuchtung zu jeder Randbedingung "eben" bzw. "Schattengrenze bei Zentralbeleuchtung" jeweils angepaßte Funktionen $f,h:M \to \mathbb{R}$ angegeben, von denen dann gezeigt wird, daß sie zusammenfallen. Mit dieser Information kann wieder das Verschwinden der Pickschen Invariante J nachgewiesen werden.
Als erstes werden eigentliche Affinsphären 1. Art betrachtet.

Satz 2 (Ebene Schattengrenzen bei Zentralbeleuchtung, $H > 0$)

Seien M eine n-dimensionale, kompakte Mannigfaltigkeit mit geschlossenem Rand ∂M, $x:M \to A_{n+1}$ eine schwach gekrümmte Affinsphäre mit $H > 0$, der Rand $x(\partial M)$ sei eben und Schattengrenze bei Zentralbeleuchtung. Dann liegt $x(M)$ auf einem Ellipsoid.
Das Lichtzentrum l liegt auf einer Geraden durch das Zentrum der Affinsphäre und den Punkt $x(p_0)$, dessen Tangentialebene parallel zur Randebene ist.

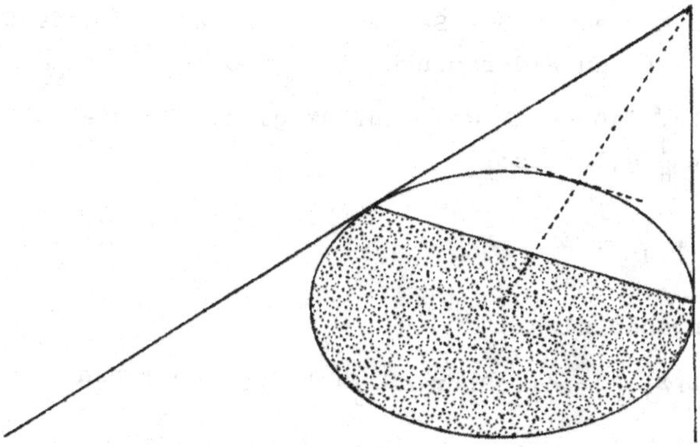

Beweis:
O.B.d.A. sei der Ursprung des A_{n+1} in das Zentrum der Affinsphäre gelegt.
(12) Hinweis: Die in dem ersten und zweiten Teil dieses Beweises gewonnenen Ergebnisse gelten auch für $H < 0$.

1. Umformulierung der Randbedingungen:

$x(\partial M)$ ist Schattengrenze bei Zentralbeleuchtung aus dem Lichtzentrum mit Ortsvektor $l \in V_{n+1}$. $l - x$ ist die Richtung der Verbindungsgeraden von l mit dem Punkt x und stimmt in beleuchteten Punkten mit der Lichtrichtung überein. Auf der Schattengrenze ist $l-x$ tangential. Für die Funktion

$$f:M \to \mathbb{R}, \quad f:= \langle X, l-x\rangle \text{ gilt also } f\big|_{\partial M} \equiv 0.$$

für die Ableitungen ergeben sich mit den Ableitungsgleichungen (2)

$$f_i = \langle X_i, l-x\rangle + \langle X, -x_i\rangle = \langle X_i, l-x\rangle = \langle X_i, l\rangle$$

$$f_{ij} = \langle X_{ij}, l-x\rangle + \langle X_i, -x_j\rangle = -A_{ij}{}^r f_r - HG_{ij} f + G_{ij} \quad \Delta f + nH\, f = n.$$

$x(\partial M)$ ist eben, $L \in V^*_{n+1}$ charakterisiere die Ebene, in der der Rand $x(\partial M)$ liegt, d.h., es gilt $\langle L,x\rangle = c =$ konst. auf ∂M.
Die Ebene des Randes $x(\partial M)$ geht nicht durch das Zentrum der Affinsphäre, denn sonst würde bei der Wahl des Ursprungs des A_{n+1} $\langle L,x\rangle \equiv 0$ auf ∂M gelten und $\langle L,x\rangle$ die Differentialgleichung $\Delta(\langle L,x\rangle) + nH\langle L,x\rangle = 0$ erfüllen. Nach dem Satz 3.3. von Schneider in [4] verschwände $\langle L,x\rangle$ dann identisch auf M.
$\langle L,x\rangle$ hat auch keine Nullstellen in inneren Punkten von M:
Ann.: $\langle L,x\rangle$ hat Nullstellen. Da $L \neq 0$, gilt in Nullstellen von $\langle L,x\rangle$ $\operatorname{grad}(\langle L,x\rangle) \neq 0$. Damit wäre $\{p \in M | \langle L,x(p)\rangle = 0\}$ eine $(n-1)$-dimensionale, geschlossene Untermannigfaltigkeit im Innern von M und Rand der n-dimensionalen Untermannigfaltigkeit $N = \{p \in M | \operatorname{sign}(c)\langle L,x(p)\rangle \leqq 0\}$ mit $c = \langle L,x\rangle\big|_{\partial M}$. Eine erneute Anwendung des Satzes 3.3 in [4] auf jede Zusammenhangskomponente liefert den Widerspruch $\langle L,x\rangle \equiv 0$ auf N.

$L \in V^*_{n+1}$ sei nun so gewählt, daß $\langle L,x\rangle = -\frac{1}{H}$ auf ∂M gilt. Für die Funktion $h:M \to \mathbb{R}$, $h:= \langle L,x\rangle + \frac{1}{H}$ gilt also

$$h\big|_{\partial M} \equiv 0 \quad \text{und} \quad \bigwedge_{p \in M} h(p) \neq \frac{1}{H} \quad ,$$

und für die Ableitungen ergeben sich

$$h_i = \langle L, x_i\rangle, \quad h_{ij} = \langle L, x_{ij}\rangle = A_{ij}{}^r h_r - HG_{ij} h + G_{ij}, \quad \Delta h + nHh = n.$$

2. Die Funktionen $f,h:M \to \mathbb{R}$ fallen zusammen:

Für die Funktion $h:M \to \mathbb{R}$ gilt $\Delta h + nHh = n$ und $h\big|_{\partial M} \equiv 0$.

Da $x(\partial M)$ Schattengrenze bei Zentralbeleuchtung ist und $x:M \to A_{n+1}$ schwach gekrümmt ist, sind die Voraussetzungen des Satzes 3.4 in [4] erfüllt und es gilt

$$h \equiv f \text{ auf } M.$$

Ein Vergleich der zweiten kovarianten Ableitungen von f und h

$$f_{ij} = -A_{ij}{}^{r}f_r + (1-Hf)G_{ij},\quad h_{ij} = A_{ij}{}^{r}h_r + (1-Hh)G_{ij}$$

ergibt

(13) $A_{ij}{}^{r}f_r \equiv 0$ auf M und

(14) $f_{ij} = (1-Hf)G_{ij}$.

3. Die Picksche Invariante verschwindet:

(15) Auf M gilt $\langle \mathrm{grad} J, \mathrm{grad}\, f\rangle = -2J(1-Hf)$:

$$J_k f^k = \frac{1}{n(n-1)} \nabla_k (A_{irs}A^{irs}) f^k = \frac{2}{n(n-1)} A^{irs}A_{irs\|k} f^k$$

$$= \frac{2}{n(n-1)} A^{irs}A_{irk\|s} f^k = \frac{2}{n(n-1)} A^{irs}[\nabla_s \underbrace{(A_{irk}f^k)}_{=0} - A_{irk}f^k{}_s]$$

$$\overset{(3)}{=} \frac{-2}{n(n-1)} A^{irs}A_{irk}G^k{}_s(1-Hf) = -2J(1-Hf) .$$

(16) Als nächstes wird gezeigt, daß $f:M \to \mathbb{R}$ das absolute Maximum auf dem Rand ∂M annimmt, daß also $f \leqq 0$ auf M gilt:

Sei dazu $p_0 \in M^0$ ein Punkt, in dem $f:M \to \mathbb{R}$ ein lokales Extremum annimmt. Da $f\big|_{\partial M} \equiv 0$ und $\bigwedge_{p\in M} f(p) \neq \frac{1}{H}$, gilt wegen der Stetigkeit $\bigwedge_{p\in M} f(p) < \frac{1}{H}$.

Insbesondere in p_0 ergibt sich:

$$f(p_0) < \frac{1}{H} \iff 1 - Hf(p_0) > 0 \iff f_{ij|p_0} \text{ ist positiv definit;}$$

damit ist jeder lokale Extremwert von f, der in inneren Punkten von M angenommen wird, ein Minimum. Da M kompakt ist, wird also das absolute Maximum auf dem Rand ∂M angenommen.

Für eigentliche Affinsphären mit $H > 0$ ergibt sich wieder mit (5) $\Delta J \geqq 0$.

Zusammenfassend kann jetzt folgende Ungleichungskette aufgestellt werden

$$0 \geqq \int_M f\Delta J \, do = \int_M (\nabla^r(fJ_r) - f^r J_r) do$$

$$= \int_{\partial M} \underbrace{f\langle \mathrm{grad}\, J, \nu\rangle d\tilde{o}}_{\overset{(15)}{=}\,0} + \int_M \underbrace{2J(1-Hf)}_{\geqq (16)} do \geqq 2\int_M J\, do \geqq 0$$

Damit erhält man $\int_M J\, do = 0$ und $J \equiv 0$ auf M, $x:M \to A_{n+1}$ liegt auf einem Ellipsoid (6).

Die Lage des Lichtzentrums ergibt sich aus der Betrachtung der Quadriken.

Versucht man den Beweis von Satz 2 auf eigentliche Affinsphären zweiter Art, $H < 0$, zu übertragen, so kommt der Unterschied $H > 0$ zu $H < 0$ wesentlich zum Tragen, wenn gezeigt werden soll: $\Delta J \geqq 0$. Die Identität für ΔJ, die zur betrachteten Ungleichung führt, lautet

$$\frac{1}{2}\, n(n-1)\Delta J = R_{ijkl}R^{ijkl} + R_{ij}R^{ij} + |\nabla A|^2 - n(n-1)(n+1)H(J + H) \quad (4).$$

Von der rechten Seite dieser Gleichung kann man für $H < 0$ nicht ohne weiteres das Vorzeichen erkennen. Mir sind auch keine Abschätzungen bekannt, mit denen man das Normquadrat des Krümmungstensors, des Ricci-Tensors und $|\nabla A|^2$ gut genug vergleichen kann.
Da aber das Ergebnis (13) $A_{ij}{}^{r}f_{r} \equiv 0$ des Beweises von Satz 2 nicht das Vorzeichen von H als Voraussetzung eingeht, kann man für $\dim M = 2$ wie in der Bemerkung 1 schließen, daß J auf M identisch verschwindet und daß $x:M \to A_3$ auf einem Hyperboloid liegt. Es gilt also der folgende

<u>Satz 3 (Ebene Schattengrenzen bei Zentralbeleuchtung, $H < 0$)</u>

Sei M eine zweidimensionale, kompakte Mannigfaltigkeit mit geschlossenem Rand ∂M. $x:M \to A_3$ sei eine schwach gekrümmte Affinsphäre mit $H < 0$, der Rand $x(\partial M)$ sei eben und Schattengrenze bei Zentralbeleuchtung, dann liegt $x(M)$ auf einem Hyperboloid.
Ist die Randbedingung für eine n-dimensionale Affinsphäre erfüllt, dann liegt das Lichtzentrum l auf einer Geraden durch das Zentrum der Affinsphäre und den Punkt $x(p_0)$, dessen Tangentialebene parallel zur Randebene ist; l liegt zwischen dem Zentrum und $x(p_0)$.

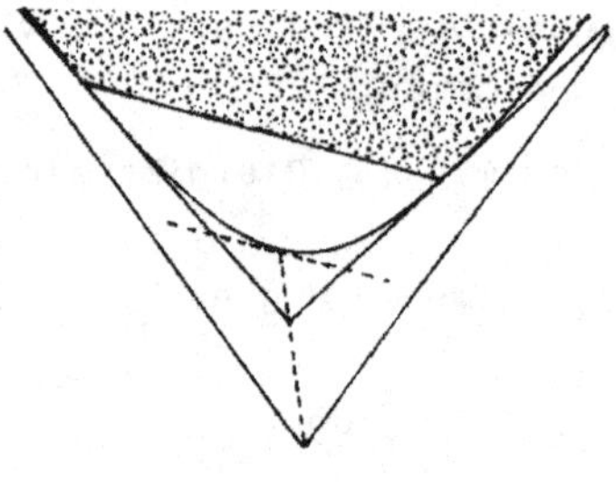

Beweis:

1. Die Picksche Invariante J verschwindet:

Seien $f,h:M \to \mathbb{R}$ wie im Beweis von Satz 2 definiert:

$$f := \langle X, l-x\rangle, \quad f\big|_{\partial M} \equiv 0$$

$$h := \langle L, x\rangle + \frac{1}{H}, \quad h\big|_{\partial M} \equiv 0, \quad \bigwedge_{p\in M} h(p) \neq \frac{1}{H}.$$

Mit dem Hinweis (12) gelten (13), (14)

$$A_{ij}{}^r f_r \equiv 0 \text{ auf } M \text{ und } f_{ij} = (1-Hf)G_{ij}\ .$$

Da dim $M = 2$, gilt wie in Bemerkung 1

$$J|\text{grad } f|^2 \equiv 0 \text{ und } J \equiv 0 \text{ auf } M.$$

Damit liegt $x(M)$ auf einem Hyperboloid.

2. Die Lage des Lichtzentrums:

Für jede Dimension n von M läßt sich aber die Lage des Lichtzentrums bestimmen.

Da $f:M \to \mathbb{R}$ stetig ist und $f\big|_{\partial M} \equiv 0 > \frac{1}{H}$ und $\bigwedge_{p\in M} f(p) \neq \frac{1}{H}$ gelten, erhält man

$$(17)\quad \bigwedge_{p\in M} f(p) > \frac{1}{H},\quad f(p) > \frac{1}{H} \Leftrightarrow 0 < 1-Hf(p)$$

Damit ist jede Hessesche von f positiv definit (14), jeder stationäre Punkt p_0 im Innern von M ist ein lokales Minimum. M ist kompakt, das absolute Maximum von f wird also auf dem Rand ∂M angenommen.

$$f\big|_{\partial M} \equiv 0 \Rightarrow \bigwedge_{p\in M} f(p) \leqq 0$$

Zusammen mit (17) gilt also

$$(18)\quad \bigwedge_{p\in M} \frac{1}{H} < f(p) \leqq 0.$$

Sei $p_0 \in M^0$ ein Punkt, in dem die Tangentialebene parallel zu der Ebene ist, in der der Rand von M liegt; p_0 ist ein stationärer Punkt von f, denn es gilt:

$$f_i\big|_{p_0} = h_i\big|_{p_0} = \langle L, x_i\big|_p\rangle = 0.$$

$$f_i\big|_{p_0} = \langle X_i\big|_{p_0}, l\rangle = 0 \text{ für } i = 1,\dots,n \Rightarrow \bigvee_{\alpha\in\mathbb{R}} l = \alpha x_{p_0}.$$

$$f(p_0) = \langle X_{p_0}, l - x_{p_0}\rangle = (\alpha-1)\langle X_{p_0}, x_{p_0}\rangle = \frac{1}{H}(1-\alpha)$$

$$\left.\begin{aligned} &\tfrac{1}{H} < f(p_0) \Rightarrow \tfrac{1}{H} < \tfrac{1}{H}(1-\alpha) \Rightarrow \alpha > 0 \\ &f(p_0) < 0 \Rightarrow \tfrac{1}{H}(1-\alpha) < 0 \Rightarrow 1 > \alpha \end{aligned}\right\} \Rightarrow 0 < \alpha < 1$$

Für das Lichtzentrum l gilt also $l = \alpha x_{p_0}$ mit $0 < \alpha < 1$.

Mit den gleichen Hilfsmitteln wie für die Fälle $H > 0$ und $H < 0$ läßt sich auch im Fall $H = 0$ zeigen, daß ebene Schattengrenzen nur bei Quadriken, d.h. hier für Paraboloide, möglich sind.

Satz 4 (Ebene Schattengrenzen bei Zentralbeleuchtung, H = O)

Sei M eine n-dimensionale, kompakte Mannigfaltigkeit mit geschlossenem Rand ∂M, $n \geqq 2$. $x: M \to A_{n+1}$ sei eine uneigentliche Affinsphäre, H = O, der Rand $x(\partial M)$ sei eben und Schattengrenze bei Zentralbeleuchtung. Dann liegt x(M) auf einem Paraboloid.
Das Lichtzentrum erhält man auf folgende Art: Man legt durch den Punkt $x(p_0)$, dessen Tangentialebene parallel zur Randebene ist, eine Gerade in Richtung der Affinnormalen. Spiegelt man den Durchstoßpunkt dieser Geraden durch die Randebene an dem Punkt $x(p_0)$, so erhält man das Lichtzentrum l.

Beweis:
O.B.d.A. sei der Ursprung des A_{n+1} in der Ebene, in der der Rand $x(\partial M)$ liegt, gewählt.

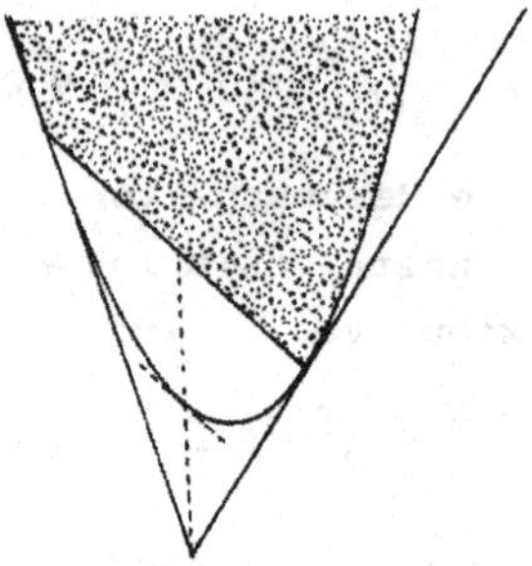

1. Umformulierung der Randbedingungen:

$x(\partial M)$ ist Schattengrenze bei Zentralbeleuchtung aus dem Lichtzentrum mit Ortsvektor $l \in V_{n+1}$, auf dem Rand ∂M gilt $\langle X, l-x\rangle = O$. Für die Funktion $f: M \to \mathbb{R}$, $f := \langle X, l-x\rangle$ erhält man also

$$f\Big|_{\partial M} \equiv O \text{ und } f_i = \langle X_i, l-x\rangle,$$
$$f_{ij} = \langle X_{ij}, l-x\rangle - \langle X_i, x_j\rangle = -A_{ij}{}^r f_r + G_{ij}\ , \quad \Delta f = n\ .$$

$x(\partial M)$ ist eben, d.h., es gibt ein $L \in V^*_{n+1}$ mit $\langle L, x\rangle = O$ auf ∂M. L sei so normiert, daß $\langle L, y\rangle = 1$ ist.
Für die Funktion $h: M \to \mathbb{R}$, $h := \langle L, x\rangle$ erhält man

$$h\Big|_{\partial M} \equiv O \text{ und } h_i = \langle L, x_i\rangle \quad h_{ij} = \langle L, x_{ij}\rangle = A_{ij}{}^r h_r + G_{ij},\ \Delta h = n.$$

2. Die Funktionen $f, h: M \to \mathbb{R}$ fallen zusammen:

Die Randbedingungen erfüllen gerade die Voraussetzungen des Eindeutigkeitssatzes 3.4 in [4], $x(\partial M)$ ist Schattengrenze bei Zentralbeleuchtung und $\Delta h = n$, damit gilt

$f \equiv h$ auf M.

(19) Konsequenzen sind $A_{ij}{}^r f_r \equiv O$ auf M und $f_{ij} = G_{ij}$ ist positiv de-

finit auf M.

3. Die Picksche Invariante J verschwindet:

f_{ij} ist stets positiv definit, damit liegt in jedem stationären Punkt von f ein lokales Minimum vor. Das absolute Maximum nimmt f auf dem Rand ∂M an.

$$\bigwedge_{p\in M} f(p) \leqq 0.$$

(20) Es gilt $\langle \operatorname{grad} J, \operatorname{grad} f\rangle = -2J$:

$$J^k f_k = \frac{2}{n(n-1)} A_{ijr} A^{ijr\|k} f_k \underset{(3)}{=} \frac{2}{n(n-1)} A_{ijr} A^{ijk\|r} f_k$$

$$= \frac{2}{n(n-1)} A_{ijr} [\nabla^r \underbrace{(A^{ijk} f_k)}_{=0} - A^{ijk} f_k{}^r] \underset{(19)}{=} -2J .$$

Für uneigentliche Affinsphären (H=O) ergibt sich mit (5)

$$\Delta J \geqq 0 .$$

Zusammenfassend ergibt sich die Ungleichungskette

$$0 \geqq \int_M f \Delta J \, do \underset{(20)}{=} \int_{\partial M} \underbrace{f}_{=0} \langle \operatorname{grad} J, \nu\rangle \, d\tilde{o} + 2\int_M J \, do \geqq 0 .$$

Mit $J \equiv 0$ und $H = 0$ liegt $x(M)$ auf einem Paraboloid.
Die Lage des Lichtzentrums ergibt sich aus der Betrachtung von Paraboloiden.

Bisher wurden Affinsphären $x:M \to A_{n+1}$ betrachtet, die gleichzeitig folgenden Randbedingungen genügten: Der Rand $x(\partial M)$ sei eben und Schattengrenze. Beide Randbedingungen sind Bedingungen an die Größen der äußeren Geometrie.
Jetzt soll jeweils eine der "äußeren" Randbedingungen ersetzt werden durch eine Randbedingung der inneren Riemannschen Geometrie: Der Rand ∂M sei eine totalgeodätische, geschlossene Untermannigfaltigkeit von M. Die so veränderten Randbedingungen charakterisieren wieder Quadriken. Der folgende Satz zeigt sogar noch mehr.

Satz 5 (Äquivalenz der äußeren Randbedingungen)

Ist M eine n-dimensionale, kompakte Mannigfaltigkeit mit totalgeodätischem, geschlossenem Rand ∂M und ist $x:M \to A_{n+1}$ eine eigentliche Affinsphäre, $H \neq 0$, $n \geqq 2$, dann sind die beiden folgenden Aussagen äquivalent:

a) Der Rand $x(\partial M)$ liegt in einer Ebene, die durch das Zentrum der Affinsphäre geht.

b) Der Rand $x(\partial M)$ ist Schattengrenze bei Parallelbeleuchtung.

Beweis: (a) ⇒ (b)

O.B.d.A. sei der Ursprung des A_{n+1} in das Zentrum der Affinsphäre gelegt.

1. Umformulierung der Randbedingungen:

Liegt der Ursprung in der Randebene, so gibt es ein $L \in V^*_{n+1}$ mit $\langle L,x\rangle = 0$ auf ∂M. Für die Funktion

$$h:M \to \mathbb{R}\,,\quad h:=\langle L,x\rangle \text{ gilt } h\big|_{\partial M} \equiv 0.$$

Entsprechend den vorangehenden Beweisen ergibt sich für die kovarianten Ableitungen von h

$$h_i = \langle L,x_i\rangle$$

(21)

$$h_{ij} = A_{ij}{}^r h_r - HG_{ij}h.$$

Ist der Rand ∂M von M eine totalgeodätische Untermannigfaltigkeit von M, so verschwindet die zweite Fundamentalform α von ∂M bzgl. M.

$\overset{M}{\nabla^2} h$ bezeichne die Hessesche von $h:M \to \mathbb{R}$ und

$\overset{\partial M}{\nabla^2} h$ bezeichne die Hessesche von $h\big|_{\partial M}: \partial M \to \mathbb{R}$.

Sind v,w Tangentialvektorfelder auf ∂M, dann gilt für Untermannigfaltigkeiten

$$\overset{M}{\nabla^2} h(v,w) = \overset{\partial M}{\nabla^2} h(v,w) - dh(\alpha(v,w)).$$

Da ∂M nun totalgeodätisch ist, folgt $\overset{M}{\nabla^2} h\big|_{T\partial M} \equiv \overset{\partial M}{\nabla^2} h$,

und zwischen $\overset{M}{\nabla^2} h\big|_{T\partial M}$ und $\overset{\partial M}{\nabla^2} h$ braucht nicht mehr unterschieden zu werden.

2. Auswertung der Randbedingungen:

Für das folgende seien auf M lokale Koordinaten $u^1,\dots,u^{n-1},u^n$ gegeben, so daß $u^1\big|_{\partial M},\dots,u^{n-1}\big|_{\partial M}$ lokale Korodinaten von ∂M sind und $\frac{\partial}{\partial u^n}\Big|_{\partial M}$ ein normiertes Normalenfeld von ∂M bzgl. M ist.

Indizes $i,j,k,l,\dots$ laufen jeweils von $1,\dots,n$ und

Indizes $\mu,\nu,\dots$ laufen jeweils von $1,\dots,n-1$.

Da $h\big|_{\partial M} \equiv 0$, ist $\overset{\partial M}{\nabla^2} h \equiv 0$, d.h.

$$(22)\quad 0 \equiv h_{\nu\mu} = A_{\nu\mu}{}^{r}h_{r} - HG_{\nu\mu}h \underset{\substack{\uparrow \\ h|_{\partial M}=0}}{=} A_{\nu\mu}{}^{r}h_{r} \text{ auf } \partial M.$$

Zu zeigen ist, daß für die Funktion $h:M \to \mathbb{R}$ die Identität $h = \langle X,a\rangle$ für einen geeigneten konstanten Vektor $a \in V_{n+1}$ gilt. Dazu wird für a der bereits von Blaschke ([1], p.215) benutzte Ansatz

$$(23)\quad a:M \to V_{n+1},\quad a:= -h^{r}x_{r} - Hhx,\ \langle X,a\rangle = h$$

gemacht. Mit einer in der globalen Differentialgeometrie häufig benutzten Integralformelmethode (vergleiche ([3], Kap.3) und ([4],Kap.3)) wird a = konst. gezeigt.

Dazu wird auf M das Vektorfeld

$$\xi^{i} := \varepsilon^{ii_2..i_n}\det(x,a,a_{i_2},x_{i_3},\ldots,x_{i_n})$$

definiert, dabei ist $\varepsilon^{i_1..i_n} = \frac{1}{\sqrt{\det G_{ij}}}\,\mathrm{sign}(i_1,..,i_n)$ der zu der Metrik G_{ij} gehörende Diskriminantentensor.

Nach den Integralsätzen gilt

$$\int_M \operatorname{div}\xi\, do = \int_{\partial M} \langle \xi,\nu\rangle\, d\tilde{o}\ .$$

3. Die Ableitung von $a = -h^{r}x_{r} - Hhx$:

Die Ableitung von a ist tangential zu M, wie die folgende Rechnung zeigt:

$$a_i = -h^{r}{}_{i}x_{r} - h^{r}x_{ri} - Hh_{i}x - Hhx_{i}$$

$$\underset{(21)(2)}{=} -A^{r}{}_{i}{}^{k}h_{k}x_{r} + HG^{r}{}_{i}hx_{r} - h^{r}A_{ri}{}^{l}x_{l} + Hh^{r}G_{ri}x - Hh_{i}x - Hhx_{i}$$

$$(24)\quad = -(2A_{i}{}^{rk}h_{k})x_{r}.$$

4. Das Randintegral $\int_{\partial M} \langle\xi,\nu\rangle\, d\tilde{o}$ verschwindet:

Für das gewählte Koordinatensystem reicht es zu zeigen, daß $\xi^{n} = 0$ auf ∂M gilt.

$$\xi^{n} = \varepsilon^{ni_2..i_n}\det(x,a,a_{i_2},x_{i_3},..,x_{i_n}).$$

$\varepsilon^{ni_2..i_n} = 0$, falls $n \in \{i_2,..,i_n\}$, darum dürfen die Indizes $i_2,..,i_n$ durch $\nu_2,..,\nu_n$ ersetzt werden, die jeweils nur noch den Laufbereich 1 bis n-1 besitzen.

$$\xi^{n} = \varepsilon^{n\nu_2..\nu_n}\det(x,a,a_{\nu_2},x_{\nu_3},..,x_{\nu_n})$$

$$\underset{(23)}{=} -h^{r}\varepsilon^{n\nu_2..\nu_n}\det(x,x_{r},a_{\nu_2},x_{\nu_3},..,x_{\nu_n})$$

Da mit $h|_{\partial M} \equiv 0$ ∂M eine Niveaufläche von h ist, auf der grad h senkrecht steht, ist $h^{\nu} = 0$, $\nu = 1,..,n-1$, auf ∂M.

$$\xi^n = -h^n \varepsilon^{n\nu_2..\nu_n} \det(x, x_n, a_{\nu_2}, x_{\nu_3}, .., x_{\nu_n})$$

$$\underset{(24)}{=} 2A_{\nu_2}{}^{rk} h_k h^n \varepsilon^{n\nu_2..\nu_n} \det(x, x_n, x_r, x_{\nu_3}, .., x_{\nu_n})$$

$\det(x, x_n, x_r, x_{\nu_3}, .., x_{\nu_n}) = 0$, falls $r = n$, darum kann r durch μ ersetzt werden.

$$\xi^n = 2A_{\nu_2}{}^{\mu k} h_k h^n \varepsilon^{n\nu_2..\nu_n} \det(x, x_n, x_\mu, x_{\nu_3}, .., x_{\nu_n}) = 0 \text{ auf } \partial M,$$

da nach (22) $A_{\nu_2}{}^{\mu k} h_k$ auf ∂M verschwindet.

5. $a = -h^r x_r - Hhx$ ist konstant:

Nach dem vierten Schritt ist $\int_M \operatorname{div} \xi do = 0$.

Berechnung von $\operatorname{div}\xi$:

$$\begin{aligned} \operatorname{div}\xi = \xi^i{}_i = \varepsilon^{ii_2..i_n} \Big[& \det(x_i, a, a_{i_2}, x_{i_3}, .., x_{i_n}) \\ & + \det(x, a_i, a_{i_2}, x_{i_3}, .., x_{i_n}) \\ & + \det(x, a, a_{i_2 i}, x_{i_3}, .., x_{i_n}) \quad (*) \\ & + \sum_{k=3}^{n} \det(x, a, a_{i_2}, x_{i_3}, .., x_{i_k i}, .., x_{i_n}) \Big] \quad (*) \end{aligned}$$

Wegen der Symmetrie der zweiten kovarianten Ableitungen a_{ij} bzw. x_{ij} zusammen mit der Schiefsymmetrie des Diskriminantentensors verschwinden die durch (*) gekennzeichneten Terme.

Betrachtung von $\varepsilon^{ii_2..i_n} \det(x_i, a, a_{i_2}, x_{i_3}, .., x_{i_n})$:

$$\det(x_i, a, a_{i_2}, x_{i_3}, .., x_{i_n}) \underset{(24)}{=} -2A_{i_2}{}^{rk} h_k \det(x_i, a, x_r, x_{i_3}, .., x_{i_n})$$

$$\underset{(1)}{=} 2A_{i_2}{}^{rk} h_k \langle X, a\rangle \varepsilon_{iri_3..i_n},$$

damit erhält man

$$\begin{aligned} & \varepsilon^{ii_2..i_n} \det(x_i, a, a_{i_2}, x_{i_3}, .., x_{i_n}) \\ & = 2\langle X, a\rangle \varepsilon^{ii_2..i_n} \varepsilon_{iri_3..i_n} A_{i_2}{}^{rk} h_k \\ & = 2\langle X, a\rangle (n-1)!\, A_r{}^{rk} h_k = 0 . \end{aligned}$$

Betrachtung von $\varepsilon^{ii_2..i_n} \det(x, a_i, a_{i_2}, x_{i_3}, .., x_{i_n})$:

$$\det(x, a_i, a_{i_2}, x_{i_3}, .., x_{i_n}) \underset{(24)}{=} 4A_i{}^{rk} h_k A_{i_2}{}^{st} h_t \det(x, x_r, x_s, x_{i_3}, .., x_{i_n})$$

$$= 4A_i{}^{rk}h_kA_{i_2}{}^{st}h_t\langle X,x\rangle\varepsilon_{rsi_3..i_n} = -4A_i{}^{rk}h_kA_{i_2}{}^{st}h_t\tfrac{1}{H}\varepsilon_{rsi_3..i_n}$$

$$\varepsilon^{ii_2..i_n}\det(x,a_i,a_{i_2},x_{i_3},..,x_{i_n})$$

$$= -4\,\tfrac{1}{H}A_i^{rk}\;h_kA_{i_2}{}^{st}h_t\varepsilon^{ii_2..i_n}\varepsilon_{rsi_3..i_n}$$

$$= -4\,\tfrac{1}{H}A_i{}^{rk}h_kA_{i_2}{}^{st}h_t(G^i{}_rG^{i_2}{}_s - G^i{}_sG^{i_2}{}_r)(n-2)!$$

$$= (n-2)!\;4\,\tfrac{1}{H}\,A_i{}^{rk}h_kA_r{}^{it}h_t = (n-2)!\;4\,\tfrac{1}{H}|A_i{}^{jk}h_k|^2\;,$$

dabei ist $|A_i{}^{jk}h_k|^2$ das Normquadrat des Tensors mit den Koeffizienten $A_i{}^{jk}h_k$.

Zusammenfassend erhält man also

$$\operatorname{div}\xi = (n-2)!\;4\,\tfrac{1}{H}\,|A_i{}^{jk}h_k|^2\quad,$$

(25)
$$\int_M |A_i{}^{jk}h_k|^2 do = 0.$$

(25) ergibt $A_i{}^{jk}h_k = 0$ auf M. Nach (24) verschwinden alle Ableitungen von a, a ist also konstant.

Damit ist nun alles gezeigt:

$$a{:}M \to V_{n+1},\; a{:}= -h^rx_r - Hhx = \text{konst.} \Rightarrow \langle X,a\rangle = h\;.$$

Da $h|_{\partial M} \equiv 0$, ist auch $\langle X,a\rangle \equiv 0$ auf ∂M, der Rand $x(\partial M)$ ist Schattengrenze bei Parallelbeleuchtung mit Licht der Richtung $a \in V_{n+1}$.

Bleibt noch zu zeigen, daß (b) ⇒ (a) gilt. In diesem Fall gilt für die Funktion $f{:}M \to \mathbb{R}$, $f{:}=\langle X,a\rangle$, $f|_{\partial M} \equiv 0$. Ist der Rand ∂M von M eine totalgeodätische Untermannigfaltigkeit, ergibt sich entsprechend dem ersten und zweiten Schritt des Beweises (a) ⇒ (b)

$$f_{\nu\mu} = A_{\nu\mu}{}^kf_k \equiv 0 \text{ auf } \partial M.$$

Zu zeigen ist diesmal, daß es einen konstanten Vektor $L \in V^*_{n+1}$ gibt, so daß $f = \langle L,x\rangle$ gilt. Der Beweis von (a) ⇒ (b) läßt sich auf den neuen Fall übertragen, wenn man die entsprechenden Ansätze

$$L{:}M \to V^*_{n+1},\quad L{:}= -f^rX_r - HfX \text{ und}$$

$$\zeta^i{:}= \varepsilon^{ii_2..i_n}\det{}^*(X,L,L_{i_2},X_{i_3},..,X_{i_n})$$

macht.

Der Beweis von (a) ⇒ (b) läßt sich auch direkt verwenden, wenn man die Immersion $X{:}M \to A^*_{n+1}$ betrachtet. Zu dieser Interpretation werden im Abschnitt 4 noch nähere Ausführungen gemacht.

Damit ist Satz 5 bewiesen.

Zusammen mit Satz 1 ergibt sich das

Korollar

Ist M eine n-dimensionale, kompakte Mannigfaltigkeit mit totalgeodätischem, geschlossenem Rand ∂M und ist $x:M \to A_{n+1}$ eine eigentliche, konvexe Affinsphäre, dann gilt:
Ist der Rand $x(\partial M)$ Schattengrenze bei Parallelbeleuchtung oder liegt $x(\partial M)$ in einer Ebene durch das Zentrum der Affinsphäre, dann ist $H > 0$ und $x(M)$ liegt auf einem "halben Ellipsoid".

Beweis:
Satz 5 und Satz 1.

4. Dualitätsbetrachtungen bei eigentlichen Affinsphären

Ist $x:M \to A_{n+1}$ eine eigentliche Affinsphäre, dann läßt sich $X:M \to A^*_{n+1}$ auch als eigentliche Affinsphäre auffassen, die auf M die gleiche Riemannsche Metrik wie $x:M \to A_{n+1}$ induziert, zu einer eigentlichen Affinsphäre existiert also eine "duale" eigentliche Affinsphäre.

Die im letzten Abschnitt 3 behandelten Randbedingungen:

a) Der Rand ist Schattengrenze

b) Der Rand liegt in einer Hyperebene

sind in dieser Sichtweise zueinander duale Bedingungen. Da x der Konormalenvektor der Tangentialebene von $X:M \to A^*_{n+1}$ ist, kann man $\langle L,x\rangle = 0$ auf ∂M wahlweise als "$x(\partial M)$ liegt in einer Ebene durch den Ursprung" und "$X(\partial M)$ ist Schattengrenze bei Parallelbeleuchtung" interpretieren.
Ebenso ergibt für $c \neq 0$ $c = \langle L,x\rangle$
$\Longleftrightarrow 0 = \langle -\frac{1}{cH}L - X, x\rangle$, daß man $\langle L,x\rangle = c$, $c \neq 0$, auf ∂M wahlweise als "$x(\partial M)$ liegt in einer Ebene, die nicht den Ursprung erhält" oder als "$X(\partial M)$ ist Schattengrenze bei Zentralbeleuchtung" interpretieren kann.
Völlig entsprechende Aussagen ergeben $\langle X,a\rangle = 0$ bzw. $\langle X,a\rangle = c$, $c \neq 0$.

Der Satz 5 aus dem Abschnitt 3 läßt sich nun auch als Satz 5' formulieren:

Satz 5'

Ist M wie in Satz 5 und ist ∂M eine totalgeodätische Untermannigfaltigkeit von M, dann ist $x(\partial M)$ genau dann Schattengrenze bei Parallelbeleuchtung, wenn $X(\partial M)$ Schattengrenze bei Parallelbeleuchtung ist.

Literaturverzeichnis

[1] BLASCHKE, W. Vorlesungen über Differentialgeometrie, Bd. II Affine Differentialgeometrie, Springer Verlag 1923.

[2] CALABI, E. Improper affine hyperspheres of convex type and generalization of a theorem by K. Jörgens, Mich. Math.J.5, 105-126 (1958; Zbl. 113, 301).

[3] HUCK, H. et al. Beweismethoden der Differentialgeometrie im Großen, Lecture Notes in Mathematics 335, Springer Verlag 1973 (Zbl. 258.53029).

[4] SCHNEIDER, R. Zur affinen Differentialgeometrie im Großen I, Math. Zeitschrift 101, 375-506 (1967; Zbl. 156, 201).

[5] SCHNEIDER, R. Zur affinen Differentialgeometrie im Großen II. Über eine Abschätzung der Pickschen Invariante auf Affinsphären, Math. Zeitschrift 102, 1-8 (1967; Zbl. 156.201).

[6] SIMON, U. The Pick invariant in equiaffine Differential Geometry, Abh. Math. Sem. Hamburg, 53, 225-228 (1983; Zbl. 508.53006).

CONFORMAL ORBITS OF ELECTROMAGNETIC RIEMANNIAN CURVATURE TENSORS
ELECTROMAGNETIC IMPLIES GRAVITATIONAL RADIATION

Hans Tilgner

Mathematisches Institut
Technische Universität Berlin

Limastr. 1, 1000 Berlin 37

Abstract

Electromagnetic curvature structures (c.s.) are defined as being bilinear in the two electromagnetic field matrices and electrovac c.s. by having the el.magn. energy-momentum as Einstein tensor. There is a one-parameter familiy of electrovac c.s. having never a component in the space of constant curvatures. Their non-Weyl component is uniquely determined by the el.magn. energy-momentum, and in general they have a Weyl component. It is shown that el.magn. implies gravitational radiation, and coversely that el.magn. gravitational radiation is induced by el.magn. radiation. A structure theory of c.s. is described with morphisms as linear conformal transformations (i.e. Lorentztransformations and dilatations) such that the above properties of c.s., and many others, are orbit properties.

Running title: EL.MAGN. RADIATION CURVATURE

Introduction

General Relativity correlates gravitational sources with curvature of a 4-dimensional pseudo-Riemannian manifold. So the basic mathematical structure is a manifold, each tangent space of which is a pseudo-orthogonal vector space whose bilinear form $\langle , \rangle$ of Minkowskian signature (+---) being the gravitational potentials. The unique torsion-free Levi-Civita-connection ∇, constructed in terms of $\langle , \rangle$ and representing the gravitational field, determines a curvature tensor, whose Riccitensor leads via Einstein's gravitational equations to the energy momentum tensor.
Levi-Civita's curvature tensor fulfills the three axioms (C.1), (C.2) and (C.3) in section 1 for vector fields on the manifold. These axioms induce the well-known symmetries of the corresponding curvature tensors over each tangent space, if one uses the index notation. Actually there is another identity, namely

$$\text{(C.2')} \qquad \langle R(x,y)z,w\rangle = \langle R(z,w)x,y\rangle \quad ,$$

which follows from the other three. More general, given (C.1), any two of the others imply the third. However, it remains to show that for each such axiomatised curvature structure C there is a pseudo-Riemannian manifold admitting it as it's Levi-Civita curvature structure.
In the following we assume as the basic mathematical structure on space-time the pseudo-Riemannian structure and a curvature structure with respect to it. This should be seen as an analogy to a Lie group (Loos' description of symmetric manifolds allows to consider more generally symmetric spaces), where the local tangent structure is the Lie algebra (for symmetric spaces the tangent structures are Lie triples which are much closer to curvature structures than Lie algebras). So the fundamental structure on space-time now is such that it's local version - which one gets by applying the tangent functor - is pseudo-orthogonal curvature described by the three C-axioms. With this view of curvature we need a structure theory as we have one for Lie algebras and -triples and generalizing those. Fortunately this structure theory exists, formulated by Singer and Thorpe, Nomizu, Kulkarni, Kowalski and others. In section 1 a short review of this structure theory is given.

Actually one proceeds in the same way as one does in the case of Lie algebras and -triples. One defines data, in terms of which the curvature structure R is descibed. The data considered are: The Ricciform ρ_R (corresponding to the Killingform of algebras), the Riccitransformation L_R, which connects the Ricciform with <,>, the curvature scalar Sc(R), which is a linear form, the sectional curvature, which, however, in the following algebraic considerations plays no role, - these data are the mathematical data - and the physical datum energy-momentum T_R, which mathematically is the Einsteintensor and only a substitute for L_R. How these data are used in the structure theory of curvature is perfectly described by Singer and Thorpe's theorem in section 1. A corresponding theorem for the physical datum energy-momentum states for the vacuum case [JES p.22]

$$R \in \operatorname{Im} W \quad \Leftrightarrow \quad T_R \text{ vanishes} \quad .$$

However, we feel there is still place for a datum which is an algebra.

A structure theory of curvature spaces needs some idea of "morphism". Given two pseudo-orthogonal vector spaces ($\mathbf{V}$,<,>) and ($\mathbf{V}'$, <,>'), a linear conformal map is a linear isomorphism $\Phi : \mathbf{V} \to \mathbf{V}'$ such that for all x,y in $\mathbf{V}$ $\langle \Phi x, \Phi y\rangle' = \lambda_\Phi \langle x,y\rangle$ for a suitable non-vanishing λ_Φ in the groundfield $\mathbf{R}$. Then for a curvature structure R on $\mathbf{V}$,

$$(\Phi^*R)(x',y') = \Phi R(\Phi^{-1}x',\Phi^{-1}y')\Phi^{-1}$$

defines a new curvature structure Φ^*R on $\mathbf{V}'$, and we have a covariant functor from the category of pseudo-orthogonal vector spaces into that of (pseudo-orthogonal) curvature spaces. Using (C.2) one proves that a linear map Φ is curvature preserving in in this sense if and only if it is linear conformal. Conformal maps were studied extensively, for instance by Kulkarni. Jordan, Ehlers, Kundt and Sachs sugguested them as a classification group for solutions of the gravitational field equations.

A better insight into the structure theory can be gained by changing the formulation a little: Given a solution R of the three C-axioms, $R(x,y)z = [x,y,z]$ defines a trilinear composition on $\mathbf{V}$, i.e. a triple, and (C.1) and (C.3) mean that the first two axioms of a Lie triple [Lo p.78] hold for [,,], whereas (c.2) means that the triple is compatible with $\langle,\rangle$. The missing data for the curvature structure R now should be a standard embedding algebra, which becomes a Lie algebra exactly for the homogeneous case [Lo p.79] and [Ti p.1120].

In section 2 the curvature morphisms are specialized to automorphisms of the space of curvature structures (which is 20-dimensional for a 4-dimensional manifold) on a given pseudo-orthogonal vector space. The pseudo-orthogonal automorphism group of this space is denoted by $O(\mathbf{V},\langle,\rangle)$, its Lie algebra by $so(\mathbf{V},\langle,\rangle)$, and the *linear conformal* group **G** is its direct product with the dilatations. Hence $G \in \mathbf{G}$ means that there is a non-vanishing λ_G in **R** such that $\langle Gx,Gy\rangle = \lambda_G^2\langle x,y\rangle$ for all x,y in **V**.

The sections 2 and 3 give the mathematical setting of the theory, in section 4 the physical basis, i.e. the field equations are formulated in a coordinate-free way. In the other sections we study special properties of curvature structures, all of them orbit properties with respect to the group **G**. They roughly may be divided into mathematical and physical ones. Among the mathematical ones, irreducibility and semi-simplicity are only mentioned. More emphasis is laid on structures which are defined by semi-simple Lie- and Jordan algebras with respect to their standard bilinear forms. In these cases an integration to a manifold with curvature is known. For Lie algebras the manifold is just the Lie group with the Killingform $\langle,\rangle$. For semi-simple Jordan algebras, the integration is even simpler, leading to the interior of a null-cone as one connectivity component of the set of all invertible elements, a so-called ω-domain [Lo II p.176]. For a symmetric curvature structure there still is a wellknown integration to a manifold. The curvature there is a pseudo-orthogonal Lie triple which is embedded in a Lie algebra, its standard embedding algebra. By Ado's theorem this Lie algebra may be taken as a matrix algebra, and the exponential series generates a connected matrix Lie group, such that the symmetric space is a homogeneous space of this group, factorized by some stability subgroup. The result in general is not simply connected. If not, it may be difficult to determine the other covering spaces, admitting the given curvature structure as the canonical one. If the result is simply connected, the discrete center of the symmetric space leads by factorization to the other covering spaces. At least one straightforeward globalization exists. Thus a symmetric pseudo-orthogonal curvature structure means integrability. In the

more general cases integration is an open problem.
Physical properties to be discussed, are Friedmann, electromagnetic and gravitational radiation curvature and combinations of these. Nomizu has constructed curvature structures in terms of two $\langle , \rangle$-self-adjoint matrices. This construction transports the structure theory of Kern$W$ curvature to the algebra of these matrices. In the same (bilinear) way we find curvature structures given in terms of two $\langle , \rangle$-skewadjoint matrices, which in four dimensions can be interpreted as the two field matrices of electromagnetism. That this interpretation makes sense can be seen from two facts. We show that these electromagnetic curvatures link electromagnetic and gravitational radiation (section 11) in the following sense: Electromagnetic induces gravitational radiation and conversely, if gravitational radiation is induced by an electromagnetic field, this field must be a radiation field. The radiation concepts are taken entirely from Lichnerowicz. Secondly, for these el.magn. curvatures we can prove Rainich's result that there is a one-parameter schar of electrovac curvature structures.
The discussion of curvature is local, i.e. in tangent spaces $\mathbf{V}$ of of pseudo-Riemannian manifolds. However, globalization is simply to take the principle $\mathbf{G}$-bundle over the manifold and to associate curvature bundles (or less orbit bundles) with respect to the $\mathbf{G}$-action described in section 2.
Up to section 8 we only have to assume $n > 2$. In the sections 8 and 9 we assume $n = 4$, though straightforeward generalizations exist. Section 10 needs an indefinite form $\langle , \rangle$, and in section 11 we specialize to 4 dimensions and Minkowski signature (+---). We define $g(x,y)z = \langle y,z\rangle x$ and frequently use that the trace of the linear map $g(x,y)$ is $\langle x,y\rangle$.
We should remark here, that the notion of a symmetric space used here, is not identical with that used in physics, for instance in Weinberg's chapter 13. There it means that space-time is a transformation (not necessarily homogeneous) space of a group. The physical reference books mainly used here, were [MTW], [SW], [W] and [F], the latter giving a curvature based representation of general relativity.

1. DECOMPOSITION OF CURVATURE SPACES

Let $so(\mathbf{V},<,>)$ be the pseudo-orthogonal Lie algebra. *A curvature structure* on $(\mathbf{V},<,>)$ is a bilinear map $R : \mathbf{V}\times\mathbf{V} \longrightarrow \text{end}\,\mathbf{V}$ such that

(C.1) $R(y,x) = -R(x,y)$,

(C.2) $R(x,y) \in so(\mathbf{V},<,>)$,

(C.3) $R(x,y)z + R(z,x)y + R(y,z)x = 0 \quad \forall\ x,y,z \in \mathbf{V}$.

[SINGER & THORPE, NOMIZU, KULKARNI, GRAY]. (C.2) can be changed to $\langle R(x,y)z,w\rangle = \langle R(z,w)x,y\rangle$. $curv(\mathbf{V},<,>)$ will be the linear space of such curvature structures. The *Ricciform* of R is the symmetric bilinear form $\rho_R(x,y) = \text{trace}(z \mapsto R(z,x)y)$, the *Riccitransformation* L_R is given by $\langle L_R x,y\rangle = \rho_R(x,y)$, the *curvature scalar* is $Sc(R) = \text{trace}\,L_R$. L_R is $<,>$-selfadjoint, i.e. in (the Jordan algebra with respect to the anticommutator) $JA(\mathbf{V},<,>) = \{A/\langle Ax,y\rangle = \langle x,Ay\rangle \quad \forall\ x,y\}$. Examples of c.s. are the *trivial* $R_o(x,y)z = \langle y,z\rangle x - \langle x,z\rangle y$ and

$$R_o^{A,B}(x,y) = \frac{1}{2}(R_o(Ax,By) + R_o(Bx,Ay)),\ \forall\ A,B \in JA(\mathbf{V},<,>).$$

The linear map $\Omega : A \longmapsto R_o^{A,id}$ from $JA(\mathbf{V},<,>)$ into the *curvature space* $curv(\mathbf{V},<,>)$ is an injection. The *Riccimap*

$$Ri : R \longmapsto \Omega(L_R) = R_o^{L_R,id}$$

is an endomorphism of $curv(\mathbf{V},<,>)$ and Sc is a linear form on it. The *Weyl*-resp. the *Einstein*-projector on $curv(\mathbf{V},<,>)$ are

$$WR = R - \frac{2}{n-2}Ri(R) + \frac{Sc(R)}{(n-1)(n-2)}R_o\ ,\quad ER = R - \frac{Sc(R)}{n(n-1)}R_o$$

with $n = \dim\mathbf{V}$. It is well known that $\rho_{WR} = 0$, $L_{WR} = 0$, $Sc(WR) = 0$, $W\circ\Omega = 0$ (especially $WR_o = 0$). Moreover $W\circ Ri = Ri\circ W = 0$ and $E\circ Ri = Ri\circ E$. $W, E, E\circ W$ and $E - W$ are projectors on $curv(\mathbf{V},<,>)$. From $id = E^{\perp} \oplus E - W \oplus W$ we get

$$curv(\mathbf{V},<,>) = \mathbf{R}\,R_o \oplus Im(E-W) \oplus Im\,W\ . \qquad (*)$$

Ω resp. $L : R \longmapsto L_R$, $L : curv(\mathbf{V},<,>) \longrightarrow JA(\mathbf{V},<,>)$ are linear bijections onto resp. from $\text{Kern}\,W = \mathbf{R}\,R_o \oplus Im(E-W)$ with $\Omega(id_{\mathbf{V}}) = R_o$ and $L_{R_o} = (n-1)id_{\mathbf{V}}$. The differentialgeometric meaning of the direct decomposition (*) is given by the THEOREM [SINGER & THORPE] : $R \in Im\,W \Leftrightarrow \rho_R$ vanishes.

$R \in \mathbb{R}\, R_o \Leftrightarrow$ the sectional curv. of R is constant,

$R \in \mathbb{R}\, R_o \oplus \operatorname{Im} W \Leftrightarrow L_R = \mathrm{id}_{\mathbb{V}}$ (*Einsteinian* case).

$R \in \operatorname{Im}(E - W) \oplus \operatorname{Im} W \Leftrightarrow \operatorname{Sc}(R) = 0$.

2. THE ACTION OF $\mathbb{G}$ AND ITS LIE ALGEBRA ON CURV($\mathbb{V}$,<,>)

is given by

$$(G\cdot R)(x,y) = GR(G^{-1}x, G^{-1}y)G^{-1}, \quad x,y \in \mathbb{V},\ G \in \mathbb{G},$$

and on JA($\mathbb{V}$,<,>) by $G\cdot A = GAG^{-1}$. From $\rho_{G\cdot R}(Gx,Gy) = \rho_R(x,y)$ we get $G\cdot L_R = \lambda_G^2 L_{G\cdot R}$, $\operatorname{Sc}(G\cdot R) = \lambda_G^{-2}\operatorname{Sc}(R)$ and $G\cdot Ri(R) = Ri(G\cdot R)$. Hence Sc is $\mathbb{G}$-equivariant and even invariant with respect to the pseudo-orthogonal subgroup.

$$G\cdot R_o^{A,B} = \lambda_G^{-2}\, R_o^{G\cdot A, G\cdot B}$$

implies $G\cdot\Omega(A) = \lambda_G^{-2}\,\Omega(G\cdot A)$ and $G\cdot R_o = \lambda_G^{-2} R_o$. The kernel of this *curvature representation* of $\mathbb{G}$ is $\mathbb{Z}_2$. Like *Ri* the projectors *W* and *E* commute with the action of $\mathbb{G}$. Hence (*) is a $\mathbb{G}$-invariant direct decomposition into $\mathbb{G}$-modules and it suffices to classify $\mathbb{G}$-orbits in the three subspaces. The Lie algebra of $\mathbb{G}$ is $\mathbb{LG} = \{Q / \langle Qx,y\rangle + \langle x,Qy\rangle = \mu_Q\langle x,y\rangle, \mu_Q \in \mathbb{R},\ \forall\, x,y\}$, its action on curv($\mathbb{V}$,<,>) is given by

$$(Q\cdot R)(x,y) = [Q,R(x,y)] - R(Qx,y) - R(x,Qy), \qquad (\Diamond)$$

and on JA($\mathbb{V}$,<,>) is $Q\cdot A = [Q,A]$. Again *Ri*, *W* and *E* commute with the action of $\mathbb{LG}$ and (*) is a decomposition into faithful $\mathbb{LG}$-modules. It is easy to check that

$$G\cdot(Q\cdot R) = (G\cdot Q)\cdot(G\cdot R), \qquad (\dagger)$$

which shows that $G\cdot Q$ annihilates $G\cdot R$ if Q annihilates R.

3. $\mathbb{G}$-TRANSFORMS OF THE ENERGY-MOMENTUM TENSOR

For given R the *gravitational equations* are

$$\rho_{ER}(x,y) - \frac{\operatorname{Sc}(R)}{n}\langle x,y\rangle = -\kappa\langle T_R x,y\rangle$$

where $T_R = -\kappa^{-1}(L_R - 2\frac{\operatorname{Sc}(R)}{n}\mathrm{id}_{\mathbb{V}}) \in$ JA($\mathbb{V}$,<,>) is the *energy-momentum matrix*. From $-\kappa$ trace $T_R = -\operatorname{Sc}(R)$ we get

$$-\rho_R(x,y) = \kappa(\langle T_R x,y\rangle - \frac{2}{n}\operatorname{trace}(T_R)\langle x,y\rangle).$$

In Kern $W = \mathbb{R}\, R_o \oplus \operatorname{Im}(E - W)$ we have $T_{\Omega(A)} = 0$ iff $A = 0$.

The transformation property of T_R under $\mathbb{G}$ resp. $\mathbb{LG}$ is

$$G\cdot T_R = \lambda_G^2\, T_{G\cdot R} \quad \text{resp.} \quad Q\cdot T = T_{Q\cdot R} + 2\mu_Q\, T_R,$$

which shows how to transport the solution of the gravitational equations on a space-time to that on another perhaps simpler but conformally equivalent space-time.

4. PSEUDO-RIEMANNIAN SYMMETRIC CURVATURE STRUCTURES

Among the homogeneous-reductive curvature structure the pseudo-riemannian *symmetric* ones are characterized by the additional axiom

$$[R(u,v),R(x,y)] = R(R(u,v)x,y) + R(x,R(u,v)y). \qquad (\Pi)$$

It is fulfilled for R_o for which it reduces to the commutation relations of $so(\mathbb{V},<,>)$ which is generated linearly by the $R_o(x,y)$. (C.1), (C.3) and (Π) are the axioms of a *Lie triple* which is *pseudo-orthogonal* from (C.2). (Π) has the equivalent form

$$R(u,v)\cdot R = 0 \quad \text{(use (C.2) and } (\Diamond) \text{ for a proof)},$$

which can be used (i) to show that symmetric curvature structures lie in symmetric orbits: $(G\cdot R(u,v))\cdot(G\cdot R) = G\cdot(R(u,v)\cdot R)$ and (ii) to characterize symmetric orbits in Kern $W = \mathrm{Im}\,\Omega$:
$0 = \Omega(A)(u,v) = \Omega(\Omega(A)(u,v)\cdot A) = \Omega([\Omega(A)(u,v),A])$
which implies $A^2 = \lambda\, id_{\mathbb{V}}$ with $\lambda \in \mathbf{R}$. In the Minkowski case there may be nilpotent A. $\mathbf{R}\, R_o$ contains three symmetric orbits of type λR_o with $\lambda \gtreqless 0$. The usual description of a symmetric space can be given in terms of the *standard embedding*: If $\mathbb{H}$ is Lie algebra spanned by the $R(x,y)$, $\mathbb{H} \oplus \mathbb{V}$ is a Lie algebra:

$$[Q \oplus x,\; P \oplus y] = [Q,P] + R(x,y) \oplus Qy - Px ,$$

and $Q + x \longmapsto Q - x$ is an involutive automorphism which decomposes $\mathbb{H} \oplus \mathbb{V}$ into the eigenspaces $\mathbb{H}$ and $\mathbb{V}$ of eigenvalues ± 1. The symmetric space then is of the form $\mathbb{F}/\mathbb{H}$ where $\mathbb{F}(\mathbb{H})$ is now a Lie group with Lie algebra $\mathbb{H} \oplus \mathbb{V}(\mathbb{H})$.

5. SYMMETRIC ORBITS DEFINED BY JORDAN ALGEBRAS

Given a Jordan algebra $\mathbb{V}$ with composition $\{,\}$,
$\{x,\{y,z\}\} - \{y,\{x,z\}\} =: R(x,y)z$ defines a Lie triple on $\mathbb{V}$.
If in addition $R(x,y) \in so(\mathbb{V},<,>)$, R is a *Jordan algebra* c.s. on $(\mathbb{V},<,>)$. Writing $L(x)y = \{x,y\}$ and $<x,y> = \mathrm{trace}\,L(\{x,y\})$,

examples are given by semisimple Jordan algebras. Especially for any $t \in V$, $\{x,y\} := \langle x,t\rangle y + \langle y,t\rangle x - \langle x,y\rangle t$ defines a Jordan algebra on $(\mathbb{V},\langle,\rangle)$ and a curvature structure in Kern $W = \operatorname{Im}\Omega$ by

$$ {}^tR_o(x,y) = \langle x,t\rangle R_o(t,y) + \langle t,y\rangle R_o(x,t) - \langle t,t\rangle R_o(x,y) $$

with $G \cdot {}^tR_o = {}^{Gt}R_o$ and ${}^tR_o = -\langle t,t\rangle \Omega(J_t)$, where $J_t z = z - 2\langle t,t\rangle^{-1}\langle t,z\rangle t$ is the *standard reflection at* t. The associated global symmetric space is the set of points outside the null-cone in $\mathbb{V}$ with respect to the symmetric composition

$$ x \,\square\, y = 2\,\frac{\langle x,y\rangle}{\langle y,y\rangle}\,x - \frac{\langle x,x\rangle}{\langle y,y\rangle}\,y\ . $$

For any Jordan algebra curvature structure the associated symmetric spaces are flat; in fact it can be identified to the open subset of invertible elements in the given Jordan algebra. A nec. and suff. condition for a Lie triple to be a Jordan algebra was derived by Jordan and Matsushita. For any $R \in \operatorname{curv}(\mathbb{V},\langle,\rangle)$ we can define ${}^tR$ as above by dropping the index zero. A nec. and suff. condition for ${}^tR \in \operatorname{curv}(\mathbb{V},\langle,\rangle)$ then in $\operatorname{Im}\Omega$ is $At = \lambda t$, and in this case ${}^t\Omega(A) = ({}^tR_o)^{A,id}$. For $G \in \mathbf{G}$, $\{\!\!\{x,y\}\!\!\} := G\{G^{-1}x, G^{-1}y\}$ defines an isomorphic J.a. on $\mathbb{V}$ and $(G \cdot R)(x,y)z = G\{G^{-1}x,\{G^{-1}y, G^{-1}z\}\} - G\{G^{-1}y,\{G^{-1}x, G^{-1}z\}\} = \{\!\!\{x,\{\!\!\{y,z\}\!\!\}\}\!\!\} - \{\!\!\{y,\{\!\!\{x,z\}\!\!\}\}\!\!\}$ shows that $G \cdot R$ again is a J.a. curvature structure. Hence there are Jordan algebraic orbits among the symmetric ones.

6. SYMMETRIC ORBITS DEFINED BY LIE ALGEBRAS, REDUCTIVE AND IRREDUCIBLE ORBITS

Analogously a Lie algebra $(\mathbb{V},[\,,])$ defines a symmetric curvature structure R by $R(x,y)z = [[x,y],z]$ if in addition (C.2) holds. Taking for $\langle,\rangle$ the Killingform, a semisimple Lie algebra defines a curvature structure by $R(x,y) = \operatorname{ad}([x,y])$. A condition for a Lie triple to be a Lie algebra seems to be unknown; it should single out the dimensions 1,3,7 for R_o, in which cases R_o indeed describes a Lie algebra. Like in 5. one proves that to be a Lie algebra is an orbit property.

Reductive **G**-orbits are defined as generalizations of symmetric ones. One has to modify (Δ) by adding on the right hand side a *torsion* term $t(x,y)$, $t : \mathbb{V} \times \mathbb{V} \longrightarrow \mathbb{V}$ being bilinear and skew. (r,t) then fulfill the axioms of the *canonical* curvature and torsion which can be deduced from the Lie algebra ones. The *pseudo-riemannian reductive* curvature structure then is given by a certain linear combination $t(t(.,.),.)$'s and $R(.,.).$'s. **G** operates on t by

$$(G\cdot t)(x,y) \;=\; G\,t(G^{-1}x, G^{-1}y)$$

and the corresponding action on the pseudo-riemannian curvature structures collects reductive curvature structure in *reductive* **G**-orbits [NOMIZU].

Semisimple orbits are characterized by a nondegenerate Ricci-form, which again is an orbit property. Clearly semisimple Lie-, Jordan algebras and Lie triples lie in semisimple orbits. In $\mathrm{Im}\,\Omega$ semisimplicity is equivalent to

$$\det\left(A + \frac{1}{n-2}\,\mathrm{trace}(A)\,\mathrm{id}_{\mathbb{V}}\right) \neq 0 \quad .$$

Irreducible orbits are characterized by an irreducible action of the linear span of the $R(x,y)$'s on **V**.

So far all properties of orbits were entirely mathematical. They may facilitate the classification of orbits, however, in general have no direct physical meaning. In the following we describe two physical orbit properties, namely the Robertson-Walker and the electromagnetic ones.

7. ROBERTSON-WALKER OR FRIEDMANN G-ORBITS

They lie in $\mathrm{Im}\,\Omega$ and are defined for any *velocity* $u \in \mathbb{V}$ by

$$A_u = \frac{(3n-4)\rho - (n-4)p}{n(n-1)(n-2)}\,\mathrm{id}_{\mathbb{V}} - \frac{2}{n-2}(\rho + p)\frac{\langle u,\cdot\rangle}{\langle u,u\rangle}u \quad \in \; JA(\mathbb{V},\langle,\rangle) \;,$$

ρ and p in $\mathbb{R}$ being *density* and *pressure* of a relativistic perfect fluid. Ricciform and energy-momentum transformation are

$$\rho_{\Omega(A_u)}(x,y) = \frac{2}{n}(\rho - p)\langle x,y\rangle - (\rho + p)\frac{\langle u,x\rangle\langle u,y\rangle}{\langle u,u\rangle} \;,$$

$$-T_{\Omega(A_u)} = \frac{4}{n}\,p\,\mathrm{id}_{\mathbb{V}} - (\rho + p)\frac{\langle u,\cdot\rangle}{\langle u,u\rangle}u \quad .$$

One proves that for the *equation of state* $(2+n)p = (2-n)\rho$

the Ricciform becomes degenerate with kernel $\mathbb{R}u$. In the other cases $\Omega(A_u)$ is semisimple. It is easy to see that

$$G\cdot A_u = A_{Gu} \qquad \text{for all } G \in \mathbb{G} ,$$

which shows that Friedmann orbits are parametrized by density and pressure, whereas the (normalized) velocity depends on the point in such an orbit. The condition that $\Omega(A_u)$ is symmetric gives besides $p = -\rho$ the same equation of state as above for the non-semisimple case. Hence the reflection

$$A_u = \frac{4\rho}{(n-2)(n+2)} \left(id_{\mathbb{V}} - 2 \frac{\langle u,\cdot\rangle}{\langle u,u\rangle} u \right)$$

describes a symmetric but not semisimple orbit.
There are three specializations for this $3p = \rho$ case ($n = 4$) according to the sign of $\langle u,u\rangle$. For the lightlike case $A_u = \text{const.}\langle u,\cdot\rangle u$ describes a homogeneous space of a six-dimensional solvable Lie group.

8. ELECTROMAGNETIC CURVATURE STRUCTURES

Recall that Maxwells equations (in arbitrary media) can be formulated in term of the two skew fieldmatrices

$$F^* = \begin{bmatrix} 0 & -\frac{1}{c}E_1 & -\frac{1}{c}E_2 & -\frac{1}{c}E_3 \\ \frac{1}{c}E_1 & 0 & B_3 & -B_2 \\ \frac{1}{c}E_2 & -B_3 & 0 & B_1 \\ \frac{1}{c}E_3 & B_2 & -B_1 & 0 \end{bmatrix}, \quad F = \begin{bmatrix} 0 & -cD_1 & -cD_2 & -cD_3 \\ cD_1 & 0 & -H_3 & H_2 \\ cD_2 & H_3 & 0 & -H_1 \\ cD_3 & -H_2 & H_1 & 0 \end{bmatrix},$$

which in general are not linearly dependent. However, if I is the matrix of $\langle , \rangle$ in a given basis, Maxwell's equations can also be formulated in terms of the matrices

$$F^* I = P \quad \text{and} \quad I F = Q \quad ,$$

which no longer are skew, but in the Lorentz Liealgebra, i.e. fulfill $\langle Px,y\rangle = -\langle x,Py\rangle$ and the same for Q. The symmetric energy-momentum tensor of F and F* is

$$T_{sym} = \begin{bmatrix} W & \frac{1}{2}\left(\frac{1}{c}\vec{E}\times\vec{H} + c\vec{D}\times\vec{B}\right) \\ \frac{1}{2}\left(\frac{1}{c}\vec{E}\times\vec{H} + c\vec{D}\times\vec{B}\right) & [\;\;] \end{bmatrix} ,$$

where $W = \frac{1}{2}\{(\vec{E},\vec{D}) + (\vec{B},\vec{H})\}$ and [ ] is the symmetric matrix

$$\begin{bmatrix} W-E_1D_1-B_1H_1 & -\frac{1}{2}(E_1D_2+B_2H_1+E_2D_1+B_1H_2) & -\frac{1}{2}(E_1D_3+B_3H_1+E_3D_1+B_1H_3) \\ & W-E_2D_2-B_2H_2 & -\frac{1}{2}(E_2D_3+B_3H_2+E_3D_2+B_2H_3) \\ & & W-E_3D_3-B_3H_3 \end{bmatrix}$$

of electromagnetic stresses. T_{sym} reduces to the ordinary energy-momentum for isotropic media [MISNER, THORNE, WHEELER p. 141]. Now we have the representation

$$\frac{1}{4}\,\mathrm{trace}(PQ)\,\mathrm{id}_4 - \{P,Q\} \;=\; I\,T_{sym} \qquad (\S)$$

for the energy-momentum tensor in terms of a traceless $JA(\mathbf{V},<,>)$ matrix and $I = \mathrm{diag}(1,-\mathrm{id}_3)$.

To find curvature structures such that the corresponding energy-momentum matrix is the electromagnetic one, we try to find curvature structures in terms of P and Q, having im mind the construction of $R_o^{A,B}$ in terms of the <,>-selfadjoint A,B. A straightforeward argumentation leaves as only elements of $\mathrm{curv}(\mathbf{V},<,>)$ which involve P and Q in a bilinear way

(i) $\mathrm{trace}(PQ)\,R_o$ which is $\Omega(\mathrm{trace}(PQ)\,\mathrm{id}_{\mathbf{V}})$,

(ii) $\Omega(\{P,Q\})$ - note that if $P,Q \in so(\mathbf{V},<,>)$ then the anticommutator of P and Q is <,>-selfadjoint - ,

(iii) $H_{P,Q}(x,y) \;=\; R_o^{P,Q}(x,y) - \langle Px,y\rangle Q - \langle Qx,y\rangle P$.

Like $R_o^{A,B}$, $H_{P,Q}$ is symmetric in P,Q. But is there something for $H_{P,Q}$ like the map Ω for $R_o^{A,B}$? Ricciform and -transformation and curvature scalar (three mathematical data) are in the three cases

(i)
$$\rho_R(x,y) = (n-1)\,\mathrm{trace}(PQ)\,\langle x,y\rangle$$
$$L_R = (n-1)\,\mathrm{trace}(PQ)\,\mathrm{id}_{\mathbf{V}}$$
$$Sc(R) = n\,(n-1)\,\mathrm{trace}(PQ)$$

(ii)
$$\rho_{\Omega(\{P,Q\})}(x,y) = \frac{1}{2}\langle((n-2)\{P,Q\} + \mathrm{trace}(PQ)\,\mathrm{id}_{\mathbf{V}})x,y\rangle$$
$$L_{\Omega(\{P,Q\})} = \frac{1}{2}((n-2)\,\{P,Q\} + \mathrm{trace}(PQ)\,\mathrm{id}_{\mathbf{V}})$$
$$Sc(\Omega(\{P,Q\})) = (n-1)\,\mathrm{trace}(PQ)$$

(iii)
$$\rho_{H_{P,Q}}(x,y) = \frac{3}{2}\left(\langle Px,Qy\rangle + \langle Qx,Py\rangle\right)$$
$$L_{H_{P,Q}} = -3\,\{P,Q\}$$
$$Sc(H_{P,Q}) = -3\,\mathrm{trace}(PQ) \quad .$$

Calculating the Einstein- and Weylprojections of $H_{P,Q}$ gives the

the direct decomposition

$$H_{P,Q} = -3\,\frac{\mathrm{trace}(PQ)}{n(n-1)}\,R_o \oplus \frac{6}{n-2}\,\Omega\left(\frac{1}{n}\mathrm{trace}(PQ)\,\mathrm{id}_{\mathbf{V}} - \{P,Q\}\right) \oplus WH_{P,Q} \quad (\oplus)$$

of the new curvature structure, which shows that it has a component in each of the three invariant subspaces.
The physical datum T_R can be calculated from the gravitational equations for the Riccitransformations

$$L_R - 2\,\frac{Sc(R)}{n}\,\mathrm{id}_{\mathbf{V}} = -\kappa\, T_R \quad .$$

We get in the three cases above

$$\text{(i)} \quad -\kappa\, T_{\mathrm{trace}(PQ) R_o} = -(n-1)\,\mathrm{trace}(PQ)\,\mathrm{id}_{\mathbf{V}} \; ,$$

$$\text{(ii)} \quad -\kappa\, T_{\Omega(\{P,Q\})} = \frac{1}{2}(n-2)\{P,Q\} - \frac{3n-4}{2n}\,\mathrm{trace}(PQ)\,\mathrm{id}_{\mathbf{V}} \; ,$$

which for $n = 4$ is $\{P,Q\} - \mathrm{trace}(PQ)\,\mathrm{id}_4$,

$$\text{(iii)} \quad -\kappa\, T_{H_{P,Q}} = -3\{P,Q\} + \frac{6}{n}\,\mathrm{trace}(PQ)\,\mathrm{id}_{\mathbf{V}} \quad .$$

We are looking now for a curvature structure

$$\alpha\,\mathrm{trace}(PQ)\,R_o + \beta\,\Omega(\{P,Q\}) + \gamma\, H_{P,Q} = H^{\alpha\beta\gamma}_{P,Q}$$

such that its energy-momentum matrix becomes the electromagnetic one, namely $\frac{1}{n}\mathrm{trace}(PQ)\,\mathrm{id}_{\mathbf{V}} - \{P,Q\}$. We get this for

$$\beta = -\frac{2}{n} - 2(n-1)\alpha \quad \text{and} \quad \gamma = -\frac{n-1}{3}\left(\frac{2}{n} + (n-2)\alpha\right) \quad .$$

Hence for $n = 4$ the one-parameter schar of curvature structures

$$H^{\alpha}_{P,Q} = \alpha\,\mathrm{trace}(PQ)\,R_o - \left(6\alpha + \frac{1}{2}\right)\Omega(\{P,Q\}) - \left(2\alpha + \frac{1}{2}\right) H_{P,Q}$$

has the energy-momentum matrix IT_{sym}. Hence we may call the $H^{\alpha}_{P,Q}$ *electrovac* (they should be divided by κ), since the only contribution to energy-momentum is given by electromagnetism.
The $H_{P,Q}$ above are the most general curvature structures constructed from the electromagnetic fields P and Q in a bilinear way. It is difficult to imagine any other construction. An example of a gravitational field, induced by an electromagnetic one, which is not electrovac is the plane gravitational wave, described by Sachs and Wu in section 7.6. They start from an electromagnetic field with a non-vanishing energy-momentum and derive a not trivial curvature tensor of Weyl type, i.e. with vanishing Ricciform and total energy-momentum. Hence there must be another contribution to the curvature whose energy-momentum cancels the electromagnetic one.

Remark: The determinaton of $H_{P,Q}$ was exactly Nomizu's construction of $R_o^{A,B}$ transported from $JA(\mathbf{V},<,>)$ to $so(\mathbf{V},<,>)$. One may ask whether there is some mixture of both procedures, namely curvature structures constructed linearly in one A and one Q. However, it turns out that contrary to the two cases above, there are only the expected types $\operatorname{trace}(AQ)R_o$ and $\Omega(Q\cdot A)$.

9. ELECTROVAC CURVATURE STRUCTURES

Suppose $n=4$. To get more insight into the structure of electrovac curvature structures we use the invariant direct decomposition ($\oplus$) to get that of $H^{\alpha}_{P,Q}$:

$$H^{\alpha}_{P,Q} = -\Omega(\tfrac{1}{4}\operatorname{trace}(PQ)\,id_4 - \{P,Q\}) \oplus -2(\alpha+\tfrac{1}{4})\,WH_{P,Q}$$
$$= -\Omega(IT_{sym}) \oplus -2(\alpha+\tfrac{1}{4})\,WH_{P,Q} .$$

We use this direct decomposition to draw some conclusions for the electrovac case. Note that

$$T_{\Omega(A)} = \operatorname{trace}(A)\,id_4 - A \qquad A \text{ in } JA(\mathbf{V},<,>)$$

shows that $T\circ\Omega$ leaves the traceless matrices and especially the energy-momentum matrix IT_{sym} invariant up to a minus sign. Singer and Thorpe's structure theorem in section 1 implies now:

THEOREM: (i) $\alpha = -\frac{1}{4}$ gives electrovac curvature structures in $Im(E-W)$; if an electrovac curvature structure is of Weyl type, its energy-momentum must vanish (which only can be the case for anisotropic media).

(ii) Electrovac curvature structures have no component in $\mathbf{R}R_o$, i.e. $Sc(H^{\alpha}_{P,Q}) = 0$; for isotropic media there always is a component in $Im(E-W)$ whence their Ricci-form does not vanish; since

$$L_{H^{\alpha}_{P,Q}} \neq \lambda\,id_4 ,$$

they are never of Einsteinian type; their sectional curvature is never constant.

That one scalar α remains arbitrary was first seen by Rainich 1925 [R p.107], who also proved the vanishing of the curvature scalar. There are three additional algebraic conditions, derived by Rainich for the electrovac case - in Ludwig's work they are (1.4), (1.2) and (1.3) respectively - namely

(c) $L_{H^{\alpha}_{P,Q}} = -IT_{sym}$ which implies $L^2_{H^{\alpha}_{P,Q}} = \omega\, id_4$

where $\omega = W^2-(p,p)$ for the Poyntingvector p in T_{sym}.
A consequence of this is

(d) $\rho_R(x,L_R y) = \frac{1}{4}\,\mathrm{trace}(L_R)\,\langle x,y\rangle$ for $R = H^{\alpha}_{P,Q}$.

In addition for any timelike t we have for this R now $\rho_R(t,t) = \langle t,L_R t\rangle = -\langle t,IT_{sym}t\rangle$ which is positive[SW p.85], i.e.

(e) $\rho_R(t,t) \geq 0$ for all timelike vectors 1 .

The proof of the second equation in (c) is given in [MTW p.481]. ω is the sum of the squares of the two invariants of the electromagnetic fields. Note that to derive the second equation in (c) we have to assume the medium to be isotropic, that is $D = \varepsilon E$ and $B = \mu H$. In this case P and Q are linearly dependent, i.e.

$$P = -\mu Q \quad .$$

These results are not exactly those of Rainich. To get $H^{\alpha}_{P,Q}$ we didn't have to restrict to isotropic media. On the other hand his conditions (which can be completed to a set of necessary and sufficient conditions for a curvature to be electrovac, the other conditions being formulated with the help of covariant derivatives acting on vector fields) are conditions on an arbitrary curvature structure, not necessarily constructed in a bilinear way in terms of the P and Q.
For an isotropic medium and $\alpha = -\frac{1}{4}$, the second equation in (c) shows that we have symmetric curvature structures of this type in $\mathrm{Im}(E-W)$. Hence there should be an integration process to a homogeneous, symmetric space-time of the form $\mathbb{G}_1/\mathbb{G}_2$, where the group $\mathbb{G}_1$ has as Lie algebra the standard embedding algebra of R and the subgroup $\mathbb{G}_2$ has as Lie algebra the inner derivations of R, compare [Lo]. Since the two invariants of the electromagnetic field may vanish for a non-trivial field, we have the possibility $\omega = 0$, i. e. L_R is nilpotent. Since this case is excluded from Rainich's conditions, we conclude that it must be the additional Rainich-conditions formulated in terms of covariant derivatives, which are not fulfilled in this case.
In the electrovac case we now can solve the gravitational equations a little easier: The entries for the setting of the problem are a mathematical one (i) the four-dimensional manifold, and a

physical one (ii) energy-momentum. To solve Einstein's equations means to find a connection ∇ and a metric $\langle,\rangle$ such that (1) ∇ is the Levi-Civita connection of $\langle,\rangle$, and (2) the curvature of ∇ gives via the gravitational equations the given energy-momentum. In the electrovac case (2) reduces now to find $H^{\alpha}_{P,Q}$, since the gravitational equations then are satisfied automatically.
To discuss the **G**-transformation behaviour of electromagnetic curvature structures, note that the linearity of $H_{P,Q}$ in P and Q implies

$$\text{(iii)} \qquad G\cdot H_{P,Q} = \lambda_G^{-2} H_{G\cdot P, G\cdot Q} = H_{\lambda_G^{-1} G\cdot P, \lambda_G^{-1} G\cdot Q} \quad .$$

In part 2 it was mentioned that the other two terms in $H^{\alpha}_{P,Q}$ transform in the same way, explicitely

$$\text{(ii)} \qquad G\circ\Omega(\{P,Q\}) = \Omega(\{\frac{1}{\lambda_G} G\cdot P, \frac{1}{\lambda_G} G\cdot Q\}) \ ,$$

$$\text{(i)} \qquad G\cdot(\mathrm{trace}(PQ)R_o) = \mathrm{trace}(PQ)\lambda_G^{-2} R_o$$

$$= \mathrm{trace}\left(\frac{1}{\lambda_G}(G\cdot P)\ \frac{1}{\lambda_G}(G\cdot Q)\right) R_o \quad .$$

This means that if $H^{\alpha\beta\gamma}_{P,Q}$ is constructed from P and Q, then $G\cdot H^{\alpha\beta\gamma}_{P,Q}$ is constructed in the same way with the same parameters α, β and γ from the transformed electromagnetic fields $\lambda_G^{-1} G\cdot P$ and $\lambda_G^{-1} G\cdot Q$. Therefore electromagnetic and especially electrovac curvature structures lie in electromagnetic and electrovac orbits, the latter being parametrized by one real constant α.

10. THE LICHNEROWICZ DEFINITION OF GRAVITATIONAL RADIATION

Let now $\langle,\rangle$ have the signature $(+-\ldots-)$. In analogy to the electromagnetic radiation concept, defined entirely in terms of the non-vanishing Poyntingvector p, Lichnerowicz [L p.45] introduced two axioms in order that the gravitational field is a *pure radiation state*: There is a nonvanishing vector $l \in \mathbf{V}$ such that

(PR.1) $R(x,y)l = 0$

(PR.2) $\langle l,x\rangle\langle y,R(v,w)z\rangle + \langle l,y\rangle\langle z,R(v,w)x\rangle + \langle l,z\rangle\langle x,R(v,w)y\rangle = 0$

for all x,y,z,v,w in $\mathbf{V}$. Using (C.2') one can bring them in the shorter form

(PR.1') $R(x,y)l = 0$

(PR.2') $\langle l,x\rangle\, R(y,z) + \langle l,y\rangle\, R(z,x) + \langle l,z\rangle\, R(x,y) = 0$.

A verification shows that λR_o never fulfills one of the axioms. For a Weyl curvature (PR.1) means that R is *of Petrov type N* [SW p.254] and l is lightlike. It is straightforeward to see, that if R satisfies one of the two axioms with respect to l, $G\circ R$ satisfies the same axiom with respect to Gl. Hence the two axioms are orbit properties. Choosing $x = l$ in (PR.2') and using (PR.1') l becomes lightlike for a nontrivial R. We give two classes of examples.
For a lightlike l, $g(l,l)$ is a traceless matrix, selfadjoint with respect to $\langle,\rangle$. Hence $\Omega(g(l,l))$ is in $\mathrm{Im}(E-W)$ and defines pure radiation with respect to l. The second class of examples is discussed in the following section.
The data of a pure radiation state were given by Lichnerowicz:
THEOREM: If R is a pure radiation state with repect to l then

(a) $\rho_R(x,y) = \tau_o\langle x,l\rangle\langle l,y\rangle$ for some real τ_o (i.e. ρ_R is degenerate, hence R not semisimple);

(b) $L_R = \tau_o g(l,l)$;

(c) $Sc(R) = 0$ (i.e. $R \in \mathrm{Im}(E-W) \oplus \mathrm{Im}\, W$);

(d) $-\kappa T_R = \tau_o g(l,l)$ (i.e. l is an eigenvector of eigenvalue 0 for L_R and T_R).

Proof: To calculate the Ricciform apply the linear transformation in (PR.2') to w and take the trace in z, and get

$$\langle l,x\rangle\langle L_R y,w\rangle = \langle l,y\rangle\langle L_R x,w\rangle ,$$

$$\langle L_R y,w\rangle\, l = \langle l,y\rangle\, L_R w \qquad \text{for all } y,w \text{ in } \mathbf{V}.$$

Choose y with $\langle l,y\rangle = 1$, then $L_R w = \delta(w)\, l$, where δ is a linear form in w. Hence

$$\rho_R(x,y) = \langle L_R x,y\rangle = \delta(x)\langle l,y\rangle = \langle x,l_o\rangle\langle l,y\rangle$$

for a suitable l_o in $\mathbf{V}$. The symmetry of ρ_R implies that l and l_o are linearly dependent. The rest of the proof follows from Singer and Thorpe's theorem in section 1. □

A pure radiation state with vanishing Ricciform, i.e. of Weyl type, was called by Lichnerowicz a *pure gravitational radiation state*.
Going back to the curvature structures in $\mathrm{Kern}W = \mathrm{Im}\,\Omega$ we prove:
THEOREM: The $\Omega(\tau_o g(l,l))$ are the only pure radiation states in $\mathrm{Kern}W$.

Proof: Writing out (PR.1) and (PR.1') for $\Omega(A)$ we get

$$\langle y,l\rangle Ax + \langle y,Al\rangle x = \langle x,Al\rangle y + \langle x,l\rangle Ay = \langle Ax,y\rangle l + \langle x,y\rangle Al$$

for all x,y. Taking the trace in the vector x we get $Al = \frac{\text{trace}A}{2-n} l$ and for $A_o = A + (2-n)^{-1}\text{trace}(A)\,\text{id}_{\mathbf{V}}$ we have

$$\langle y,l\rangle A_o x = \langle x,l\rangle A_o = \langle A_o x,y\rangle l \quad .$$

Choosing y with $\langle l,y\rangle = 1$ we have $A_o x = \langle A_o y,x\rangle l$, i.e. with $A_o y = l_o$ also $\langle x,l\rangle l_o = \langle x,l_o\rangle l$; whence l_o and l are linearly dependent. From $A_o x = \tau_o \langle l,x\rangle l$ for a suitable τ_o, now

$$A = \tau_o g(l,l) - \frac{\text{trace } A}{2-n}\,\text{id}_{\mathbf{V}} \quad .$$

Introducing this in the first equations

$$\text{trace}(A)\langle y,l\rangle x = \text{trace}(A)\langle x,l\rangle y = \text{trace}(A)\langle x,y\rangle l$$

for all x,y implies trace $A = 0$. □

Given $G \in \mathbf{G}$ and $l \in \mathbf{V}$ we have

$$G\cdot\Omega(g(l,l)) = \lambda_G^{-2}\Omega(Gg(l,l)G^{-1}) = \lambda_G^{-4}\Omega(Gl,Gl) \quad .$$

If $\Omega(A)$ is a pure radiation state with respect to l then $G\cdot\Omega(A)$ is a pure radiation state with respect to $\lambda_G^{-2}Gl$. Since **G** acts transitively on the nullcone (without zeroelement), there are exactly two nontrivial radiation $\mathbb{G}$-orbits in $\text{Im}(E-W)$ according to the sign of τ_o.

From the degeneracy of a pure radiation state Ricciform we can exclude a Lie algebra curvature from the radiation state curvature orbits: In fact the Ricciform of the (necessarily semisimple) Lie algebra is just the non-degenerate Killingform <,>.

11. ELECTROMAGNETIC RADIATION IMPLIES GRAVITATIONAL RADIATION

A second type of examples for pure radiation states was given by Lichnerowicz: Given a lightlike l we choose a q in **V** with $\langle l,q\rangle = 0$ but $\langle q,q\rangle \neq 0$, which for the chosen dimensions and signatures always is possible [Gb p.268]. Define

$$Q = g(q,l) - g(l,q) \quad , \qquad P = -\mu Q \ ,$$

μ being the magnetic permeability of the isotropic medium. Q and P are in so(**V**,<,>). Then $H_{P,Q}$ fulfills the two PR-axioms with respect to l and

$$Q^2 = - \langle q,q\rangle \; g(1,1) \; . \qquad (\P)$$

Therefore trace(PQ) $= 2\{(\vec{B},\vec{H}) - (\vec{E},\vec{D})\} = 0$ and the energy-momentum becomes

$$IT_{sym} = -\{P,Q\} = \mu \, Q^2 = - \mu \langle q,q\rangle \; g(1,1) \quad ,$$

which is of the type considered in the first theorem in the preceeding section. Lichnerowicz has shown that this example is equivalent to saying that there is a (necessarily lightlike) vector l such that for all x,y,z in $\mathbb{W}$

(EMR.1) $\quad Ql = 0 \; ,$

(EMR.2) $\quad \langle l,x\rangle\langle y,Qz\rangle + \langle l,y\rangle\langle z,Qx\rangle + \langle l,z\rangle\langle x,Qy\rangle = 0 \quad .$

These axioms of *electromagnetic radiation* in turn are equivalent to a nonvanishing Pointingvector $p = \frac{1}{2}\{\frac{1}{c}\vec{E} \times \vec{H} + c\vec{D} \times \vec{B}\}$ in T_{sym}, which is the physical description of electromagnetic radiation. From the axioms one sees the analogy of the two sorts of radiation. (EMR.2) can be simplified to

(EMR.2') $\quad -\langle l,x\rangle Qy + \langle l,y\rangle Qx + \langle x,Qy\rangle l = 0 \quad .$

From (¶) and the two examples for pure radiation discussed so far one proves:

THEOREM: For the above Q and $P = -\mu Q$ the electromagnetic curvature structures $\beta\Omega(\{P,Q\}) + \gamma H_{P,Q}$ and especially the electrovac $H^{\alpha}_{P,Q}$ fulfill the two PR-axioms with respect to l.

It means that electromagnetic radiation always leads to gravitational radiation, and that the two types of radiation are linked not only by analogy. As in the preceeding section we may invert the question: If an electromagnetic curvature structure corresponds to a pure radiation state, is it then induced by electromagnetic radiation? In other words, if the electromagnetic $H^{\alpha\beta\gamma}_{P,Q}$ fulfills the two PR-axioms, do the P,Q fulfill the two EMR-axioms? For an isotropic medium with $P = -\mu Q$, the proof consists in showing that if a general electromagnetic curvature structure fulfills either (PR.1) or (PR.2) then l must be an eigenvector of eigenvalue 0 of IT_{sym}. Lichnerowicz [L] then has proven the rest.

12. THE RELATION TO THE PETROV-CLASSIFICATION

The basic fact of the above is that all concepts fit into the structure theory, i.e. are orbit properties. Certainly there are more orbit properties, like isotropy and perhaps black and white hole states. It seems to be unknown what condition one has to impose on energy-momentum to be "physical" [G]. In [SW] various such conditions for the energy-momentum tensor are described. However, whatever such a condition is, it should be valid for the whole orbit in question. Presumably such an energy-momentum condition excludes one of the two nontrivial **G**-orbits in $\mathbf{RR}_0$, but not one of the two radiation orbits in Im(E-W).

A mathematical meaning of curvature **G**-orbits may come from the following guess. Given two pseudo-Riemannian manifolds (of equal dimension and signature) such that the levi-Civita curvature section $R(X,Y) = \nabla_X\nabla_Y - \nabla_Y\nabla_X - \nabla_{[X,Y]}$ lies for both manifolds in all points in the same orbit. Then there is a conformal cover between the both manifolds. Such a result would allow to define *pure* manifolds as those whose Levi-Civita section of its curvature bundle is entirely in one orbit. However, pseudo-Riemannian manifolds and especially space-time need not be of this pure type. But the question arises, whether one can desompose space--time into pure portions.

So the first problem is to classify the curvature **G**-orbits. Since the **G**-action commutes with Ω this is for Kern$W$ the classification of **G**-orbits in the Jordan algebra JA(**V**,<,>). It was done for instance in [KSMH chap.5]. The classification in Im W seems to be more involved. We may look at Ω as a kind of Lueg ins Land onto the plane ImΩ. Is there something like this for the $\frac{1}{12}n(n+1)(n+2)(n-3)$-dimensional plane Im$W$ of Weyl curvatures? It would considerably facilitate the classification by transporting it elswhere.

Here the Petrov classification comes in. In [Gy], [ST], [S] and [T] curvature structures are defined in a different but equivalent way. This equivalence is described for instance in [N] and [T]. In the Grassmann algebra over **V** (in order to stay in a pseudo-orthogonal category one could try the Clifford algebra instead) one considers the $\frac{1}{2}n(n-1)$-dimensional space of polynomials of second power Λ^2, the socalled bivectors, generated by the $x \wedge y$

for x,y in **V**. On Λ^2 a symmetric bilinearform is given by

$$\langle\!\langle x \wedge y, z \wedge w \rangle\!\rangle = \langle x,z\rangle\langle y,w\rangle - \langle x,w\rangle\langle y,z\rangle$$

which has the signature (+++---) in the four-dimensional physical case. Given a curvature structure R one defines

$$\langle\!\langle x \wedge y, \tilde{R}(z \wedge w) \rangle\!\rangle = \langle x, R(z,w)y\rangle$$

to get a $\langle\!\langle , \rangle\!\rangle$-selfadjoint linear transformation on Λ^2. From

$$(R_o^{A,B})^\sim(z \wedge w) = \frac{1}{2}\,(\,Az \wedge Bw + Bz \wedge Aw\,)$$

$$(\Omega(A))^\sim(z \wedge w) = \frac{1}{2}\,(\,Az \wedge w + z \wedge Aw\,)$$

we see that R_o is mapped onto the identity on Λ^2.
Petrov's classification now is the reduction of the $\langle\!\langle , \rangle\!\rangle$-selfadjoint matrices $\tilde{R}$ into Jordan normal forms, which is done by generalized eigenvectors and eigenvalues. In general different $Gl(\Lambda^2,\mathbb{R})$-orbits may have the same Jordan normal form. Therefore taking the even smaller subgroup $\mathbb{G}$ a reduction of this structure group we see that in general Petrov classes split into several $\mathbb{G}$-orbits. The action of G on $\tilde{R}$ should be such that the map $R \mapsto \tilde{R}$ is **G**-equivariant, i.e.

$$(G \cdot R)^\sim = G \cdot \tilde{R} \quad .$$

It remains to show that the **G**-orbits for n = 4 and the Lorentz signature are those described in [JES p.25] and [JEK p.42]. In [CS] orbits in the space of curvature structures with respect to the action of the proper orthochronous Lorentz group are considered.

13. ADDITIONAL PROBLEMS

(1) To complete the structure theory for curvature triples one should try to find an algebra datum for a curvature structure R, which allows to state whether R is isotropic, homogeneous or has one of the other properties described in section 4 in [Ti]. Since for homogeneous R which are not symmetric, torsion and an additional structure map come in, the standard embedding algebra of Lie triples may fail to achieve this. Then Lister's structure theory for Lie triples [Li] should be generalized. Note that for curvature triples the third axiom of a Lie triple is dropped but pseu-

do-orthogonality of the triple comes in.

(2) For each **G**-orbit find a pseudo-Riemannian manifold, admitting one element in the orbit as its Levi-Civita curvature section. Its physical datum T_R then gives some idea on the physics involved. Clearly it would be desirable to find the simply connected unicover, since the other covering spaces can be determined from it.

(3) We should like to define space-times, having their Levi-Civita curvature section in one orbit in all points, as *cosmological models*, excluding those with an unphysical energy-momentum complex. On any such cosmological model one has to study classical and quantum physics. In Irving Segal's language any such model would be a *variant of special relativity*. Actually any of those models could be a better approximation to actual space-time than Minkowski space, at least somewhere in space-time.

(4) The physical space-time is not of that pure orbit type. However, some space-time regions may be. So one may ask whether there are boundaries between these regions.

Acknowledgement: The author is indepted to the referee for pointing out that the guess in section 12 might be related to the equivalence problem, described for riemannian manifolds by Nomizu and Yano [NY],
and to Frau Döring for carefully typing the manuscript.

REFERENCES

[CS] C.D. Collinson, R. Shaw, *The Rainich Conditions for Neutrino Fields*, Int.J.Theor.Phys. 6, 347-357 (1972)

[F] T. Frankel, *Gravitaitonal Curvature*, Freeman & Co, San Francisco (1979)

[G] R. Geroch, *Building things in general relativity*, Gen.Rel.Grav. 13, 37-41 (1981)

[Gy] A. Gray, *Invariants of Curvature Operators in Four-Dimensional Riemannian Manifolds*, in Proceedings of the 13th Biennial Seminar of the Canadian Mathematical Congress, Halifax (1971)

[Gb] W.H. Greub, *Linear Algebra*, Springer-Verlag, Berlin (1967)

[JM] P. Jordan, S. Matsushita, *Zur Theorie der Lie-Tripel-Algebren*, Akad.Wiss. Mainz, Abh. math.naturwissenschaftlicher Klasse 1,123-134 (1967)

[JEK] P. Jordan, J. Ehlers, W. Kundt, *Strenge Lösungen der Feldgleichungen der allgemeinen Relativitätstheorie*, Akad.Wiss.Mainz, Abh. math. naturwissenschaftlicher Klasse 2, 1-85 (1960)

[JES] P. Jordan, J. Ehlers, K. Sachs, *Beiträge zur Theorie der reinen Gravitationsstrahlung. Strenge Lösungen der Feldgleichungen der allgemeinen Relativitätstheorie*, Akad.Wiss.Mainz, Abh. math. naturwissenschaftlicher Klasse 1, 1-62 (1961)

[K] O. Kowalski, *Partial Curvature Structures and Conformal Transformations*, Jour.Diff.Geom. 8, 53-70 (1973)

[KSMH] D. Kramer, H. Stephani, M. MacCallum, E. Herlt, *Exact Solutions of Einstein's Field equations*, Cambridge Univ. Press, Cambridge (1980)

[Ku] R.S. Kulkarni, *Curvature and Metric*, Ann.Math. 91, 311-331 (1970)

[Ku] R.S. Kulkarni, *Curvature Structures and Conformal Transformations*, Jour. Diff.Geom. 4, 53-70 (1970)

[L] A. Lichnerowicz, *Ondes et radiations électromagnetiques et gravitationelles en relativité générale*, Annali di Mathematica Pura e. Applica 4, 1-95 (1960)

[Li] W.G. Lister, *A structure Theory of Lie Triples*, Trans.Am.Math.Soc. 72, 217-242 (1972)

[Lo] O. Loos, *Symmetric Spaces I,II*, Benjamin, N.Y. (1969)

[Lu] G. Ludwig, *Geometricdynamics of Electromagnetic Fields in the Newman-Penrose Formalism*, Commun.math.Phys. 17, 98-108 (1970)

[MTW] C.W. Misner, K.S. Thorne, J.A. Wheeler, *Gravitation*, Freeman and Co., San Francisco (1971)

[N] K. Nomizu, *Invariant Affine Connections on Homogeneous Spaces*, Am.Jour. Math. 76, 33-65 (1954)

[N] K. Nomizu, *The Decomposition of Generalized Curvature Tensor Fields*, in *Differential Geometry, Papers in Honour of K. Yano*, Kinukuniya, Tokyo, 335-345 (1972)

[NY] K. Nomizu, K. Yano, *Some results related to the equivalence problem in Riemannian Geometry*, in Proc. of the US-Japan Seminar in Differential Geometry, Kyoto, Japan, 95-100 (1965)

[R] G.Y. Rainich, *Electrodynamics in the General Relativity Theory*, Trans. Am. Math.Soc. 27, 106-136 (1925)

[SW] R.K. Sachs, H. Wu, *General Relativity for Mathematicians*, Springer-Verlag, Berlin (1977)

[Se] I.E. Segal, *Mathematical Cosmology and Extra-Galactic Astronomy*, Academic Press, N.Y (1977)

[ST] I.M. Singer, J.A. Thorpe, *The Curvature of 4-Dimensional Einstein Spaces*, in *Papers in honou of K. Kodaira*, Princeton Univ. Press, Princeton (1968)

[Sö] F. Söler, *r-Manifolds and Gravitational Radiation*, Gen.Rel.Grav. 13, 37-41 (1981)

[S] A. Stehney, *Principal Null Directions without Spinors*, Jour.Math.Phys. 17, 1793-1796

[T] J.A. Thorpe, *Curvature and the the Petrov Canonical Forms*, Jour.Math.Phys. 10, 1-6 (1969)

[Ti] H. Tilgner, *The Group Structure of Pseudo-Riemannian Curvature Spaces*, Jour.Math.Phys. 19, 1118-1125 (1978)

[W] S. Weinberg, *Gravitation and Cosmology*, Wiley, N.Y. (1972)